STATA GRAPHICS
REFERENCE MANUAL
RELEASE 8

A Stata Press Publication
STATA CORPORATION
College Station, Texas

Stata Press, 4905 Lakeway Drive, College Station, Texas 77845

The suggested citation for this software is

StataCorp. 2003. *Stata Statistical Software: Release 8.0*. College Station, TX: Stata Corporation.

Table of Contents

Title

> **graph intro** — Introduction to graphics manual

Remarks

Remarks are presented under the headings

Suggested reading order
What's new
A quick tour
Using the menus

Suggested reading order

We recommend that you read the entries in this manual in the following order:

Read *A quick tour* below and then ...

Entry	description
[G] **graph**	Overview of the `graph` command
[G] **graph twoway**	Overview of the `graph twoway` command
[G] **graph twoway scatter**	Overview of the `graph twoway scatter` command

When reading those sections, follow references to other entries that interest you. They will take you to such useful topics as

Entry	description
[G] *marker_label_options*	Options for specifying marker labels
[G] *by_option*	Option for repeating graph command
[G] *title_options*	Options for specifying titles
[G] *legend_option*	Option for specifying legend

We could list many, many more, but you will find them on your own. Follow the references that interest you and ignore the rest. Afterwards, you will have a working knowledge of twoway graphs. To fill that out, glance at each of

Entry	description
[G] **graph twoway line**	Overview of the `graph twoway line` command
[G] **graph twoway connected**	Overview of the `graph twoway connected` command
etc.	

Turn to [G] **graph twoway**, which lists all the different `graph twoway` plottypes, and browse the manual entry for each.

Now is the time to understand schemes. Schemes have a great effect on how graphs look and, anyway, you may want to specify a different scheme before printing your graphs.

1

Entry	description
[G] **schemes**	Schemes and what they do
[G] **set printcolor**	Set how colors are treated when graphs are printed
[G] **graph print**	Printing graphs the easy way
[G] **graph export**	Exporting graphs to other file formats

Now you are an expert on the `graph twoway` command, and you can even print the graphs it produces.

Turn to learning about the other types of graphs:

Entry	description
[G] **graph matrix**	Scatterplot matrices
[G] **graph bar**	Bar and dot charts
[G] **graph box**	Box plots
[G] **graph pie**	Pie charts

Finally, turn to learning tricks of the trade. See, if you have not already,

Entry	description
[G] **graph save**	Saving graphs to disk
[G] **graph use**	Redisplaying graphs from disk
[G] **graph describe**	Finding out what is in a .gph file
[G] *name_option*	How to name a graph in memory
[G] **graph display**	Redisplaying a named graph
[G] **graph dir**	Obtaining directory of named graphs
[G] **graph rename**	Renaming a named graph
[G] **graph copy**	Copying a named graph
[G] **graph drop**	Eliminating graphs in memory
[P] **discard**	Clearing memory

What's new

This section is intended for users of previous versions of Stata. If that is not you, skip to *A quick tour* below.

So what's new? Everything. There is not one little bit that is not new, even if it seems familiar.

Before you panic, let us tell you that all the old graphics are still in Stata. If you type

    . graph7 ...

or

    . gr7 ...

you will be back to using the old graph command. Moreover, the old graph command is still invoked under version control; see [P] **version**. If you set your version to 7.0 or earlier, graph does not mean what is defined in this manual; it means what it used to mean, which means that old do-files and ado-files continue to work.

One new feature requires some adjustment. What used to be called symbols are now called markers, and marker symbols are the shapes of the markers. Thus, you no longer specify the symbol() or s() option, you specify the msymbol() or ms() option. In addition, the old s(.) for specifying the dot symbol is now ms(p) (p stands for point). ms(.) means to use the default.

A quick tour

graph is easy to use:

```
. sysuse auto, clear
(1978 Automobile Data)
. graph twoway scatter mpg weight
```

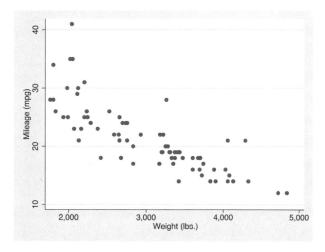

All the commands documented in this manual begin with the word graph, but in many instances the graph is optional. You could get the same graph by typing

```
. twoway scatter mpg weight
```

and, in the case of scatter, you could omit the twoway, too:

```
. scatter mpg weight
```

We, however, will continue to type twoway to emphasize when the graphs we are demonstrating are in the twoway family.

Twoway graphs can be combined with by():

(Continued on next page)

```
. twoway scatter mpg weight, by(foreign)
```

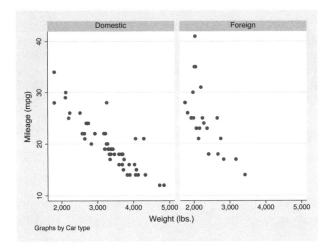

Graphs in the `twoway` family can also be overlaid. The members of the `twoway` family are called *plottypes*; `scatter` is a plottype, and another plottype is `lfit`, which calculates the linear prediction and plots it as a line chart. When we want one plottype overlaid on another, we combine the commands, putting `||` in between:

```
. twoway scatter mpg weight || lfit mpg weight
```

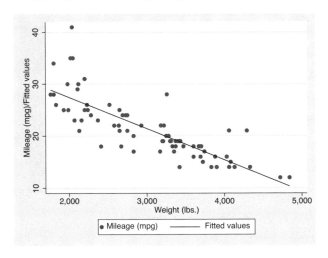

Another notation for this is called the ()-binding notation:

```
. twoway (scatter mpg weight) (lfit mpg weight)
```

It does not matter which notation you use.

Overlaying can be combined with by(). This time, we will substitute `qfitci` for `lfit`. `qfitci` plots the prediction based on a quadratic regression, and it adds a confidence interval. We will add the confidence interval based on the standard error of the forecast:

```
. twoway (qfitci mpg weight, stdf) (scatter mpg weight), by(foreign)
```

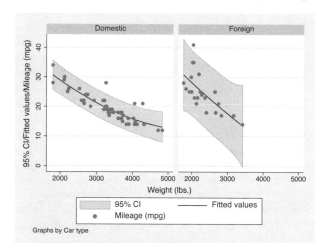

We used the ()-binding notation just because it makes it easier to see what modifies what:

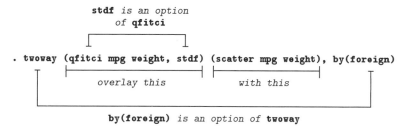

We could just as well have typed this command using the ||-separator notation,

```
. twoway qfitci mpg weight, stdf || scatter mpg weight ||, by(foreign)
```

and, as a matter of fact, we do not have to separate the twoway option by(foreign) (or any other twoway option) from the qfitci and scatter options, so we can type

```
. twoway qfitci mpg weight, stdf || scatter mpg weight, by(foreign)
```

or even

```
. twoway qfitci mpg weight, stdf by(foreign) || scatter mpg weight
```

All of these syntax issues are discussed in [G] **graph twoway**. In our opinion, the ()-binding notation is easier to read, but the ||-separator notation is easier to type. You will see us using both.

It was not an accident that we put qfitci first and scatter second. qfitci shades an area and, had we done it the other way around, that shading would have been put right on top of our scattered points and erased (or at least hidden) them!

Plots of different types or the same type may be overlaid:

```
. sysuse uslifeexp, clear
(U.S. life expectancy, 1900-1999)
. twoway line le_wm year || line le_bm year
```

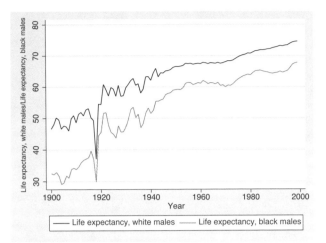

Here is a rather fancy version of the same graph:

```
. generate diff = le_wm - le_bm
. twoway line le_wm year, yaxis(1 2) xaxis(1 2)
        || line le_bm year
        || line diff  year
        || lfit diff  year
        ||,
            ytitle( "",           axis(2) )
            xtitle( "",           axis(2) )
            xlabel( 1918,         axis(2) )
            ylabel( 0(5)20,       axis(2) gmin angle(horizontal) )
            ylabel( 0 20(10)80,           gmax angle(horizontal) )
            ytitle( "Life expectancy at birth (years)" )
            title( "White and black life expectancy" )
            subtitle( "USA, 1900-1999" )
            note( "Source: National Vital Statistics, Vol 50, No. 6"
                  "(1918 dip caused by 1918 Influenza Pandemic)" )
            legend( label(1 "White males") label(2 "Black males") )
```

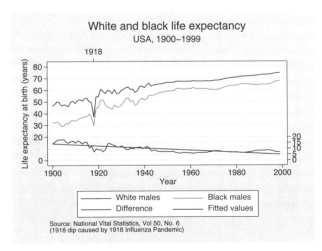

There are a lot of options on this command! Strip away the obvious ones, such as `title()`, `subtitle()`, and `note()`, and you are left with

```
. twoway line le_wm year, yaxis(1 2) xaxis(1 2)
        || line le_bm year
        || line diff  year
        || lfit diff  year
        ||,
           ytitle( "",           axis(2) )
           xtitle( "",           axis(2) )
           xlabel( 1918,         axis(2) )
           ylabel( 0(5)20,       axis(2) gmin angle(horizontal) )
           ylabel( 0 20(10)80,          gmax angle(horizontal) )
           legend( label(1 "White males") label(2 "Black males") )
```

Let's take the longest option first:

```
           ylabel( 0(5)20,       axis(2) gmin angle(horizontal) )
```

The first thing to note is that options have options:

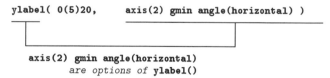

are options of **ylabel()**

Now look back at our graph. It has two *y* axes, one on the right and a second on the left. What

```
           ylabel( 0(5)20,       axis(2) gmin angle(horizontal) )
```

did was cause the right axis—axis(2)—to have labels at 0, 5, 10, 15, and 20—0(5)20. gmin forced the grid line at 0 because, by default, graph does not like to draw grid lines too close to the axis. angle(horizontal) turned the 0, 5, 10, 15, and 20 to be horizontal rather than, as usual, vertical.

You can now guess what

```
           ylabel( 0 20(10)80,          gmax angle(horizontal) )
```

did. It labeled the left *y* axis—axis(1) in the jargon—but we did not have to specify an axis(1) suboption since that is what ylabel() assumes. The purpose of

```
           xlabel( 1918,         axis(2) )
```

is now obvious, too. That labeled a value on the second *x* axis.

So now we are left with

```
. twoway line le_wm year, yaxis(1 2) xaxis(1 2)
        || line le_bm year
        || line diff  year
        || lfit diff  year
        ||,
           ytitle( "",           axis(2) )
           xtitle( "",           axis(2) )
           legend( label(1 "White males") label(2 "Black males") )
```

Options ytitle() and xtitle() specify the axis titles. We did not want titles on the second axes, so we got rid of them. The legend() option,

```
           legend( label(1 "White males") label(2 "Black males") )
```

merely respecified the text to be used for the first two keys. By default, legend() uses the variable label, which in this case would be the labels of variables le_wm and le_bm. In our dataset those labels are "Life expectancy, white males" and "Life expectancy, black males". It was not necessary—and undesirable—to repeat "Life expectancy", so we specified an option to change the label. It was either that or change the variable label.

So now we are left with

```
. twoway line le_wm year, yaxis(1 2) xaxis(1 2)
        || line le_bm year
        || line diff  year
        || lfit diff  year
```

and that is almost perfectly understandable. The yaxis() and xaxis() options are what caused the creation of two y and two x axes rather than, as usual, one.

Understand how we arrived at

```
. twoway line le_wm year, yaxis(1 2) xaxis(1 2)
        || line le_bm year
        || line diff  year
        || lfit diff  year
        ||,
           ytitle( "",          axis(2) )
           xtitle( "",          axis(2) )
           xlabel( 1918,        axis(2) )
           ylabel( 0(5)20,      axis(2) gmin angle(horizontal) )
           ylabel( 0 20(10)80,          gmax angle(horizontal) )
           ytitle( "Life expectancy at birth (years)" )
           title( "White and black life expectancy" )
           subtitle( "USA, 1900-1999" )
           note( "Source: National Vital Statistics, Vol 50, No. 6"
                 "(1918 dip caused by 1918 Influenza Pandemic)" )
           legend( label(1 "White males") label(2 "Black males") )
```

We started with the first graph we showed you,

```
. twoway line le_wm year || line le_bm year
```

and then, to emphasize the comparison of life expectancy for whites and blacks, we added the difference,

```
. generate diff = le_wm - le_bm
. twoway line le_wm year,
        || line le_bm year
        || line diff  year
```

and then, to emphasize the linear trend in the difference, we added "lfit diff year",

```
. twoway line le_wm year,
        || line le_bm year
        || line diff  year,
        || lfit diff  year
```

and then we added options to make the graph look more like we wanted. The options we introduced one at a time. Rather fun, really. As our command grew, we switched to using the Do-file Editor. There, we could add an option and hit the **do** button to see where we were. Since the command was so long, when we opened the do-file editor, we typed on the first line

```
#delimit ;
```

and we typed on the last line

```
;
```

and then we typed our ever growing command in between.

While we are on the subject of life expectancy, using another dataset, we drew

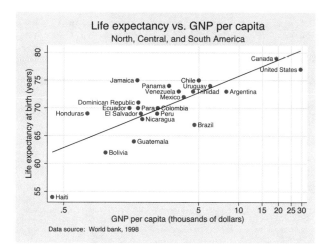

How we did that is explained in [G] *marker_label_options*. Along the same lines is

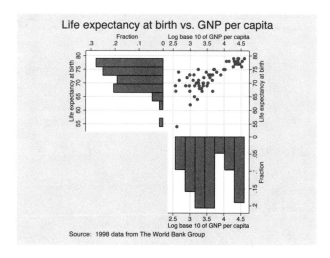

which we drew by separately drawing three rather easy graphs

```
. twoway scatter lexp loggnp,
        yscale(alt) xscale(alt)
        xlabel(, grid gmax)          saving(yx)
. twoway histogram lexp, fraction
        xscale(alt reverse) horiz     saving(hy)
. twoway histogram loggnp, fraction
        yscale(alt reverse)
        ylabel(,nogrid)
        xlabel(,grid gmax)           saving(hx)
```

and then combining them into one:

```
. graph combine hy.gph yx.gph hx.gph,
        hole(3)
        imargin(0 0 0 0) grapharea(margin(l 22 r 22))
        title("Life expectancy at birth vs. GNP per capita")
        note("Source:  1998 data from The World Bank Group")
```

See [G] **graph combine** for more information.

Returning to our tour, twoway, by() can produce graphs that look like this:

```
. sysuse auto, clear
(1978 Automobile Data)
. scatter mpg weight, by(foreign, total row(1))
```

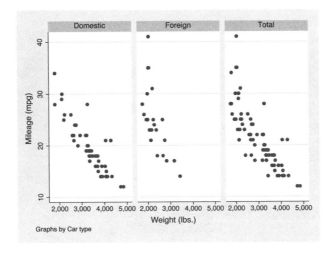

or like this

```
. scatter mpg weight, by(foreign, total col(1))
```

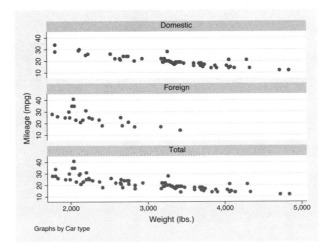

or like this

```
. scatter mpg weight, by(foreign, total)
```

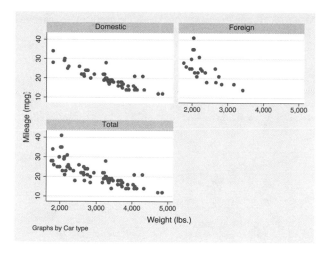

See [G] *by_option*.

There are lots of plottypes within the `twoway` family, including areas, bars, spikes, dropped lines, and dots. Just to illustrate a couple:

```
. sysuse sp500, clear
(S&P 500)
. replace volume = volume/1000
. twoway
        rspike hi low date ||
        line   close date ||
        bar    volume date, barw(.25) yaxis(2) ||
  in 1/57
, yscale(axis(1) r(900 1400))
  yscale(axis(2) r(  9    45))
  ytitle("                         Price -- High, Low, Close")
  ytitle(" Volume (millions)", axis(2) astext just(left))
  legend(off)
  subtitle("S&P 500", margin(b+2.5))
  note("Source:  Yahoo!Finance and Commodity Systems, Inc.")
```

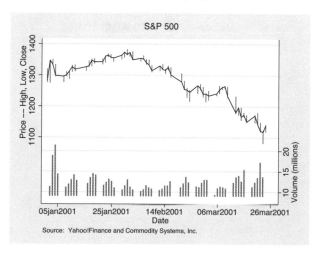

The above is explained in [G] **graph twoway rspike**.

Moving outside the `twoway` family, `graph` can draw scatterplot matrices, box plots, pie charts, and bar and dot plots. Here's an example of each:

Scatterplot matrix:

```
. sysuse lifeexp, clear
(Life expectancy, 1998)
. generate lgnppc = ln(gnppc)
. gr matrix popgr lexp lgnppc safe, maxes(ylab(#4, grid) xlab(#4, grid))
```

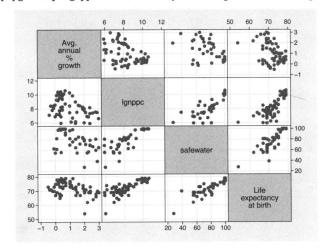

See [G] **graph matrix**.

Box plot:

```
. sysuse bplong, clear
(fictional blood-pressure data)
. graph box bp, over(when) over(sex)
        ytitle("Systolic blood pressure")
        title("Response to Treatment, by Sex")
        subtitle("(120 Preoperative Patients)" " ")
        note("Source:  Fictional Drug Trial, Stata Corporation, 2003")
```

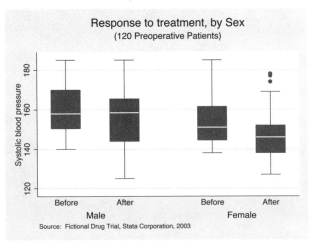

See [G] **graph box**.

Pie chart:

```
. graph pie sales marketing research development,
        plabel(_all name, size(*1.5) color(white))
        legend(off)
        plotregion(lstyle(none))
        title("Expenditures, XYZ Corp.")
        subtitle("2002")
        note("Source:  2002 Financial Report (fictional data)")
```

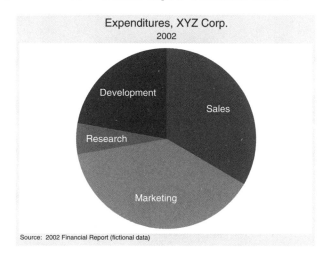

See [G] **graph pie**.

Vertical and horizontal bar charts:

```
. sysuse nlsw88, clear
(NLSW, 1988 extract)
```

```
. graph bar (mean) wage,
            over( smsa, descend gap(-30) )
            over( married )
            over( collgrad, rclabel(0 "Not college graduate"
                                    1 "College graduate"   ) )
            ytitle("")
            title("Average Hourly Wage, 1988, Women Aged 34-46")
            subtitle("by College Graduation, Marital Status,
                and SMSA residence")
            note("Source:  1988 data from NLS, U.S. Dept of Labor,
                Bureau of Labor Statistics")
```

(Continued on next page)

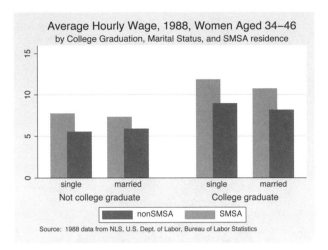

```
. sysuse educ99gdp, clear
(Education and GDP)

. generate total = private + public

. graph hbar (asis) public private,
            over(country, sort(total) descending)
            stack
            title("Spending on tertiary education as % of GDP,
                1999", span position(11) )
            subtitle(" ")
            note("Source:  OECD, Education at a Glance 2002", span)
```

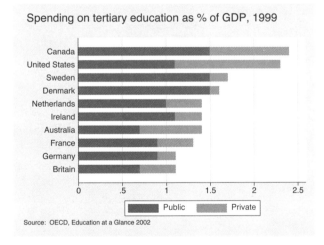

See [G] **graph bar**.

Dot chart:

```
. graph dot (mean) wage,
        over(occ, sort(1))
        by(collgrad,
            title("Average hourly wage, 1988, women aged 34-46", span)
            subtitle(" ")
            note("Source:  1988 data from NLS, U.S. Dept. of Labor,
                Bureau of Labor Statistics", span)
        )
```

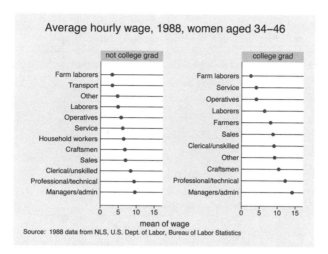

See [G] **graph dot**.

Have fun. Follow our advice in the *Suggested reading order* above; turn to [G] **graph**, then [G] **graph twoway**, and then to [G] **graph twoway scatter**.

Using the menus

In addition to the command line interface, most of graph's features can be accessed via menus. To start, load a dataset, pull down **Graphics**, and select what interests you.

When you have finished filling in the dialog box (do not forget to click on the tabs—lots of useful features are hidden there!), rather than click **OK**, click **Submit**. This way, once the graph appears, you can easily modify it and click **Submit** again.

Feel free to experiment. Clicking **Submit** (or **OK**) never hurts; if you have left a required field blank, you will be told. The dialog boxes make it easy to spot what you can change.

Also See

Complementary: [G] **graph**, [G] **graph other**

Title

> *added_line_options* — Options for adding lines to twoway graphs

Syntax

The *added_line_options* are

added_line_options	description
<u>y</u>line(*linearg*)	add horizontal lines at specified y values
<u>x</u>line(*linearg*)	add vertical lines at specified x values

yline() and xline() are *merged-implicit*; see [G] **repeated options** and see *Interpretation of repeated options* below.

where *linearg* is

  numlist [, *suboptions*]

For a description of *numlist*, see [U] **14.1.8 numlist**. The *suboptions* are

suboptions	description
<u>ax</u>is(#)	which axis to use, $1 \le \# \le 9$
style(*addedlinestyle*)	overall style of added line
[<u>no</u>]extend	extend line through plot region's margins
<u>lst</u>yle(*linestyle*)	overall style of line
<u>lp</u>attern(*linepatternstyle*)	whether solid, dashed, etc.
<u>lw</u>idth(*linewidthstyle*)	thickness of line
<u>lc</u>olor(*colorstyle*)	color of line

See [G] *addedlinestyle*, [G] *linestyle*, [G] *linepatternstyle*, [G] *linewidthstyle*, and [G] *colorstyle*.

Description

yline() and xline() are used with twoway to add lines to the plot region.

Options

yline(*linearg*) and xline(*linearg*) specify the y and x values where lines should be added to the plot.

Suboptions

axis(*#*) is for use only when multiple y or x axes are being used (see [G] *axis_selection_options*). axis() specifies to which axis the yline() or xline() is to be applied.

style(*addedlinestyle*) specifies the overall style of the added line, which includes [no]extend and lstyle(*linestyle*) documented below. The [no]extend and lstyle() options allow you to change the added line's attributes individually, but style() is the starting point.

You need not specify style() just because there is something that you want to change and, in fact, most people seldom specify the style() option. You specify style() when another style exists that is exactly what you desire or when another style would allow you to specify fewer changes to obtain what you want.

extend and noextend specify whether the line should extend through the plot region's margin and so touch the axis; see [G] *region_options*. Usually noextend is the default and extend is the option, but that is determined by the overall style() and, of course, the scheme; see [G] **schemes**.

lstyle(*linestyle*), lpattern(*linepatternstyle*), lwidth(*linewidthstyle*), and lcolor(*colorstyle*) specify the look of the line; see [G] **graph twoway line**. lstyle() can be of particular use:

If you wish to create a line with the same look as the lines used to draw axes, specify lstyle(foreground).

If you wish to create a line with the same look as the lines used to draw grid lines, specify lstyle(grid).

Remarks

yline() and xline() add lines where specified. If, however, your interest is in obtaining grid lines, see the grid option in [G] *axis_label_options*.

Remarks are presented under the headings

> *Typical use*
> *Interpretation of repeated options*

Typical use

yline() or xline() are typically used to add reference values:

```
. scatter yvar xvar, yline(10)
. scatter yvar year, xline(1944 1989)
```

If we wanted to give the line in the first example the same look as used to draw an axis, we would specify

```
. scatter yvar xvar, yline(10, lstyle(foreground))
```

If we wanted to give the lines used in the second example the same look as used to draw grids, we would specify

```
. scatter yvar year, xline(1944 1989, lstyle(grid))
```

Interpretation of repeated options

Options `yline()` and `xline()` may be repeated and each is executed separately. Thus, different styles can be used for different lines on the same graph:

```
. scatter yvar year, xline(1944) xline(1989, lwidth(3))
```

Also See

Complementary: [G] *addedlinestyle*, [G] *colorstyle*, [G] *linestyle*, [G] *linepatternstyle*,
                        [G] *linewidthstyle*

Title

added_text_option — Option for adding text to twoway graphs

Syntax

The *added_text_option* is

added_text_option	description
text(*text_arg*)	add text at specified y x to graph

The above option is *merged-implicit*; see [G] **repeated options**.

where *text_arg* is

loc_and_text [loc_and_text ...] [, textoptions]

and *loc_and_text* is

$#_y$ $#_x$ "*text*" ["*text*" ...]

and where *textoptions* are

textoptions	description
yaxis(#)	how to interpret $#_y$
xaxis(#)	how to interpret $#_x$
placement(*compassdirstyle*)	where to locate relative to $#_y$ $#_x$
textbox_options	look of text

See [G] *compassdirstyle* and [G] *textbox_options*. Note that placement() is also a textbox option, but ignore the description of placement() found there in favor of the one below.

Description

text() adds the specified text to the specified location in the plot region.

Options

text(*text_arg*) specifies the location and text to be displayed.

Suboptions

yaxis(#) and xaxis(#) specify how $#_y$ and $#_x$ are to be interpreted when there are multiple y and x axis scales; see [G] *axis_selection_options*.

In the usual case, there is one y axis and one x axis, so options yaxis() and xaxis() are not specified. $#_y$ is specified in units of the y scale and $#_x$ in units of the x scale.

19

In the multiple-axis case, specify yaxis(#) and/or xaxis(#) to specify which units you wish to use. yaxis(1) and xaxis(1) are the defaults.

placement(*compassdirstyle*) specifies where the textbox is to be displayed relative to $\#_y$ $\#_x$. The default is usually placement(center). The default is controlled both by the scheme and by the *textbox_option* tstyle(*textboxstyle*); see [G] **schemes** and [G] *textbox_options*. The available choices are

placement()	location of text
c	centered on the point, vertically and horizontally
n	above the point, centered
ne	upper-right corner on the point
e	right of the point, vertically centered
se	lower-right corner on the point
s	below point, centered
sw	lower-left corner on the point
w	left of the point, vertically centered
nw	upper-left corner on the point

north	*northwest northeast*
west X *east*	X
south	*southwest southeast*

You can see [G] *compassdirstyle*, but that will just give you synonyms for c, n, ne, . . . , nw.

textbox_options specifies the look of the text; see [G] *textbox_options*.

Remarks

Remarks are presented under the headings

Typical use
Advanced use
Use of the textbox option width()

Typical use

text() is used for placing annotations on graphs. One example is the labeling of outliers. For instance, we type

```
. sysuse auto, clear
(1978 Automobile Data)
. twoway qfitci mpg weight, stdf || scatter mpg weight
(graph omitted)
```

and we notice four outliers. First, we find the outliers by typing

```
. quietly regress mpg weight
. predict hat
. predict s, stdf
. generate upper = hat + 1.96*s
```

```
. list make mpg weight if mpg>upper
```

	make	mpg	weight
13.	Cad. Seville	21	4,290
42.	Plym. Arrow	28	3,260
57.	Datsun 210	35	2,020
66.	Subaru	35	2,050
71.	VW Diesel	41	2,040

Now we can remake the graph and label the outliers:

```
. twoway qfitci  mpg weight, stdf ||
       scatter mpg weight, ms(O)
               text(41 2040 "VW Diesel", place(e))
               text(28 3260 "Plymouth Arrow", place(e))
               text(35 2050 "Datsun 210 and Subaru", place(e))
```

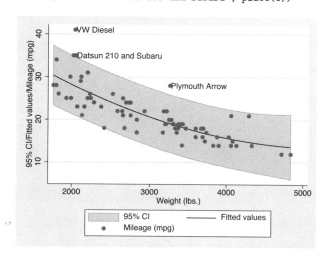

Advanced use

Another common use of *text* is to add an explanatory box of text inside the graph:

```
. sysuse uslifeexp, clear
(U.S. life expectancy, 1900-1999)
```

(*Continued on next page*)

```
. twoway line  le year ||
        fpfit le year ||
 , ytitle("Life Expectancy, years")
   xlabel(1900 1918 1940(20)2000)
   title("Life Expectancy at Birth")
   subtitle("U.S., 1900-1999")
   note("Source:  National Vital Statistics Report, Vol. 50 No. 6")
   legend(off)
   text( 48.5 1923
        "The 1918 Influenza Pandemic was the worst epidemic"
        "known in the U.S."
        "More citizens died than in all combat deaths of the"
        "20th century."
        , place(se) box just(left) margin(l+4 t+1 b+1) width(85) )
```

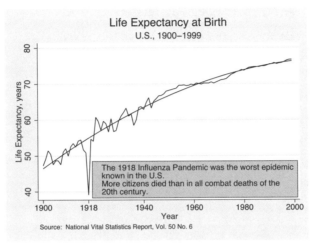

The only thing to note in the above command is the `text()` option:

```
text( 48.5 1923
        "The 1918 Influenza Pandemic was the worst epidemic"
        "known in the U.S."
        "More citizens died than in all combat deaths of the"
        "20th century."
        , place(se) box just(left) margin(l+4 t+1 b+1) width(85) )
```

and, in particular, we want to draw your eye to the location of the text and the suboptions:

```
text( 48.5 1923
        ...
        , place(se) box just(left) margin(l+4 t+1 b+1) width(85) )
```

We placed the text at $y = 48.5$, $x = 1923$, `place(se)`, meaning the lower-right corner of the box was located at $y = 48.5$, $x = 1923$.

The other suboptions, `box just(left) margin(l+4 t+1 b+1) width(85)`, are *textbox_options*. We specified `box` to draw a border around the textbox, and we specified `just(left)`—an abbreviation for `justification(left)`—so that the text was left-justified inside the box.

`margin(l+4 t+1 b+1)` made the text in the box look better. On the left we added 4%, and on the top and bottom we added 1%; see [G] ***textbox_options*** and [G] ***relativesize***.

`width(85)` was specified to solve the problem described below.

Use of the textbox option width()

Let us look at the results of the above command omitting the `width()` suboption. What you would see on your screen—or in a printout—might look virtually identical to the version we just drew, or it might look like this

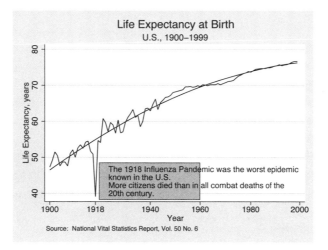

or like this:

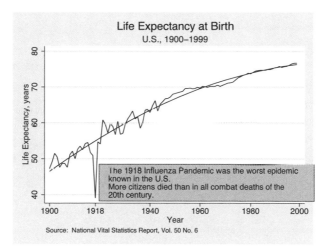

That is, Stata might make the textbox too narrow or too wide. In the above illustrations we have exaggerated the extent of the problem, but it is not uncommon for the box to run a little narrow or a little wide. Moreover, with respect to this one problem, how the graph appears on your screen is no guarantee as to how it will appear when printed.

This problem arises because Stata uses an approximation formula to determine the width of the text. This approximation is good for some fonts and poorer for others.

When the problem arises, use the *textbox_option* `width(`*relativesize*`)` to work around it. `width()` overrides Stata's calculation. In fact, we drew the two examples above by purposely misstating the `width()`. In the first case, we specified `width(40)`, and in the second, `width(95)`.

Getting the `width()` right is a matter of trial and error. The correct width will nearly always be between 0 and 100.

Corresponding to `width`(*relativesize*), there is also the *textbox_option* `height`(*relativesize*), but Stata never gets the height incorrect.

Also See

Complementary: [G] *compassdirstyle*, [G] *textbox_options*

Title

addedlinestyle — Choices for overall look of added lines

Syntax

addedlinestyle may be

addedlinestyle	description
default	determined by scheme
extended	extends through plot region margins
unextended	does not extend through margins

Other *addedlinestyles* may be available; type

    . graph query addedlinestyle

to obtain the full list installed on your computer.

Description

Added lines are those added by the *added_line_options*. *addedlinestyle* specifies the overall look of those lines. See [G] **added_line_options**.

Remarks

Remarks are presented under the headings

> *What is an added line?*
> *What is an addedlinestyle?*
> *You do not need to specify an addedlinestyle*

What is an added line?

Added lines are lines added by the *added_line_options* that extend across the plot region and perhaps across the plot region's margins, too.

What is an addedlinestyle?

Added lines are defined by two attributes:

1. whether the lines extend into the plot region's margin, and

2. the style of the lines, which includes the lines' thickness, color, and whether solid, dashed, etc.; see [G] **linestyle**.

The *addedlinestyle* specifics both of these attributes.

25

You do not need to specify an addedlinestyle

The *addedlinestyle* is specified in the options

yline(..., style(*addedlinestyle*) ...) xline(..., style(*addedlinestyle*) ...)

Correspondingly, there are other yline() and xline() suboptions that allow you to specify the individual attributes; see [G] *added_line_options*.

You specify the *addedlinestyle* when a style exists that is exactly what you desire or when another style would allow you to specify fewer changes to obtain what you want.

Also See

Complementary: [G] *added_line_options*

Title

advanced_options — Rarely specified options for use with graph twoway

Syntax

The *advanced_options* are

title_options	description
<u>yvarl</u>abel(*quoted_strings*)	respecify y-variable labels
<u>xvarl</u>abel(*quoted_string*)	respecify x-variable label
<u>yvarf</u>ormat(*%fmt* [...])	respecify y-variable formats
<u>xvarf</u>ormat(*%fmt*)	respecify x-variable format
recast(*newplottype*)	treat plot as *newplottype*

The above options are *rightmost*; see [G] **repeated options**.

where *quoted_string* is one quoted string and *quoted_strings* are one or more quoted strings such as

"*plot 1 label*"

"*plot 1 label*" "*plot 2 label*"

and where *newplottype* may be

newplottype	description
<u>sc</u>atter	treat as graph twoway scatter
<u>li</u>ne	treat as graph twoway line
<u>c</u>onnected	treat as graph twoway connected
bar	treat as graph twoway bar
spike	treat as graph twoway spike
dropline	treat as graph twoway dropline
dot	treat as graph twoway dot
rarea	treat as graph twoway rarea
rbar	treat as graph twoway rbar
rspike	treat as graph twoway rspike
rcap	treat as graph twoway rcap
rcapsym	treat as graph twoway rcapsym
rline	treat as graph twoway rline
rconnected	treat as graph twoway rconnected
rscatter	treat as graph twoway rscatter

Note: *newplottypes* in the second group (rarea though rscatter) may be recast only among themselves.

Description

The *advanced_options* are not so much advanced as they are difficult to explain and are rarely used. They are also invaluable when you need them.

Options

yvarlabel(*quoted_strings*) and xvarlabel(*quoted_string*) specify strings that are to be treated as if they were the variable labels of the first, second, ..., y variables and of the x variable.

yvarformat(*%fmt*) and xvarformat(*%fmt*) specify display formats that are to be treated as if they were the display formats of the first, second, ..., y variables and of the x variable.

recast(*newplottype*) specifies the new plottype to which the original graph twoway *plottype* command is to be recast.

Remarks

Remarks are presented under the headings

> Use of yvarlabel() and xvarlabel()
> Use of yvarformat() and xvarformat()
> Use of recast()

Use of yvarlabel() and xvarlabel()

When you type, for instance,

    . scatter mpg weight

the axes are titled using the variable labels of mpg and weight or, if the variables have no variable labels, using the names of the variables themselves. Options yvarlabel() and xvarlabel() allow you to specify strings that will be used in preference to both the variable label and the name.

    . scatter mpg weight, yvarl("Miles per gallon")

would label the y axis "Miles per gallon" (omitting the quotes) regardless of how variable mpg was labeled. Similarly,

    . scatter mpg weight, xvarl("Weight in pounds")

would label the x axis "Weight in pounds" regardless of how variable weight was labeled.

Obviously, you could specify both options.

In neither case will the actual variable label be changed. Options yvarlabel() and xvarlabel() treat the specified strings as if they were the variable labels. yvarlabel() and xvarlabel() arc quite literal in this treatment. If you specified, for instance, xvarlabel(""), the variable label would be treated as if it were nonexistent, and thus the variable name would be used to title the x axis.

What makes these two options "advanced" is that they affect not only the way axes are titled, but they literally substitute the specified strings for the variable labels wherever the variable label might be used. Variable labels are also used, for instance, in the construction of legends (see [G] *legend_option*).

Use of yvarformat() and xvarformat()

Options `yvarformat()` and `xvarformat()` work much like `yvarlabel()` and `xvarlabel()` except that, rather than overriding the variable labels, they override the variable formats. If you type

```
. scatter mpg weight, yvarformat(%9.2f)
```

the values on the y axis will be labeled 10.00, 20.00, 30.00, and 40.00 rather than 10, 20, 30, and 40.

Use of recast()

`scatter`, `line`, `histogram`, ... —the word that appears directly after `graph twoway`—is called a *plottype*. Plottypes come in two flavors: *base plottypes* and *derived plottypes*.

Base plottypes plot the data as given according to some style. `scatter` and `line` are examples of base plottypes.

Derived plottypes do not plot the data as given, but instead derive something from the data and then plot that according to one of the base plottypes. `histogram` is an example of a derived plottype. It derives from the data the values for the frequencies at certain x ranges, and then it plots that derived data using the base plottype `graph twoway bar`. `lfit` is another example of a derived plottype. It takes the data, fits a linear regression, and then passes that result along to `graph twoway line`.

`recast()` is useful when using derived plottypes. It specifies that the data are to be derived just as they would be ordinarily, but rather than passing the derived data to the default base plottype for plotting, they are passed to the specified base plottype.

For instance, if we typed

```
. twoway lfit mpg weight, pred(resid)
```

we would obtain a graph of the residuals as a line plot because the `lfit` plottype produces line plots. If we typed

```
. twoway lfit mpg weight, pred(resid) recast(scatter)
```

we would obtain a scatterplot of the residuals. `graph twoway lfit` would use `graph twoway scatter` rather than `graph twoway line` to plot the data it derives.

`recast(`*newplottype*`)` may be used with both derived and base plottypes, although it is most useful when combined with derived plots.

❏ **Technical Note**

The syntax diagram shown for `scatter` in [G] **graph twoway scatter**, although extensive, is incomplete, and so are all the other plottype syntax diagrams shown in this manual.

Consider what would happen if you specified

```
. scatter ... , ... recast(bar)
```

You would be specifying that `scatter` be treated as a `bar`. Results would be exactly the same as if you typed

```
. bar ... , ...
```

but let's ignore that and pretend that you typed the `recast()` version. What if you wanted to specify the look of the bars? You could type

```
. scatter ... , ... bar_options recast(bar)
```

That is, `scatter` allows *bar_options*, even though they do not appear in `scatter`'s syntax diagram. Similarly, `graph twoway bar` allows all of `scatter`'s options, even though they do not appears in bar's syntax diagram because you might type

> . bar ... , ... *scatter_options* recast(scatter)

The same is true for all other pairs of base plottypes, with the result that all base plottypes allow all base plottype options. The emphasis here is on base: the derived plottypes do not allow in this sharing.

If you use a base plottype without `recast()` and if you specify irrelevant options from other base types, that is not an error, but the irrelevant options are ignored. In the syntax diagrams for the base plottypes, we have listed only the options that matter under the assumption that you do not specify `recast`.

❏

Also See

Complementary: [G] **graph twoway**

Title

> *alignmentstyle* — Choices for vertical alignment of text

Syntax

alignmentstyles may be

alignmentstyle	description
baseline	bottom of textbox = baseline of letters
bottom	bottom of textbox = bottom of letters
middle	middle of textbox = middle of letters
top	top of textbox = top of letters

Other *alignmentstyles* may be available; type

        . graph query alignmentstyle

to obtain the full list installed on your computer.

Description

See [G] ***textbox_options*** for a description of textboxes. *alignmentstyle* specifies how the text is vertically aligned in a textbox. Think of the textbox as being horizontal even if it is vertical when specifying this option.

alignmentstyle is specified inside options such as the alignment() suboption of title() (see [G] ***title_options***):

        . graph ..., title("My title", alignment(*alignmentstyle*)) ...

In some cases, an *alignmentstylelist* is allowed. An *alignmentstylelist* is a sequence of *alignmentstyles* separated by spaces. Shorthands are allowed to make specifying the list easier; see [G] ***stylelists***.

Remarks

Think of the text as being horizontal even if it is not and think of the textbox as containing one line, such as

Hpqgxyz

alignment() specifies how the bottom of the textbox aligns with the bottom of the text.

alignment(baseline) specifies that the bottom of the textbox should be the baseline of the letters in the box. That would result in something like

....Hpqgxyz....

where dots represent the bottom of the textbox. Periods in most fonts are located on the baseline of letters, and note how the letters p, q, and g extend below the baseline.

alignment(`bottom`) specifies that the bottom of the textbox should be the bottom of the letters, which would be below the dots in the above example, lining up with the lowest part of the p, q, and g.

alignment(`middle`) specifies that the middle of the textbox should line up with the middle of a capital H. This is useful when you want to align text with a line.

alignment(`top`) specifies that the top of the textbox should line up with the top of a capital H.

Also See

Complementary: [G] *textbox_options*; [G] *justificationstyle*

Title

> *anglestyle* — Choices for the angle at which text is displayed

Syntax

anglestyle may be

anglestyle	description
horizontal	horizontal; reads left to right
vertical	vertical; reads bottom to top
rvertical	vertical; reads top to bottom
rhorizontal	horizontal; upside down
0	0 degrees; same as horizontal
45	45 degrees
90	90 degrees; same as vertical
180	180 degrees; same as rhorizontal
270 or -90	270 degrees; same as rvertical
#	# degrees; whatever you desire; # may be positive or negative

Note: Under Unix, only angles 0, 90, 180, and 270 display correctly on the screen. Angles are correctly rendered when printed.

Other *anglestyles* may be available; type

   . graph query anglestyle

to obtain the full list installed on your computer. If other *anglestyles* do exist, they are merely words attached to numeric values.

Description

anglestyle specifies the angle at which text is to be displayed.

Remarks

anglestyle is specified inside options such as the marker-label option mlabangle() (see [G] *marker_label_options*),

   . graph ..., ... mlabel(...) mlabangle(*anglestylelist*) ...

or the axis-label suboption angle() (scc [G] *axis_label_options*):

   . graph ..., ... ylabel(..., angle(*anglestyle*) ...) ...

In the case of mlabangle(), an *anglestylelist* is allowed. An *anglestylelist* is a sequence of *anglestyles* separated by spaces. Shorthands are allowed to make specifying the list easier; see [G] *stylelists*.

Also See

Complementary: [G] *marker_label_options*

Title

area_options — Options for specifying the look of special areas

Syntax

The *area_options* are

area_options	description
style(*areastyle*)	overall look of area
color(*colorstyle*)	outline and fill color
fcolor(*colorstyle*)	fill color
lstyle(*linestyle*)	overall look of outline
lcolor(*colorstyle*)	outline color
lwidth(*linewidthstyle*)	thickness of outline
lpattern(*linepatternstyle*)	whether outline solid, dashed, etc.

See [G] *areastyle*, [G] *colorstyle*, [G] *linestyle*, [G] *linewidthstyle*, and [G] *linepatternstyle*.

All options are *merged-implicit*; see [G] **repeated options**.

Description

The *area_options* determine the look of, for instance, the "rectangles" used by graph dot.

Options

style(*areastyle*) specifies the overall look of the area. The options listed below allow you to change each attribute, but style() provides the starting point.

You need not specify style() just because there is something you want to change. You specify style() when another style exists that is exactly what you desire or when another style would allow you to specify fewer changes to obtain what you want.

See [G] *areastyle* for a list of available area styles.

color(*colorstyle*) specifies a single color to be used both to outline the shape of the area and to fill its interior. See [G] *colorstyle* for a list of color choices.

fcolor(*colorstyle*) specifies the color to be used to fill the interior of the area. See [G] *colorstyle* for a list of color choices.

lstyle(*linestyle*) specifies the overall style of the line used to outline the area, which includes its pattern (solid, dashed, etc.), its thickness, and its color. The other options listed below allow you to change the line's attributes, but lstyle() is the starting point. See [G] *linestyle* for a list of choices.

lcolor(*colorstyle*) specifies the color to be used to outline the area. See [G] *colorstyle* for a list of color choices.

lwidth(*linewidthstyle*) specifies the thickness of the line to be used to outline the area. See
[G] ***linewidthstyle*** for a list of choices.

lpattern(*linepatternstyle*) specifies whether the line used to outline the area is solid, dashed, etc.
See [G] ***linepatternstyle*** for a list of pattern choices.

Remarks

If you specify graph dot's linetype(rectangle) option, the dot chart will be drawn with
rectangles substituted for the dots. In that case, the *area_options* determine the look of the rectangle.
The *area_options* are specified inside graph dot's rectangles() option:

    . graph dot ..., ... linetype(rectangle) rectangles(*area_options*) ...

If, for instance, you wanted to make the rectangles green, you could specify

    . graph dot ..., ... linetype(rectangle) rectangles(color(green)) ...

Also See

Complementary: [G] **graph dot**

Title

> *areastyle* — Choices for look of regions

Syntax

areastyle may be

areastyle	description
background	determined by scheme
foreground	determined by scheme
outline	foreground outline with no fill
histogram	default used for bars of histograms
ci	default used for confidence interval
ci2	default used for second confidence interval
none	no outline and no background color
p1–p15	used by first plot, second plot, ...

Other *areastyles* may be available; type

    . graph query areastyle

to obtain the full list installed on your computer.

Description

The shape of the area is determined by context. The *areastyle* determines whether the area is outlined and filled and, if so, how and in what color.

Remarks

areastyle is used to determine the look of

1. The entire region on which the graph appears
   (see option style(*areastyle*) in [G] *region_options*).

2. The look of bars
   (see option bstyle(*areastyle*) in [G] **graph bar** or [G] **graph twoway bar**).

3. The look of an area filled under a curve
   (see option bstyle() in [G] **graph twoway area** or [G] **graph twoway rarea**).

For an example of the use of the *areastyle* none, see *Suppressing the border around the plot region* in [G] *region_options*.

Also See

Complementary: [G] *region_options*; [G] **graph bar**, [G] **graph pie**, [G] **graph twoway area**, [G] **graph twoway bar**, [G] **graph twoway rarea**

Title

axis_label_options — Options for specifying axis labels

Syntax

axis_label_options are a subset of axis_options; see [G] **axis_options**. axis_label_options control the placement and the look of ticks and labels on an axis. The axis_label_options are

axis_label_options	description
{ y | x }label(rule_or_values)	major ticks plus labels
{ y | x }tick(rule_or_values)	major ticks only
{ y | x }mlabel(rule_or_values)	minor ticks plus labels
{ y | x }mtick(rule_or_values)	minor ticks only

The above options are *merged-explicit*; see [G] **repeated options**.

where *rule_or_values* is defined as

$$\big[\, rule \,\big] \; \big[\, numlist \; \big[\, "label" \; \big[\, numlist \; \big[\, "label" \; \big[\, \ldots \,\big]\big]\big]\big]\big] \; \big[\, ,\; suboptions \,\big]$$

where either *rule* or *numlist* must be specified and both may be specified. *rule* may be

rule	example	description
##	#6	approximately 6 nice values
###	##10	$10 - 1 = 9$ values between major ticks; allowed with mlabel() and mtick() only
#(#)#	-4(.5)3	specified range: -4 to 3 in steps of .5
minmax	minmax	minimum and maximum values
none	none	label no values
.	.	skip the rule

and where *numlist* is as described in [U] **14.1.8 numlist**,

and where *suboptions* are

(Continued on next page)

suboptions	description
<u>axis</u>(#)	which axis, $1 \leq \# \leq 9$
add	combine options
[<u>no</u>]ticks	suppress ticks
[<u>no</u>]labels	suppress labels
<u>val</u>uelabel	label values using first variable's value label
format(%*fmt*)	format values per %*fmt*
angle(*anglestyle*)	angle the labels
<u>alt</u>ernate	offset adjacent labels
tstyle(*tickstyle*)	labels & ticks: overall style
labgap(*relativesize*)	labels: margin between tick and label
labstyle(*textstyle*)	labels: overall style
<u>labs</u>ize(*textsizestyle*)	labels: size of text
<u>labc</u>olor(*colorstyle*)	labels: color of text
<u>tl</u>ength(*relativesize*)	ticks: length
<u>tp</u>osition(<u>outside</u> | <u>cr</u>ossing | <u>in</u>side)	ticks: position/direction
tlstyle(*linestyle*)	ticks: linestyle of
<u>tlw</u>idth(*linewidthstyle*)	ticks: thickness of line
<u>tlc</u>olor(*colorstyle*)	ticks: color of line
tlpattern(*linepatternstyle*)	ticks: line pattern of line
[no]grid	grid: whether to include
[no]gmin	grid: whether grid line at minimum
[no]gmax	grid: whether grid line at maximum
<u>gsty</u>le(*gridstyle*)	grid: overall style
[<u>no</u>]<u>ge</u>xtend	grid: extend into plot region margin
glstyle(*linestyle*)	grid: linestyle of
glwidth(*linewidthstyle*)	grid: thickness of line
glcolor(*colorstyle*)	grid: color of line
glpattern(*linepatternstyle*)	grid: line pattern of line

See [G] ***anglestyle***, [G] ***relativesize***, [G] ***textstyle***, [G] ***textsizestyle***, [G] ***colorstyle***, [G] ***tickstyle***, [G] ***linestyle***, [G] ***linewidthstyle***, [G] ***linepatternstyle***, and [G] ***gridstyle***.

Description

axis_label_options control the placement and the look of ticks and labels on an axis.

Options

ylabel(*rule_or_values*) and xlabel(*rule_or_values*) specify the major values to be labeled and ticked along the axis. For instance, to label the values 0, 5, 10, ..., 25 along the x axis, specify xlabel(0(5)25).

ytick(*rule_or_values*) and xtick(*rule_or_values*) specify the major values to be ticked but not labeled along the axis. For instance, to tick the values 0, 5, 10, ..., 25 along the x axis, specify xtick(0(5)25).

ymlabel(*rule_or_values*) and xmlabel(*rule_or_values*) specify minor values to be labeled and ticked along the axis.

ymtick(*rule_or_values*) and xmtick(*rule_or_values*) specify minor values to be ticked along the axis.

Suboptions

axis(#) specify to which scale this axis belongs and is specified when dealing with multiple x or y axes; see [G] *axis_selection_options*.

add specifies that what is specified is to be added to any previous xlabel(), ylabel(), xtick(), ..., or ymtick() option previously specified. See *Interpretation of repeated options* below.

noticks and ticks suppress/force the drawing of ticks. ticks is the usual default, so noticks makes $\{$ y | x $\}$label() and $\{$ y | x $\}$mlabel() display the labels only.

nolabels and labels suppress/force the display of the labels. labels is the usual default so nolabels turns $\{$ y | x $\}$label() into $\{$ y | x $\}$tick() and $\{$ y | x $\}$mlabel() into $\{$ y | x $\}$mtick(). Why anyone would want to do this is difficult to imagine.

valuelabel specifies that values should be mapped through the first y variable's value label (y*() options) or the x variable's value label (x*() options). Consider the command "scatter yvar xvar" and assume that xvar has been previously given a value label:

```
. label define cat 1 "Low" 2 "Med" 3 "Hi"
. label values xvar cat
```

Then

```
. scatter yvar xvar, xlabel(1 2 3, valuelabel)
```

would, rather than putting the numbers 1, 2, and 3, put the words Low, Med, and Hi on the x axis. It would have the same effect as

```
. scatter yvar xvar, xlabel(1 "Low" 2 "Med" 3 "Hi")
```

format(*%fmt*) specifies how numeric values on the axes should be formatted. The default format() is obtained from the variables specified with the graph command, which for ylabel(), ytick(), ymlabel(), and ymtick() usually means the first y variable, and for xlabel(), ..., xmtick(), means the x variable. For instance, in

```
. scatter y1var y2var xvar
```

the default format for the y axis would be y1var's format, and the default for the x axis would be xvar's format.

You may specify the format() suboption (or any suboption) without specifying values if you want the default labeling presented differently. For instance,

```
. scatter y1var y2var xvar, ylabel(,format(%9.2fc))
```

would present default labeling of the y axis but the numbers would be formatted with the %9.2fc format. Note carefully the comma in front of format. Inside the ylabel() option, we are specifying suboptions only.

angle(*anglestyle*) causes the labels to be presented at an angle. See [G] **anglestyle**.

alternate causes adjacent labels to be offset from one another and is useful when lots of values are being labeled. For instance, rather than obtaining

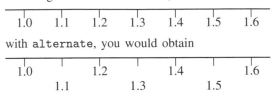

with alternate, you would obtain

tstyle() specifies the overall look of ticks and labels. The options documented below will allow you to change each attribute of a tick and its label, but the *tickstyle* specifies the starting point.

You need not specify tstyle() just because there is something you want to change about the look of ticks. You specify tstyle() when another style exists that is exactly what you desire or when another style would allow you to specify fewer changes to obtain what you want.

labgap(), labstyle(), labsize(), and labcolor() specify details about how the labels are presented.

tlength(*relativesize*) specifies the overall length of the ticks.

tposition(outside | crossing | inside) specifies whether the ticks are to extend outside (from the axis out, the usual default), are to extend crossing (crossing the axis line, extending in and out), or are to extend inside (from the axis into the plot region).

tlstyle(), tlwidth(), tlcolor(), tlpattern() specify other details about the look of the ticks. Ticks are just lines. See [G] **lines** for more information.

grid (and nogrid) specify whether grid lines are to be drawn across the plot region in addition to whatever else is specified in the $\{\,y\,|\,x\,\}\,\lceil m \rceil$label() or $\{\,y\,|\,x\,\}\,\lceil m \rceil$tick() option in which grid or nogrid appears. Typically, nogrid is the default and grid is the option for all except ylabel(), where things are reversed and grid is the default and nogrid is the option. (Which is the default and which is the option is controlled by the scheme; see [G] **schemes**.

For instance, specifying option

ylabel(, nogrid)

would suppress the grid lines in the y direction and specifying

xlabel(, grid)

would add them in the x. Specifying

xlabel(0(1)10, grid)

would place major labels, major ticks, and grid lines at $x = 0, 1, 2, \ldots, 10$.

$\lceil no \rceil$gmin and $\lceil no \rceil$gmax are relevant only if grid is in effect (because grid is the default and nogrid was not specified or because grid was specified). $\lceil no \rceil$gmin and $\lceil no \rceil$gmax specify whether grid lines are to be drawn at the minimum and maximum values. Consider

    . scatter yvar xvar, xlabel(0(1)10, grid)

Clearly the values 0, 1, ..., 10 are to be ticked and labeled, and clearly, grid lines should be drawn at 1, 2, ..., 9; but should grid lines be drawn at 0 and 10? If 0 and 10 are at the edge of the plot region, you probably do not want grid lines there. They will be too close to the axis and border of the graph.

What you want will differ from graph to graph and the `graph` command tries to be smart, which is to say, neither `gmin` nor `nogmin` (and neither `gmax` nor `nogmax`) are the defaults: The default is for `graph` to decide which looks best; the options force the decision one way or the other.

If `graph` decided to suppress the grids at the extremes and you wanted them, you could type

    . scatter yvar xvar, xlabel(0(1)10, grid gmin gmax)

`gstyle(`*gridstyle*`)` specifies the overall style of the grid lines, including whether the lines extend beyond the plot region and into the plot region's margins, along with the style, color, width, and pattern of the lines themselves. The options that follow allow you to change each attribute, but the *gridstyle* provides the starting point.

You need not specify `gstyle()` just because there is something you want to change. You specify `gstyle()` when another style exists that is exactly what you desire or when another style would allow you to specify fewer changes to obtain what you want.

`gextend` and `nogextend` specify whether the grid lines should extend beyond the plot region and pass through the plot region's margins; see [G] *region_options*. The default is determined by the `gstyle()` and scheme, but in most cases, `nogextend` is the default and `gextend` is the option.

`glstyle()`, `glwidth()`, `glcolor()`, `glpattern()` specify other details about the look of the grid. Grids are just lines. See [G] **lines** for more information. Of these options, `glpattern()` is of particular interest because, with it, you can make the grid lines dashed.

Remarks

axis_label_options are a subset of *axis_options*; see [G] **axis_options** for an overview. The other appearance options are

axis_scale_options	(see [G] **axis_scale_options**)
axis_title_options	(see [G] **axis_title_options**)

Remarks are presented under the headings

> Default labeling and ticking
> Controlling the labeling and ticking
> Adding extra ticks
> Adding minor labels and ticks
> Adding grid lines
> Suppressing grid lines
> Substituting text for labels
>
> Appendix: Details of syntax
> > Suboptions without rules, numlists, or labels
> > Rules
> > Rules and numlists
> > Rules and numlists and labels
> > Interpretation of repeated options

Default labeling and ticking

By default, approximately five values are labeled and ticked on each axis. For example, in

```
. sysuse auto
(1978 Automobile Data)
. scatter mpg weight
```

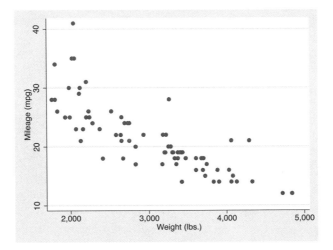

four values are labeled on each axis because choosing five would have required widening the scale too much.

Controlling the labeling and ticking

We would obtain the same results as we did in the above example if we typed,

```
. scatter mpg weight, ylabel(#5) xlabel(#5)
```

Options `ylabel()` and `xlabel()` specify the values to be labeled and ticked and `#5` specifies that Stata choose approximately five values for us. If we wanted lots of values labeled, we might type

```
. scatter mpg weight, ylabel(#10) xlabel(#10)
```

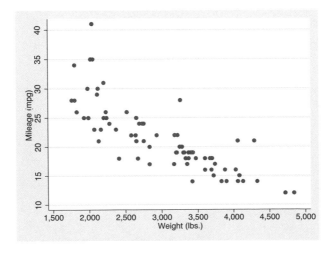

As with #5, #10 was not taken too seriously; we obtained 7 labels on the y axis and 8 on the x.

Alternatively, we can specify precisely the values we want labeled by specifying #(#)# or by specifying a list of numbers:

```
. scatter mpg weight, ylabel(10(5)45)
                      xlabel(1500 2000 3000 4000 4500 5000)
```

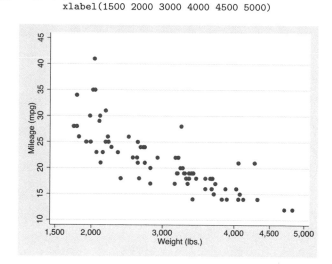

In option `ylabel()` we specified the rule 10(5)45, which means 10 to 45 in steps of 5. In option `xlabel()`, we typed out the values to be labeled.

Adding extra ticks

Options `ylabel()` and `xlabel()` draw ticks plus labels. Options `ytick()` and `xtick()` draw ticks only, so you can do things like

```
. scatter mpg weight, ytick(#10) xtick(#15)
```

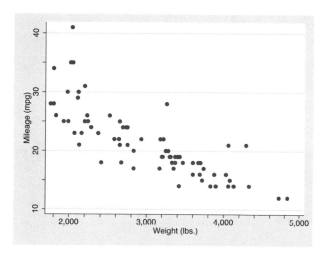

Of course, as with `ylabel()` and `xlabel()`, you can specify the exact values you want ticked.

Adding minor labels and ticks

Minor ticks and minor labels are smaller than regular ticks and regular labels. Options `ymlabel()` and `xmlabel()` allow for placement of minor ticks-plus-labels and `ymtick()` and `xmtick()` allow for placement of minor ticks without labels. When using minor ticks and labels, in addition to the usual syntax of #5 to mean approximately 5 values, 10(5)45 to mean 10 to 45 in steps of 5, and a list of numbers, there is an additional syntax: ##5. ##5 means $5 - 1 = 4$ minor ticks or labels between the major ticks. Why ##5 means 4 and not 5 we will explain momentarily.

The graph below is intended more for demonstration than an example of a good-looking graph:

```
. scatter mpg weight, ymlabel(##5) xmtick(##10)
```

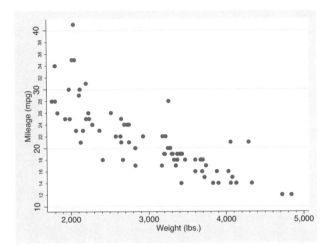

##5 means 4 ticks and ##10 means 9 ticks because most people think in reciprocals: They say to themselves, "I want to tick the fourths so I want 4 ticks between", or "I want to tick the tenths so I want 10 ticks between". They think incorrectly. They should think that if they want fourths, they want $4 - 1 = 3$ ticks between, or if they want tenths, they want $10 - 1 = 9$ ticks between. Stata subtracts one so that they can think—and correctly—when they want fourths that they want ##4 ticks between and that when they want tenths they want ##10 ticks between.

Adding grid lines

To obtain grid lines, specify the `grid` suboption of `ylabel()`, `xlabel()`, `ymlabel()`, or `xmlabel()`. `grid` specifies that, in addition to whatever else the option would normally do, grid lines are to be drawn at the same values. Note that in the example below,

```
. sysuse uslifeexp
(U.S. life expectancy, 1900-1999)
```

. line le year, xlabel(,grid)

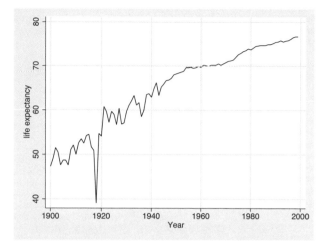

we specify `xlabel(,grid)`, omitting any mention of the specific values to use. Thus, `xlabel()` did what it does ordinarily (labeled approximately 5 nice values) and it drew grid lines at those same values.

Of course, we could have specified the values to be labeled and gridded:

```
. line le year, xlabel(#10, grid)
. line le year, xlabel(1900(10)2000, grid)
. line le year, xlabel(1900 1918 1940(20)2000, grid)
```

The `grid` suboption is usually specified with `xlabel()` (and with `ylabel()` if, given the scheme, `grid` is not the default), but it may be specified with any of the *axis_label_options*. In the example below, we "borrow" `ymtick()` and `xmtick()`, specify `grid` to make them draw grids, and also specify `style(none)` to make the ticks themselves invisible:

```
. sysuse auto, clear
(1978 Automobile Data)
. scatter mpg weight, ymtick(#20, grid tstyle(none))
                      xmtick(#20, grid tstyle(none))
```

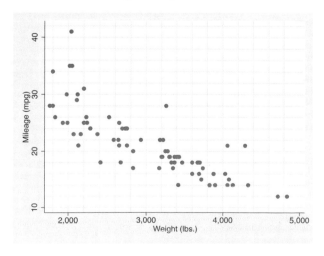

If you look carefully at the graph above, you will find no grid line was drawn at $x = 5,000$. Stata suppresses grid lines when they get too close to the axes or borders of the graph. If you want to force Stata to draw them anyway, you can specify the `gmin` and `gmax` options:

```
. scatter mpg weight, ymtick(#20, grid tstyle(none))
                      xmtick(#20, grid tstyle(none) gmax)
```

Suppressing grid lines

Some commands, and option `ylabel()`, usually draw grids by default. For instance, in the following, results are as if you specified `ylabel(,grid)`:

```
. sysuse auto, clear
(1978 Automobile Data)
. scatter mpg weight, by(foreign)
```

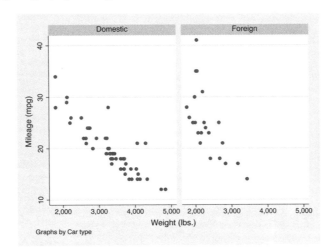

To suppress the grid lines, specify `ylabel(,nogrid)`:

```
. scatter mpg weight, by(foreign) ylabel(,nogrid)
```

Substituting text for labels

In addition to specifying explicitly the values to be labeled by specifying things such as `ylabel(10(10)50)` or `ylabel(10 20 30 40 50)`, you can specify text to be substituted for the label. If you type

```
. graph ... , ... ylabel(10 20 30 "mean" 40 50)
```

The values 10, 20, ..., 50 will be labeled, just as you would expect, but for the middle value, rather than the text "30" appearing, the text "mean" (without the quotes) would appear.

In the advanced example below we specify

```
xlabel(1 "J" 2 "F" 3 "M" 4 "A" 5 "M" 6 "J" 7 "J" 8 "A" 9 "S" 10 "O" 11 "N" 12 "D")
```

so that rather than seeing the numbers 1, 2, ..., 12 (which are month numbers), we see J, F, ..., D; and we specify

```
ylabel(12321 "12,321 (mean)", axis(2) angle(0))
```

so that we label 12321 but, rather than seeing 12321, we see "12,321 (mean)". The axis(2) option puts the label on the second y axis (see [G] *axis_selection_options*) and angle(0) makes the text appear horizontally rather than vertically (see *Options* above):

```
. sysuse sp500, clear
(S&P 500)

. generate month = month(date)

. sort month

. by month: egen lo = min(volume)

. by month: egen hi = max(volume)

. format lo hi %10.0gc

. summarize volume
```

Variable	Obs	Mean	Std. Dev.	Min	Max
volume	248	12320.68	2585.929	4103	23308.3

```
. by month: keep if _n==_N
(236 observations deleted)

. twoway rcap lo hi month,
          xlabel(1 "J"  2 "F"  3 "M"  4 "A"  5 "M"  6 "J"
                 7 "J"  8 "A"  9 "S" 10 "O" 11 "N" 12 "D")
          xtitle("Month of 2001")
          ytitle("High and Low Volume")
          yaxis(1 2) ylabel(12321 "12,321 (mean)", axis(2) angle(0))
          ytitle("", axis(2))
          yline(12321, lstyle(foreground))
          msize(*2)
          title("Volume of the S&P 500", margin(b+2.5))
          note("Source:  Yahoo!Finance and Commodity Systems Inc.")
```

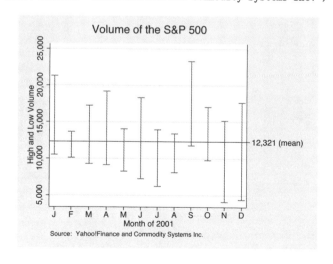

Appendix: Details of syntax

Suboptions without rules, numlists, or labels

What may appear in each of the options $\{\,y\,|\,x\,\}\{\,\texttt{label}\,|\,\texttt{tick}\,|\,\texttt{mlabel}\,|\,\texttt{mtick}\,\}()$ is a rule or numlist followed by suboptions:

$$\big[\,rule\,\big]\ \big[\,numlist\ \big[\,\texttt{"}label\texttt{"}\ \big[\,numlist\ \big[\,\texttt{"}label\texttt{"}\ \big[\,\ldots\,\big]\,\big]\,\big]\,\big]\,\big]\ \big[\,,\ suboptions\,\big]$$

Note that *rule*, *numlist*, and *label* are optional. Remove those and you are left with

, *suboptions*

That is to say, the options $\{\,\texttt{y}\,|\,\texttt{x}\,\}\{\,\texttt{label}\,|\,\texttt{tick}\,|\,\texttt{mlabel}\,|\,\texttt{mtick}\,\}()$ may be specified with just suboptions and, in fact, they are often specified that way. If you want default labeling of the y axis and x axis, but you want grid lines in the x direction as well as the y, specify

. `scatter` *yvar xvar* , `xlabel(,grid)`

When you do not specify the first part—the *rule*, *numlist*, and *label*—you are saying that you do not want that part to change. You are saying that you merely wish to change how the *rule*, *numlist*, and *label* are displayed.

Of course, you may specify more than one suboption. You might type

. `scatter` *yvar xvar*, `xlabel(,grid format(%9.2f))`

if, in addition to grid lines, you wanted the numbers presented on the x axis to be presented in a `%9.2f` format.

Rules

What may appear in each of the axis-label options is a rule or numlist

$$\big[\,rule\,\big]\ \big[\,numlist\ \big[\,\texttt{"}label\texttt{"}\ \big[\,numlist\ \big[\,\texttt{"}label\texttt{"}\ \big[\,\ldots\,\big]\,\big]\,\big]\,\big]\,\big]\ \big[\,,\ suboptions\,\big]$$

where either *rule* or *numlist* must be specified and both may be specified. Let us ignore the `"`*label*`"` part right now. Then the above simplifies to

$$\big[\,rule\,\big]\ \big[\,numlist\,\big]\ \big[\,,\ suboptions\,\big]$$

where *rule* or *numlist* must be specified, both may be specified, and most often you will simply specify the *rule*. *rule* may be any of the following:

rule	example	description
`##`	`#6`	6 nice values
`###`	`##10`	$10 - 1 = 9$ values between major ticks; allowed with `mlabel()` and `mtick()` only
`#(#)#`	`-4(.5)3`	specified range: -4 to 3 in steps of .5
`minmax`	`minmax`	minimum and maximum values
`none`	`none`	label no values
`.`	`.`	skip the rule

The most commonly specified rules are `##` and `###`.

Specifying `##` says to choose # nice values. Specifying `#5` says to choose five nice values, `#6` means to choose six, and so on. If you specify `ylabel(#5)`, then five values will be labeled (on the y axis). If you also specify `ymtick(#10)`, then ten minor ticks will also be placed on the axis. Actually, `ylabel(#5)` and `ymtick(#10)` will result in approximately five labels and ten minor ticks because the choose-a-nice-number routine will change your choice a little if, in its opinion, that would result in a nicer overall result. You may not agree with the routine about what is nice and, in that case, the `#(#)#` rule will let you specify exactly what you want, assuming that what you want are evenly spaced labels and numbers.

is allowed only with the {y|x}mlabel() and {y|x}mtick() options—the options that result in minor ticks. ### says to put #−1 minor ticks between the major ticks. ##5 would put four and ##10 would put nine. In this case, # is taken seriously, at least after subtraction, and you are given exactly what you request.

#(#)# can be used with major or minor labels and ticks. It says to label the first number specified, increment by the second number, and keep labeling as long as the result is less than or equal to the last number specified. ylabel(1(1)10) would label (and tick) the values 1, 2, ..., 10. ymtick(1(.5)10) would put minor ticks at 1, 1.5, 2, 2.5, ..., 10. It would be perfectly okay to specify both of those options. When specifying rules, minor ticks and labels will check what is specified for major ticks and labels and remove the intersection so as not to overprint. The results will be just as if you specified ymtick(1.5(1)9.5).

The rule minmax specifies that you want the minimum and maximum. ylabel(minmax) would label only the minimum and maximum.

Rule none means precisely that: the rule that results in no labels and no ticks.

Rule . only makes sense when add is specified, although it is allowed at other times and then . means the same as none.

Rules and numlists

Following the *rule*—or instead of it—you can specify a *numlist*. A numlist is a list of numbers: "1 2 5 7" (without the quotes) is a numlist. Other shorthands are allowed (see [U] **14.1.8 numlist**) and, in fact, one of *numlist*'s syntaxes looks just like a *rule*: #(#)#. It has the same meaning, too.

There is, however, a subtle distinction between, for example,

ylabel(1(1)10) (a *rule*) and ylabel(none 1(1)10) (a *numlist*)

Rules are more efficient. Visually, however, there is no difference.

Use numlists when the values that you wish to label or to tick are unequally spaced,

ylabel(none 1 2 5 7)

or when there is one or a few extra values that you want to label or to tick:

ylabel(1(1)10 3.5 7.5)

Rules and numlists and labels

Numlists serve an additional purpose—you can specify text that is to be substituted for the value to be labeled. For instance,

ylabel(1(1)10 3.5 "Low" 7.5 "Hi")

says to label 1, 2, ..., 10 (that is the *rule* part) and in addition, to label the special values 3.5 and 7.5. Rather than actually printing "3.5" and "7.5" next to the ticks at 3.5 and 7.5, however, graph will instead print the words "Low" and "Hi".

Interpretation of repeated options

Each of the axis-label options may be specified more than once in the same command. If you do that and you do not specify suboption add, it is the rightmost of each that is honored. If you specify suboption add, then the option just specified and the previous options are merged; see [G] **repeated options**.

Also See

Complementary: [G] *axis_options*; [G] *axis_scale_options*, [G] *axis_title_options*

Title

axis_options — Options for specifying numeric axes	

Syntax

The *axis_options* are

axis_scale_options	description
{ y	x }scale(*axis_description*)	log scales, range, appearance

See [G] ***axis_scale_options***.

axis_label_options	description
{ y	x }label(*rule_or_values*)	major ticks plus labels
{ y	x }tick(*rule_or_values*)	major ticks only
{ y	x }mlabel(*rule_or_values*)	minor ticks plus labels
{ y	x }mtick(*rule_or_values*)	minor ticks only

(also allows control of grid lines; see [G] ***axis_label_options***)

axis_title_options	description
{ y	x }title(*axis_title*)	specify axis title

See [G] ***axis_title_options***.

Description

Axes are the graphical elements that indicate the scale.

Options

yscale() and xscale() specify how the *y* and *x* axes are scaled (arithmetic, log, reversed), the range of the axes, and the look of the lines that are the axes. See [G] ***axis_scale_options***.

ylabel(), ytick(), ymlabel(), ymtick(), and xlabel(), ..., xmtick() specify how the axes should be labeled and ticked. These options allow you to control the placement of major and minor ticks and labels. In addition, these options allow you to add or to suppress grid lines on your graphs. See [G] ***axis_label_options***.

ytitle() and xtitle() specify the titles to appear next to the axes. See [G] ***axis_title_options***.

Remarks

Numeric axes are allowed with graph twoway and graph matrix and are allowed for one of the axes of graph barchart, graph dotchart, and graph boxplot. How the numeric axes look is affected by the *axis_options*.

Remarks are presented under the headings

> Use of axis-appearance options with graph twoway
> Multiple y and x scales
> Axis on the left, axis on the right?

Use of axis-appearance options with graph twoway

When you type

```
. scatter yvar xvar
```

the resulting graph will have y and x axes. How the axes look will be determined by the scheme; see [G] **schemes**. The *axis_options* allow you to modify the look of the axes in terms of whether the y axis is on the left or on the right, whether the x axis is on the bottom or on the top, the number of major and minor ticks that appear on each axis, the values that are labeled, and the titles that appear along each.

For instance, you might type

```
. scatter yvar xvar, ylabel(#6) ymtick(##10) ytitle("values of y") xlabel(#6)
> xmtick(##10) xtitle("values of x")
```

to draw a graph of `yvar` versus `xvar`, putting on each axis approximately 6 labels and major ticks, 10 minor ticks between major ticks, and labeling the y axis "values of y" and the x axis "values of x".

```
. scatter yvar xvar, ylabel(0(5)30) ymtick(0(1)30) ytitle("values of y")
> xlabel(0(10)100) xmtick(0(5)100) xtitle("values of x")
```

would draw the same graph, putting major ticks on the y axis at 0, 5, 10, ..., 30 and minor ticks at every integer over the same range, and putting major ticks on the x axis at 0, 10, ..., 100 and minor ticks at every 5 units over the same range.

The way we have illustrated it, it appears that the axis options are options of `scatter`, but that is not so. In this case, they are options of `twoway`, and the "right" way to write the last command is

```
. twoway (scatter yvar xvar), ylabel(0(5)30) ymtick(0(1)30) ytitle("values of y")
> xlabel(0(10)100) xmtick(0(5)100) xtitle("values of x")
```

The parentheses around `(scatter yvar xvar)` and the placing of the axis-appearance options outside the parentheses makes clear that the options are aimed at `twoway` rather than at `scatter`. Whether you use the ||-separator notation or the ()-binding notation makes no difference, but it is important to understand that there is only one set of axes, especially when you type more complicated commands, such as

```
. twoway (scatter yvar xvar)
         (scatter y2var x2var)
                 , ylabel(0(5)30) ymtick(0(1)30) ytitle("values of y")
                 xlabel(0(10)100) xmtick(0(5)100) xtitle("values of x")
```

There is one set of axes in the above, and it just so happens that both `yvar` versus `xvar` and `y2var` versus `x2var` appear on it. You are free to type the above command how you please, such as

```
. scatter yvar   xvar ||
    scatter y2var x2var ||,
          ylabel(0(5)30) ymtick(0(1)30) ytitle("values of y")
          xlabel(0(10)100) xmtick(0(5)100) xtitle("values of x")
```

or

```
. scatter yvar   xvar ||
    scatter y2var x2var, ylabel(0(5)30) ymtick(0(1)30)
                        ytitle("values of y") xlabel(0(10)100)
                        xmtick(0(5)100) xtitle("values of x")
```

or

```
. scatter yvar xvar, ylabel(0(5)30) ymtick(0(1)30)
                        ytitle("values of y") xlabel(0(10)100)
                        xmtick(0(5)100) xtitle("values of x") ||
    scatter y2var x2var
```

All of the above result in the same graph, even though the last makes it appear that the axis options are associated with just the first `scatter`, and the next to the last makes it appear that they are associated with just the second. However you type it, the command is really `twoway`. `twoway` draws twoway graphs, twoway graphs have one set of axes (or one set per by-group), and all the plots that appear on the twoway graph share that set.

Multiple y and x scales

Actually, a twoway graph can have more than one set of axes. Consider the command:

```
. twoway (scatter yvar xvar) (scatter y2var x2var, yaxis(2))
```

The above graphs `yvar` versus `xvar` and `y2var` versus `x2var`, but two y scales are provided. The first (which will appear on the left) applies to `yvar` and the second (which will appear on the right) applies to `y2var`. The `yaxis(2)` option says that the y axis of the specified scatter is to appear on the second y scale.

See [G] *axis_selection_options*.

Axis on the left, axis on the right?

When there is only one y scale, whether the axis appears on the left or the right is determined by the scheme; see [G] **schemes**. The default scheme puts the y axis on the left, but the scheme that mirrors the style used by *The Economist* puts it on the right:

```
scatter yvar xvar, scheme(economist)
```

Specifying `scheme(economist)` will change other things about the appearance of the graph, too. If you just want to move the y axis to the right, you can type

```
scatter yvar xvar, yscale(alt)
```

As explained in *axis_scale_options*, `yscale(alt)` switches the axis from one side to the other, so if you typed

```
scatter yvar xvar, scheme(economist) yscale(alt)
```

you would get *The Economist* scheme but with the y axis on the left.

`xscale(alt)` switches the x axis from the bottom to the top, or the top to the bottom; see [G] *axis_scale_options*.

Also See

Complementary: [G] *axis_label_options*, [G] *axis_scale_options*, [G] *axis_title_options*;
[G] *axis_selection_options*; [G] *region_options*

Title

> *axis_scale_options* — Options for specifying axis scale, range, and look

Syntax

axis_scale_options are a subset of *axis_options*; see [G] **axis_options**. The *axis_scale_options* are

axis_scale_options	description
yscale(*axis_suboptions*)	how y axis looks
xscale(*axis_suboptions*)	how x axis looks

The above options are *merged-implicit*; see [G] **repeated options**.

where *axis_suboptions* are

axis_suboptions	description
axis(#)	which axis to modify; $1 \leq \# \leq 9$
[no]log	use logarithmic scale
[no]reverse	reverse scale to run from max to min
range(*numlist*)	expand range of axis
off and on	suppress/force display of axis
fill	allocate space for axis even if off
alt	move axis from left to right or from top to bottom
fextend	extend axis line through plot region and plot region's margin
extend	extend axis line through plot region
noextend	do not extend axis line at all
noline	do not even draw axis line
line	force drawing of axis line
titlegap(*relativesize*)	margin between axis title and tick labels
outergap(*relativesize*)	margin outside of axis title
lstyle(*linestyle*)	overall style of axis line
lcolor(*colorstyle*)	color of axis line
lwidth(*linewidthstyle*)	thickness of axis line
lpattern(*linepatternstyle*)	whether axis solid, dashed, etc.

See [U] **14.1.8 numlist**, [G] *relativesize*, [G] *linestyle*, [G] *colorstyle*, [G] *linewidthstyle*, and [G] *linepatternstyle*.

Description

The *axis_scale_options* determine how axes are scaled (arithmetic, log, reversed), the range of the axes, and the look of the lines that are the axes.

Options

yscale(*axis_suboptions*) and xscale(*axis_suboptions*) specify the look of the y and x axes. Inside the parentheses, you specify *axis_suboptions*.

yscale() and xscale() suboptions

axis(*#*) specifies to which scale this axis belongs and is specified when dealing with multiple y or x axes; see [G] *axis_selection_options*.

log and nolog specify whether the scale should be logarithmic or arithmetic. nolog is the usual default, and thus log is the option. See *Obtaining log scales* under *Remarks* below.

reverse and noreverse specify whether the scale should run from the maximum to the minimum or from the minimum to the maximum. noreverse is the usual default, and thus reverse is the option. See *Obtaining reversed scales* under *Remarks* below.

range(*numlist*) specifies that the axis is to be expanded to include the numbers specified. Missing values, if specified, are ignored. See *Specifying the range of a scale* under *Remarks* below.

off and on suppress or force the display of the axis. on is the default and off the option. See *Suppressing the axes* under *Remarks* below.

fill goes with off and is seldom specified. If you turned an axis off but still wanted the space to be allocated for the axis, you could specify fill.

alt specifies that if the axis is by default on the left, it is to be on the right; if it is by default on the bottom, it is to be on the top. The following would draw a scatterplot with the y axis on the right:

    . scatter yvar xvar, yscale(alt)

fextend, extend, noextend, line, and noline determine how much of the line representing the axis is to be drawn. They are alternatives.

noline specifies that the line is not to be drawn at all. The axis is there, ticks and labels will appear, but the line that is the axis itself will not be drawn.

line is the opposite of noline, for use if the axis line somehow got turned off.

noextend specifies that the axis line should not extend beyond the range of the axis. Pretend that the axis extends from -1 to $+20$. With noextend, the axis line begins at -1 and ends at $+20$.

extend specifies that the line should be longer than that and extend all the way across the plot region. For instance, -1 and $+20$ might be the extent of the axis, but the scale might extend from -5 to $+25$, with the range $[-5, -1)$ and $(20, 25]$ being unlabeled on the axis. With extend, the axis line begins at -5 and ends at 25.

fextend specifies that the line should be longer than that and extend across the plot region and across the plot region's margins. For a definition of the plot region's margins, see [G] *region_options*. If the plot region has no margins (which would be rare), then fextend means the same as extend. If the plot region does have margins, extend would result in the y and x axes not meeting. With fextend, they touch.

fextend is the default with most schemes.

titlegap(*relativesize*) specifies the margin to be inserted between the axis title and the axis' tick labels.

outergap(*relativesize*) specifies the margin to be inserted outside of the axis title.

lstyle(*linestyle*), lcolor(*colorstyle*), lwidth(*linewidthstyle*), and lpattern(*linepatternstyle*) determine the overall look of the line that is the axis; see [G] **lines**.

Remarks

Remarks are presented under the headings

> *Use of the yscale() and xscale()*
> *Specifying the range of a scale*
> *Obtaining log scales*
> *Obtaining reversed scales*
> *Suppressing the axes*

axis_scale_options are a subset of *axis_options*; see [G] ***axis_options*** for an overview. The other appearance options are

axis_label_options	(see [G] ***axis_label_options***)
axis_title_options	(see [G] ***axis_title_options***)

Use of the yscale() and xscale()

yscale() and xscale() specify the look of the y and x axes. Inside the parentheses, you specify *axis_suboptions*, for example:

    . twoway (scatter ...) ... , yscale(range(0 10) titlegap(1))

Note that yscale() and xscale() may be abbreviated ysc() and xsc(), suboption range() may be abbreviated r(), and titlegap() may be abbreviated titleg():

    . twoway (scatter ...) ... , ysc(r(0 10) titleg(1))

Multiple yscale() and xscale() may be specified on the same command and results of the multiple options will be combined. Thus, the above command could also be specified

    . twoway (scatter ...) ... , ysc(r(0 10)) ysc(titleg(1))

Suboptions may also be specified more than once, either within a single yscale() or xscale() option, or across multiple options, and it will be the rightmost suboption that takes effect. In the following command, titlegap() will be 2, and range() 0 and 10:

    . twoway (scatter ...) ... , ysc(r(0 10)) ysc(titleg(1)) ysc(titleg(2))

Specifying the range of a scale

To specify the range of a scale, specify the $\{\,y\mid x\,\}$scale(range(*numlist*)) option. The option specifies that the axis is to be expanded to include the numbers specified.

Consider the graph

    . scatter *yvar* *xvar*

Assume that it resulted in a graph where the y axis varied over 1 to 100 and assume that, given the nature of the y variable, it would be more natural if the range of the axis were expanded to go from 0 to 100. You could type

. scatter *yvar xvar*, ysc(r(0))

Similarly, if the range without the yscale(range()) option went from 1 to 99 and you wanted it to go from 0 to 100, you could type

. scatter *yvar xvar*, ysc(r(0 100))

If the range without yscale(range()) went from 0 to 99 and you wanted it to go from 0 to 100, you could type

. scatter *yvar xvar*, ysc(r(100))

Specifying missing for a value leaves the current minimum or maximum unchanged; specifying a nonmissing value changes the range, but only if the specified value is outside the value that would otherwise have been chosen. range() never narrows the scale of an axis or causes data to be omitted from the plot. If you wanted to graph yvar versus xvar for the subset of xvar values between 10 and 50, typing

. scatter *yvar xvar*, xsc(r(10 50))

would not suffice. You need to type

. scatter *yvar xvar* if *xvar* >=10 & *xvar* <=50

Obtaining log scales

To obtain log scales specify the $\{\,y\,|\,x\,\}$scale(log) option. Ordinarily when you draw a graph, you obtain arithmetic scales:

. sysuse lifeexp, clear
(Life expectancy, 1998)
. scatter lexp gnppc

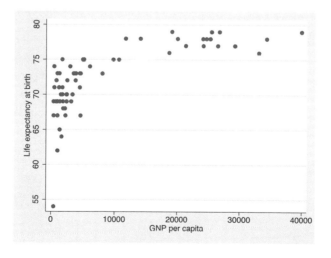

To obtain the same graph with a log x scale, we type

```
. scatter lexp gnppc, xscale(log)
```

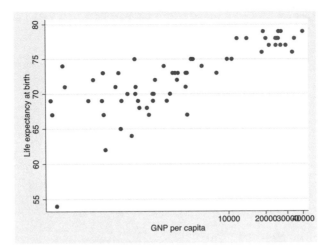

We obtain the same graph as if we typed

```
. generate log_gnppc = log(gnppc)
. scatter lexp log_gnppc
```

The difference concerns the labeling of the axis. When we specify $\{\,y\,|\,x\,\}$scale(log), the axis is labeled in natural units. In this case, the overprinting of the 30,000 and 40,000 is unfortunate, but we could fix that by dividing gnppc by 1,000.

Obtaining reversed scales

To obtain reversed scales—scales that run from high to low—specify the $\{\,y\,|\,x\,\}$scale(reverse) option:

```
. sysuse auto, clear
(1978 Automobile Data)
. scatter mpg weight, yscale(rev)
```

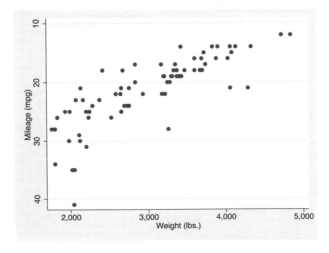

Suppressing the axes

There are two ways to suppress the axes. The first is to turn them off completely, which means that the line that is the axis is suppressed, along with all of its ticking, labeling, and titling. The second is to simply suppress the line that is the axis while leaving the ticking, labeling, and titling in place.

The first is done by $\{\,y\,|\,x\,\}$scale(off) and the second by $\{\,y\,|\,x\,\}$scale(noline). In addition, you will probably need to specify the plotregion(style(none)) option; see [G] *region_options*.

The axes and the border around the plot region are right on top of each other. Specifying plotregion(style(none)) will do away with the border and reveal the axes to us:

```
. sysuse auto, clear
(1978 Automobile Data)
. scatter mpg weight, plotregion(style(none))
```

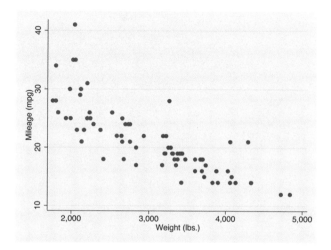

To eliminate the axes in their entirety, type

```
. scatter mpg weight, plotregion(style(none))
            yscale(off) xscale(off)
```

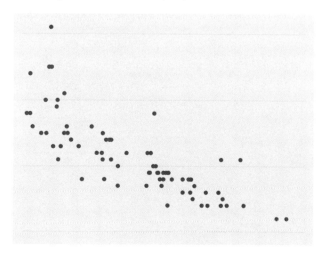

To eliminate the lines that are the axes while leaving in place the labeling, ticking, and titling, type

```
. scatter mpg weight, plotregion(style(none))
                      yscale(noline) xscale(noline)
```

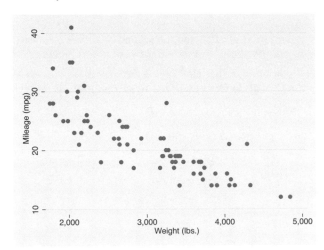

Rather than $\{y \mid x\}$scale(noline), you may specify $\{y \mid x\}$scale(lstyle(noline)) or $\{y \mid x\}$scale(lstyle(none)). They all mean the same thing.

Also See

Complementary: [G] *axis_options*; [G] *axis_label_options*, [G] *axis_title_options*;
                        [G] *region_options*

Title

axis_selection_options — Options for specifying on which axes a variable appears

Syntax

The *axis_selection_options* arc

axis_selection_options	description
yaxis(# [# ...])	which y axis to use, $1 \leq \# \leq 9$
xaxis(# [# ...])	which x axis to use, $1 \leq \# \leq 9$

yaxis() and xaxis() are *unique*; see [G] **repeated options**.

These options are allowed with any of the plotstyles (scatter, line, etc.) allowed by twoway.

Description

The *axis_selection_options* determine on which y and x axis (or axes) the variable is to appear.

Options

yaxis(# [# ...]) and xaxis(# [# ...]) specify the y or x axis to be used. The default is yaxis(1) and xaxis(1).

Typically yaxis() and xaxis() are treated as if their syntax is yaxis(#) and xaxis(#)—that is, just one number is specified. In fact, however, more than one number may be specified, and specifying a second is sometimes useful with yaxis(). The first y axis appears on the left, and the second (if there is a second) appears on the right. Specifying yaxis(1 2) allows you to force there being two identical y axes. You could use the one on the left in the usual way and the one on the right to label special values.

Remarks

Options yaxis() and xaxis() are used when you wish to create a single graph with multiple axes. These options are specified with twoway's scatter, line, etc., to specify which axis is to be used for each individual plot.

Remarks are presented under the headings

> Usual case: one set of axes
> Special case: multiple axes due to multiple scales
> yaxis(1) and xaxis(1) are the defaults
> Notation style is irrelevant
> yaxis() and xaxis() are plot options
> Specifying the other axes options with multiple axes
> Each plot may have at most one x and one y axis
> Special casc: multiple axes with a shared scale

Usual case: one set of axes

Normally, when you construct a `twoway` graph with more than one plot, as in

. scatter y1 y2 x

or equivalently,

. twoway (scatter y1 x) (scatter y2 x)

the two plots share common axes for y and for x.

Special case: multiple axes due to multiple scales

In some cases, you want the two y variables graphed on separate scales. In that case, you type

. twoway (scatter gnp year, c(l) yaxis(1))
         (scatter r   year, c(l) yaxis(2))

yaxis(1) specified on the first `scatter` says, "This scatter is to appear on the first y axis." yaxis(2) specified on the second `scatter` says, "This scatter is to appear on the second y axis."

The result is that two y axes will be constructed. The one on the left will correspond to gnp and the one on the right to r. If we had two x axes instead, one would appear on the bottom and one on the top:

. twoway (scatter year gnp, c(l) xaxis(1))
         (scatter year r,   c(l) xaxis(2))

You are not limited to having just two y axes or two x axes. You could have two of each,

. twoway (scatter y1var x1var, c(l) yaxis(1) xaxis(1))
         (scatter y2var x2var, c(l) yaxis(2) xaxis(2))

You may have up to nine y and nine x axes, although graphs become pretty well unreadable by that point. When there are three or more y axes (x axes), the axes are stacked up on the left (on the bottom). In any case, you specify yaxis(#) and xaxis(#) to specify which axis applies to which plot.

Also note that you may reuse axes:

. twoway (scatter gnp year, c(l) yaxis(1))
         (scatter nnp year, c(l) yaxis(1))
         (scatter r   year, c(l) yaxis(2))
         (scatter r2 year, c(l) yaxis(2))

The above graph has two y axes; one on the left and one on the right. The left axis is used for gnp and nnp; the right axis is used for r and r2.

The order in which we type the plots is not significant; the following would result in the same graph,

. twoway (scatter gnp year, c(l) yaxis(1))
         (scatter r   year, c(l) yaxis(2))
         (scatter nnp year, c(l) yaxis(1))
         (scatter r2 year, c(l) yaxis(2))

except that the symbols, colors, and linestyles associated with each plot would change.

yaxis(1) and xaxis(1) are the defaults

In our first multiple-axis example,

```
. twoway (scatter gnp year, c(l) yaxis(1))
         (scatter r   year, c(l) yaxis(2))
```

xaxis(1) is assumed because we did not specify otherwise. The command is interpreted as if we had typed,

```
. twoway (scatter gnp year, c(l) yaxis(1) xaxis(1))
         (scatter r   year, c(l) yaxis(2) xaxis(1))
```

Because yaxis(1) is the default, you need not bother to type it. Similarly, because xaxis(1) is the default, you could omit typing it, too:

```
. twoway (scatter gnp year, c(l))
         (scatter r   year, c(l) yaxis(2))
```

Notation style is irrelevant

Whether you use the ()-binding notation or the ||-separator notation never matters. You could just as well type

```
. scatter gnp year, c(l) || scatter r year, c(l) yaxis(2)
```

yaxis() and xaxis() are plot options

It is important to appreciate that, unlike all the other axis options, yaxis() and xaxis() are options of the individual plots and not of twoway itself. You may not type

```
. scatter gnp year, c(l) || scatter r year, c(l) ||, yaxis(2)
```

because twoway would have no way of knowing whether you wanted yaxis(2) to apply to the first or to the second scatter. While it is true that how the axes appear are a property of twoway—see [G] *axis_options*—which axes are used for which plots is a property of the plots themselves.

For instance, options ylabel() and xlabel() are options that specify the major ticking and labeling of an axis (see [G] *axis_label_options*). If you want the x axis to have ten ticks with labels, you can type

```
. scatter gnp year, c(l) ||
  scatter r   year, c(l) yaxis(2) ||, xlabel(#10)
```

and indeed you are "supposed" to type it that way to illustrate your deep understanding that xlabel() is a twoway option. Nonetheless, you may type

```
. scatter gnp year, c(l) ||
  scatter r   year, c(l) yaxis(2) xlabel(#10)
```

or

```
. scatter gnp year, c(l) xlabel(#10) ||
  scatter r   year, c(l) yaxis(2)
```

because twoway can reach inside the individual plots and pull out options intended for it. What twoway cannot do is redistribute options specified explicitly as twoway back to the individual plots.

Specifying the other axes options with multiple axes

Continuing with our example,

```
. scatter gnp year, c(l) ||
  scatter r   year, c(l) yaxis(2) ||
  , xlabel(#10)
```

say you also wanted ten ticks with labels on the first y axis and eight on the second. You type

```
. scatter gnp year, c(l) ||
  scatter r   year, c(l) yaxis(2) ||
  , xlabel(#10)  ylabel(#10, axis(1))  xlabel(#8, axis(2))
```

Each of the other axis options (see [G] *axis_options*) itself has an `axis(#)` option that specifies to which axis the option applies. When you do not specify that suboption, `axis(1)` is assumed.

As always, even though the other axis options are options of `twoway`, you can let them run together with the options of individual plots:

```
. scatter gnp year, c(l) ||
  scatter r   year, c(l) yaxis(2) xlabel(#10) ylabel(#10, axis(1))
                    ylabel(#8, axis(2))
```

Each plot may have at most one x and one y scale

Each `scatter`, `line`, `connected`, etc.—i.e., each plot—is allowed to have only one y scale and one x scale, so you may not type the shorthand

```
. scatter gnp r year, c(l l) yaxis(1 2)
```

to put `gnp` on one axis and `r` on another. In fact, `yaxis(1 2)` is not an error—we will get to that in the next section—but it will not put `gnp` on one axis and `r` on another. To do that, you must type

```
. twoway (scatter gnp year, c(l) yaxis(1))
         (scatter r   year, c(l) yaxis(2))
```

which, of course, you may type as

```
. scatter gnp year, c(l) yaxis(1) || scatter r year, c(l) yaxis(2)
```

The overall graph may have multiple scales, but the individual plots that appear in it may not.

Special case: multiple axes with a shared scale

It is sometimes useful to have multiple axes just so you have extra places to label special values. Consider graphing blood pressure versus concentration of some drug:

```
. scatter bp concentration
```

Perhaps you would like to add a line at $bp = 120$ and label that value specially. One thing you might do is

```
. scatter bp concentration, yaxis(1 2) ylabel(120, axis(2))
```

The `ylabel(120, axis(2))` part is explained in [G] *axis_label_options*; it caused the second axis to have the value 120 labeled. It is the option `yaxis(1 2)` that caused there to be a second axis, which we could label. When you specify `yaxis()` (or `xaxis()`) with more than one number, you are specifying that the axes be created sharing the same scale.

To better understand what `yaxis(1 2)` does, compare the results of

```
. scatter bp concentration
```

with

```
. scatter bp concentration, yaxis(1 2)
```

In the first graph, there is one y axis and it is on the left. In the second graph, there are two y axes, one on the left and one on the right, and they are labeled identically.

Now compare

```
. scatter bp concentration
```

with

```
. scatter bp concentration, xaxis(1 2)
```

In the first graph, there is one x axis and it is on the bottom. In the second graph, there are two x axes, one on the bottom and one on the top, and they are labeled identically.

Finally, try

```
. scatter bp concentration, yaxis(1 2) xaxis(1 2)
```

In this graph, there are two y and two x axes: left and right, and top and bottom.

Also See

Complementary: [G] *axis_label_options*, [G] *axis_options*, [G] *axis_scale_options*, [G] *axis_title_options*

Title

---

axis_title_options — Options for specifying axis titles

Syntax

axis_title_options are a subset of *axis_options*; see [G] **axis_options**. *axis_title_options* control the titling of an axis. The *axis_title_options* are

axis_title_options	description
ytitle(*axis_title*)	specify y axis title
xtitle(*axis_title*)	specify x axis title

The above options are *merged-explicit*; see [G] **repeated options**.

where *axis_title* is

"*string*" ["*string*" [...]] [, *suboptions*]

and *suboptions* are

suboptions	description
axis(#)	which axis, $1 \leq \# \leq 9$
prefix	combine options
suffix	combine options
textbox_options	control details of text appearance
	see [G] *textbox_options*

Description

axis_title_options specify the titles to appear on axes.

Options

ytitle(*axis_title*) and xtitle(*axis_title*) specify the titles to appear on the y and x axes.

Suboptions

axis(#) specifies to which axis this title belongs and is specified when dealing with multiple y axes or multiple x axes; see [G] *axis_selection_options*.

prefix and suffix specify that what is specified in this option is to be added to any previous xtitle() or ytitle() options previously specified. See *Interpretation of repeated options* below.

textbox_options specifies the look of the text. See [G] *textbox_options*.

68

Remarks

axis_title_options are a subset of *axis_options*; see [G] *axis_options* for an overview. The other appearance options are

axis_scale_options	(see [G] *axis_scale_options*)
axis_label_options	(see [G] *axis_label_options*)

Remarks are presented under the headings

Default axis titles
Overriding default titles
Specifying multiline titles
Suppressing axis titles
Interpretation of repeated options
Titles with multiple y axes or multiple x axes

Default axis titles

Even if you do not specify the `ytitle()` or `xtitle()` options, in most cases axes will be titled. In those cases, { y | x }`title()` changes the title. If an axis is not titled, specifying { y | x }`title()` adds a title.

Default titles are obtained using the corresponding variable's variable label or, if it does not have a label, using its name. For instance, in

    . twoway scatter yvar xvar

the default title for the y axis will be obtained from variable yvar and the default title for the x axis will be obtained from xvar. In some cases, the plottype substitutes a different title; for instance,

    . twoway lfit yvar xvar

labels the y axis "Fitted values" regardless of the name or variable label associated with variable yvar.

If multiple variables are associated with the same axis, their individual titles (variable label, variable name, or as substituted) are joined, with a slash (/) in between. For instance, in

    . twoway scatter y1var xvar || line y2var xvar || lfit y1var xvar

the y axis will be titled

y1var_title/*y2var_title*/`Fitted values`

When many plots are overlaid, this often results in titles that run off the end of the graph.

Overriding default titles

You may specify the title to appear on the y axis using `ytitle()` and the title to appear on the x axis using `xtitle()`. You specify the text—surrounded by double quotes—inside the option:

`ytitle("My y title")`

`xtitle("My x title")`

In the case of `scatter`, the command might read

    . scatter yvar xvar, ytitle("Price") xtitle("Quantity")

Specifying multiline titles

Titles may include more than one line. Lines are specified one after the other, each enclosed in double quotes:

```
ytitle("First line" "Second line")
```

```
xtitle("First line" "Second line" "Third line")
```

Suppressing axis titles

To eliminate an axis title altogether, specify $\{$ y $|$ x $\}$title("").

To eliminate the title on a second, third, ..., axis, specify $\{$ y $|$ x $\}$title("", axis(#)). See *Titles with multiple y or multiple x axes* below.

Interpretation of repeated options

xtitle() and ytitle() may be specified more than once in the same command. When you do that, it is the rightmost one that takes effect.

See *Interpretation of repeated options* in [G] *axis_label_options*. Multiple ytitle() and xtitle() options work the same way. The twist in the case of the title options is that you specify whether the extra information is to be prefixed or suffixed onto what came before.

For instance, pretend that sts graph produced the x axis title "analysis time". If you typed

```
. sts graph, xtitle("My new title")
```

the title you specified would replace that. If you typed

```
. sts graph, xtitle("in days", suffix)
```

the x axis title would be (first line) "analysis time" (second line) "in days". If you typed

```
. sts graph, xtitle("Time to failure", prefix)
```

the x axis title would be (first line) "Time to failure" (second line) "analysis time".

Titles with multiple y or multiple x axes

When you have more than one y or x axis (see [G] *axis_selection_options*), remember to specify the axis(#) suboption to indicate to which axis you are referring.

Also See

Complementary: [G] *axis_options*; [G] *axis_label_options*, [G] *axis_scale_options*

Title

> *barlook_options* — Options for setting the look of bars

Syntax

The *barlook_options* are

barlook_options	description
bstyle(*areastyle*)	overall look of bars
bcolor(*colorstyle*)	outline and fill color
bfcolor(*colorstyle*)	fill color
blstyle(*linestyle*)	overall look of outline
blcolor(*colorstyle*)	outline color
blwidth(*linewidthstyle*)	thickness of outline
blpattern(*linepatternstyle*)	whether outline solid, dashed, etc.

See [G] *areastyle*, [G] *colorstyle*, [G] *linestyle*, [G] *linewidthstyle*, and [G] *linepatternstyle*.

All options are *merged-implicit*; see [G] **repeated options**.

Description

The *barlook_options* determine the look of bars produced by graph bar, graph hbar, graph twoway bar, and graph twoway hbar.

Options

bstyle(*areastyle*) specifies the look of the bar. The options listed below allow you to change each attribute, but bstyle() provides the starting point.

You need not specify bstyle() just because there is something you want to change. You specify bstyle() when another style exists that is exactly what you desire or when another style would allow you to specify fewer changes to obtain what you want.

See [G] *areastyle* for a list of available area styles.

bcolor(*colorstyle*) specifies a single color to be used both to outline the shape of the bar and to fill its interior. See [G] *colorstyle* for a list of color choices.

bfcolor(*colorstyle*) specifies the color to be used to fill the interior of the bar. See [G] *colorstyle* for a list of color choices.

blstyle(*linestyle*) specifies the overall style of the line used to outline the bar, which includes its pattern (solid, dashed, etc.), its thickness, and its color. The other options listed below allow you to change the line's attributes, but blstyle() is the starting point. See [G] *linestyle* for a list of choices.

blcolor(*colorstyle*) specifies the color to be used to outline the bar. See [G] *colorstyle* for a list of color choices.

blwidth(*linewidthstyle*) specifies the thickness of the line to be used to outline the bar. See [G] *linewidthstyle* for a list of choices.

blpattern(*linepatternstyle*) specifies whether the line used to outline the bar is solid, dashed, etc. See [G] *linepatternstyle* for a list of pattern choices.

Remarks

The *barlook_options* are allowed inside graph bar's and graph hbar's option bar(#, *barlook_options*), as in

```
. graph bar yvar1 yvar2, bar(1, bcolor(green)) bar(2, bcolor(red))
```

The above would set the bar associated with *yvar1* to green and the bar associated with *yvar2* to red; see [G] **graph bar**.

barlook_options are also allowed as options with graph twoway bar and graph twoway rbar, as in

```
. graph twoway bar yvar xvar, bcolor(green)
```

The above would set all the bars (which are located at *xvar* and extend to *yvar*) to be green; see [G] **graph twoway bar** and [G] **graph twoway rbar**.

Also See

Complementary: [G] *areastyle*, [G] *colorstyle*, [G] *linestyle*, [G] *linepatternstyle*,
                  [G] *linewidthstyle*; [G] **graph bar**, [G] **graph twoway bar**,
                  [G] **graph twoway rbar**

Title

> *blabel_option* — Option for labeling bars

Syntax

<u>grap</u>h { bar | hbar } ..., ... <u>blabe</u>l(*what* [, *where_and_how*]) ...

where *what* is defined

what	description
none	no label; default
bar	label is bar height
total	label is cumulative bar height
name	label is name of *yvar*
group	label is first over() group

and *where_and_how* are defined

where_and_how	description
<u>p</u>osition(outside | inside | base | center)	where to locate label on bar
gap(*relativesize*)	distance from position
format(%*fmt*)	format if bar or total
textbox_options	look of label

See [G] **relativesize** and [G] **textbox_options**.

Description

Option blabel() is for use with graph bar and graph hbar. It adds a label on top of or inside each bar.

Options

blabel(*what*, *where_and_how*) specifies the label and where it is to be located relative to the bar. *where_and_how* is optional and documented under *Suboptions for use with blabel()* below. *what* specifies the contents of the label.

blabel(bar) specifies that the label be the height of the bar. In

. graph bar (mean) empcost, over(division) blabel(bar)

the labels would be the mean employee cost.

blabel(total) specifies that the label be the cumulative height of the bar. blabel(total) is for use with graph bar's stack option. In

. graph bar (sum) cost1 cost2, stack over(group) blabel(total)

73

the labels would be the total height of the stacked bar, which is to say, the sum of costs. In addition, the cost1 part of the stack bar would be labeled with its height.

blabel(name) specifies that the label be the name of the *yvar*. In

```
. graph bar (mean) y1 y2 y3 y4, blabel(name)
```

The labels would be "mean of y1", "mean of y2", ..., "mean of y4". Usually, you would also want to suppress the legend in this case and so would type

```
. graph bar (mean) y1 y2 y3 y4, blabel(name) legend(off)
```

blabel(group) specifies that the label be the name of the first over() group. In

```
. graph bar cost, over(division) over(year) blabel(group)
```

the labels would be the name of the divisions. Usually, you would also want to suppress the appearance of the division labels on the axis:

```
. graph bar cost, over(division, axis(off)) over(year) blabel(group)
```

Suboptions for use with blabel()

position() specifies where the label is to appear.

position(outside) is the default. The label appears just above the bar (graph bar), or just to its right (graph hbar).

position(inside) specifies the label is to appear inside the bar, at the top (graph bar), or at its rightmost extent (graph hbar).

position(base) specifies the label is to appear inside the bar, at the bar's base; at the bottom of the bar (graph bar), or at the left of the bar (graph hbar).

position(center) specifies the label is to appear inside the bar, at its center.

gap(*relativesize*) specifies a distance by which the label is to be offset from its location (outside, inside, base, or center). The default is usually gap(1.7). Note that the gap() may be positive or negative, and note that you can specify, for instance, gap(*1.2) and gap(*.8) to increase or decrease the gap by 20%; see [G] *relativesize*.

format(*%fmt*) is for use with blabel(bar) and blabel(total); it specifies the display format to be used to format the height value. See [R] **format**.

textbox_options are any of the options allowed with a textbox. Important options include size(), which determines the size of the text, box, which draws a box around the text, and color(), which determines the color of the text. See [G] *textbox_options*.

Remarks

blabel() serves two purposes: to increase the information content of the chart (blabel(bar) and blabel(total)), or to change how bars are labeled (blabel(name) and blabel(group)).

Remarks are presented under the headings

Increasing the information content
Changing how bars are labeled

Increasing the information content

Under the heading *Multiple bars* in [G] **graph bar** we drew the following graph:

```
. graph bar (mean) tempjuly tempjan, over(region)
        bargap(-30)
        legend( label(1 "July") label(2 "January") )
        ytitle("Degrees Fahrenheit")
        title("Average July and January temperatures")
        subtitle("by regions of the United States")
        note("Source:  U.S. Census Bureau, U.S. Dept. of Commerce")
```

To the above, we now add

```
blabel(bar, position(inside) format(%9.1f))
```

which will add the average temperature to the bar, position the average inside the bar, at the top, and format its value using %9.1f:

```
. graph bar (mean) tempjuly tempjan, over(region)
        bargap(-30)
        legend( label(1 "July") label(2 "January") )
        ytitle("Degrees Fahrenheit")
        title("Average July and January temperatures")
        subtitle("by regions of the United States")
        note("Source:  U.S. Census Bureau, U.S. Dept. of Commerce")
        blabel(bar, position(inside) format(%9.1f) color(white))     ← new
```

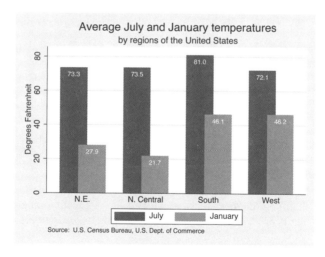

We also specified the *textbox_option* color(white) to change the color of the text; see [G] *textbox_options*. Dark text on a dark bar would have blended too much.

Changing how bars are labeled

Placing the labels on the bars works especially well with horizontal bar charts:

```
. sysuse nlsw88, clear
(NLSW, 1988 extract)
```

```
. graph hbar (mean) wage,
        over( occ, axis(off) sort(1) )
        blabel( group, position(base) color(bg) )
        ytitle( "" )
        by( union,
            title("Average Hourly Wage, 1988, Women Aged 34-46")
            note("Source:  1988 data from NLS, U.S. Dept. of Labor,
                Bureau of Labor Statistics")
        )
```

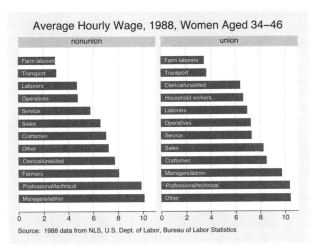

What makes moving the labels from the axis to the bars work so well in this case is that it saves so much horizontal space.

In the above command, note the first two option lines:

```
            over( occ, axis(off) sort(1) )
        blabel( group, position(base) color(bg) )
```

blabel(group) puts the occupation labels on top of the bars and suboption position(base) located the labels at the base of each bar. We specified over(,axis(off)) to prevent the labels from appearing on the axis. Let us run though all the options:

over(occ, axis(off) sort(1))
> Specified the chart be done over occupation, that the occupation labels not be shown on the axis, and that the bars be sorted by the first (and only) *yvar*, namely (mean) wage.

ytitle("")
> Specified that the title on the numerical y axis (the horizontal axis in this horizontal case), be suppressed.

by(union, title(...) note(...))
> Specified the entire graph be repeated by values of variable union, and specified the title and note to be added to the overall graph. Had we specified the title() and note() options outside the by(), they would have been placed on each graph.

Also See

Complementary: [G] **graph bar**

Title

> *by_option* — Option for repeating graph command

Syntax

The *by_option* is

by_option	description
by(*varlist*[, *suboptions*])	repeat for by-groups

by() is *unique*; see [G] **repeated options**.

where *suboptions* are

suboptions	description
total	add total group
missing	add missing group(s)
colfirst	display down columns
rows(#) | cols(#)	display in # rows or # columns
holes(*numlist*)	positions to leave blank
iscale([*]#)	size of text and markers
compact	synonym for style(compact)
style(*bystyle*)	overall style of presentation
[no]edgelabel	label x axes of edges
[no]rescale	separate y and x scales for each group
[no]yrescale	separate y scale for each group
[no]xrescale	separate x scale for each group
[no]iyaxes	show individual y axes
[no]ixaxes	show individual x axes
[no]iytick	show individual y axes ticks
[no]ixtick	show individual x axes ticks
[no]iylabel	show individual y axes labels
[no]ixlabel	show individual x axes labels
[no]iytitle	show individual y axes titles
[no]ixtitle	show individual x axes titles
imargin(*marginstyle*)	margin between graphs
title_options	overall titles
region_options	overall outlining, shading, and aspect

See [U] **14.1.8 numlist**, [G] *bystyle*, [G] *marginstyle*, [G] *title_options*, and [G] *region_options*; note that the *title_options* and *region_options* on the command on which by() is appended will become the titles and regions for the individual by-groups.

Description

Option by() draws separate plots within a single graph. *varlist* may be a numeric or a string variable.

Options

by(*varlist*[, *suboptions*]) specifies that the graph command is to be repeated for each unique set of values of *varlist* and that the resulting individual graphs are to be arrayed into a single graph.

Suboptions

total specifies that, in addition to the graphs for each by-group, a graph is to be added for all by-groups combined.

missing specifies that, in addition to the graphs for each by-group, graph(s) are to be added for missing values of *varlist*. Missing is defined as ., .a, ..., .z for numeric variables and "" for string variables.

colfirst specifies that the individual graphs are to be arrayed down the columns rather than across the rows. That is, if there were four groups, the graphs would be displayed

default	colfirst
1 2	1 3
3 4	2 4

rows(#) and cols(#) are alternatives. They specify that the resulting graphs are to be arrayed as # rows and however many columns are necessary, or as # columns and however many rows are necessary. The default is

cols(*c*), *c* = ceiling(sqrt(*G*))

where *G* is the total number of graphs to be presented and ceiling() is the function that rounds nonintegers up to the next integer. For instance, if four graphs are to be displayed, the result will be presented in a 2×2 array. If five graphs are to be displayed, the result will be presented as a 2×3 array because ceiling(sqrt(5))==3.

cols(#) may be specified as larger or smaller than *c*; *r* will be the number of rows implied by *c*. Similarly, rows(#) may be specified as larger or smaller than *r*.

holes(*numlist*) specifies which positions in the array are to be left unfilled. Consider drawing a graph with three groups and assume that the three graphs are being displayed in a 2×2 array. By default the first group will appear in the graph at (1,1), the second in the graph at (1,2), and the third in the graph at (2,1). Nothing will be displayed in the (2,2) position.

Specifying holes(3) would cause position (2,1) to be left blank, so the third group would appear in (2,2).

The numbers that you specify in holes() correspond to the position number,

```
1   2      1   2   3      1   2   3   4      1   2   3   4   5
3   4      4   5   6      5   6   7   8      6   7   8   9  10
           7   8   9      9  10  11  12     11  12  13  14  15
                        12  14  15  16     16  17  18  19  20
                                           21  22  23  24  25    etc.
```

The above is the numbering when `colfirst` is not specified. If `colfirst` is specified, the positions are transposed:

```
1  3     1  4  7     1  5   9  13     1   6  11  16  21
2  4     2  5  8     2  6  10  14     2   7  12  17  22
         3  6  9     3  7  11  15     3   8  13  18  23
                     4  8  12  16     4   9  14  19  24
                                     5  10  15  20  25     etc.
```

`iscale(#)` and `iscale(*#)` specify a size adjustment (multiplier) to be used to scale the text and markers.

By default, `iscale()` gets smaller and smaller the larger is G, the number of by-groups and hence the number of graphs presented. The default is parameterized as a multiplier $f(G)$— $0 < f(G) < 1$, $f'(G) < 0$—that is used to multiply `msize()`, $\{y\,|\,x\}$ `label(,labsize())`, and the like. The size of everything except the overall titles, subtitles, captions, and notes is affected by `iscale()`.

If you specify `iscale(#)`, the number you specify is substituted for $f(G)$. `iscale(1)` means text and markers should appear at the same size as they would were each graph drawn separately. `iscale(.5)` displays text and markers at half that size. We recommend you specify a number between 0 and 1, but you are free to specify numbers larger than 1.

If you specify `iscale(*#)`, the number you specify is multiplied by $f(G)$ and that product is used to scale text and markers. `iscale(*1)` is the default. `iscale(*1.2)` means text and markers should appear 20% larger than `graph, by()` would usually choose. `iscale(*.8)` would make them 20% smaller.

`compact` is a synonym for `style(compact)`. It makes no difference which you type. See the description of the `style()` option below and see *By-styles* under *Remarks*.

`style(`*bystyle*`)` specifies the overall look of the by-graphs. The style determines whether individual graphs have their own axes and labels or if instead the axes and labels are shared across graphs arrayed in the same row or in the same column, how close the graphs are to be placed to each other, etc. The other options documented below will allow you to change the way the results are displayed, but the *bystyle* specifies the starting point.

You need not specify `style()` just because there is something you want to change. You specify `style()` when another style exists that is exactly what you desire or when another style would allow you to specify fewer changes to obtain what you want.

See [G] *bystyle* for a list of by-style choices. The suboptions listed below modify the by-style:

`edgelabel` and `noedgelabel` specify whether the last graphs of a column that do not appear in the last row are to have their x axes labeled. See *Labeling the edges* under *Remarks* below.

`rescale`, `yrescale`, and `xrescale` (and `norescale`, `noyrescale`, and `noxrescale`) specify that the scales of each graph be allowed to differ (or forced to be the same). Whether X or noX is the default is determined by `style()`.

Usually, noX is the default and `rescale`, `yrescale`, and `xrescale` are the options. By default, all the graphs will share the same scaling for their axes. Specifying `yrescale` will allow the y scales to differ across the graphs, specifying `xrescale` will allow the x scales to differ, and specifying `rescale` is equivalent to specifying `yrescale` and `xrescale`.

`iyaxes` and `ixaxes` (and `noiyaxes` and `noixaxes`) specify whether the y axes and x axes are to be displayed with each individual graph. The default with most styles and schemes is to place y axes on the leftmost graph of each row and to place x axes on the bottommost graph of each column. The y and x axes include the default ticks and labels but exclude the axes titles.

iytick and ixtick (and noiytick and noixtick) are seldom specified. If you specified iyaxis and then wanted to suppress the ticks, you could also specify noiytick. In the rare event where specifying iyaxis did not result in the ticks being displayed (because of how the style or scheme works), specifying iytick would cause the ticks to be displayed.

iylabel and ixlabel (and noiylabel and noixlabel) are seldom specified. If you specified iyaxis and then wanted to suppress the axes labels, you could also specify noiylabel. In the rare event where specifying iyaxis did not result in the labels being displayed (because of how the style or scheme works), specifying iylabel would cause the labels to be displayed.

iytitle and ixtitle (and noiytitle and noixtitle) are seldom specified. If you specified iyaxis and then wanted to add the *y*-axes titles (which would make the graph very busy), you could also specify iytitle. In the rare event where specifying iyaxis resulted in the titles being displayed (because of how the style of scheme works), specifying noiytitle would suppress displaying the title.

imargin(*marginstyle*) specifies the margins between the individual graphs.

Remarks

Remarks are presented under the headings

> *Typical use*
> *Placement of graphs*
> *Treatment of titles*
> *by() uses subtitle() within graph*
> *Placement of the subtitle()*
> *by() uses the overall note()*
> *Use of legends with by()*
> *By-styles*
> *Labeling the edges*
> *Specifying separate scales for the separate plots*

Typical use

One often has data that divides into different groups—person data where the persons are male or female or in different age categories (or both), country data where the countries can be categorized into different regions of the world, or, as below, automobile data where the cars are foreign or domestic. If you type

```
. scatter mpg weight
```

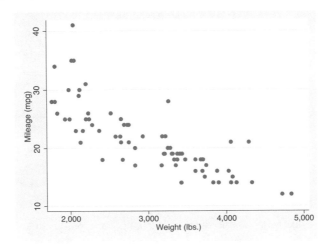

you obtain a scatterplot of mpg versus weight. If you add by(foreign) as an option, you obtain two graphs, one for each value of foreign:

. scatter mpg weight, by(foreign)

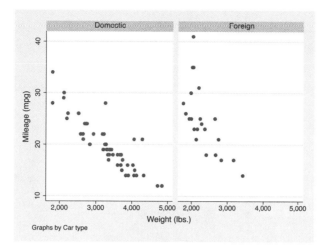

If you add the total suboption, another graph will be added representing the overall total:

. scatter mpg weight, by(foreign, total)

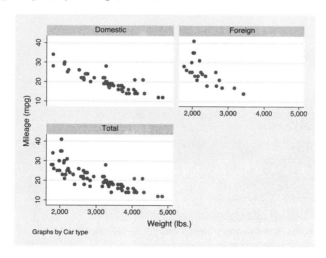

In this case, there were three graphs to be presented and by() chose to display them in a 2×2 array, leaving the last position empty.

Placement of graphs

By default, by() places the graphs in a rectangular $R \times C$ array and leaves empty the positions at the end:

Number of graphs	array dimension	positions left empty
1	1×1	
2	1×2	
3	2×2	4=(2,2)
4	2×2	
5	2×3	6=(3,3)
6	2×3	
7	3×3	8=(3,2) 9=(3,3)
8	3×3	9=(3,3)
9	3×3	
10	3×4	11=(3,3) 12=(3,4)
11	3×4	12=(3,4)
12	3×4	
13	4×4	14=(4,2) 15=(4,3) 16=(4,4)
14	4×4	15=(4,3) 16=(4,4)
15	4×4	16=(4,4)
16	4×4	
17	4×5	18=(4,3) 19=(4,4) 20=(4,5)
18	4×5	19=(4,4) 20=(4,5)
19	4×5	20=(4,5)
20	4×5	
21	5×5	22=(5,2) 23=(5,3) 24=(5,4) 25=(5,5)
22	5×5	23=(5,3) 24=(5,4) 25=(5,5)
23	5×5	24=(5,4) 25=(5,5)
24	5×5	25=(5,5)
25	5×5	
etc.		

Options `rows()`, `cols()`, and `holes()` allow you to control this behavior.

You may specify either `rows()` or `cols()`, but not both. In the previous section, we drew

    . scatter mpg weight, by(foreign, total)

and had 3 graphs displayed in a 2×2 array with a hole at 4. We could draw the graph in a 1×3 array by specifying either `rows(1)` or `cols(3)`,

(Continued on next page)

```
. scatter mpg weight, by(foreign, total rows(1))
```

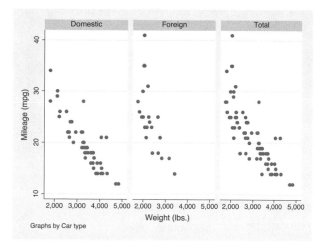

or we could stay with the 2 × 2 array and move the hole to 3,

```
. scatter mpg weight, by(foreign, total holes(3))
```

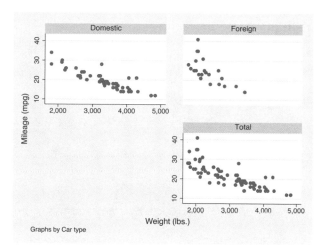

Treatment of titles

Were you to type,

```
. scatter yvar xvar, title("My title") by(catvar)
```

"My title" will be repeated above each individual graph. by() repeats the entire graph command and then arrays the results.

To specify titles for the entire graph, specify the *title_options*—see [G] ***title_options***—inside the by() option:

```
. scatter yvar xvar, by(catvar, title("My title"))
```

by() uses the subtitle() with graph

by() labels each of the individual graphs using the subtitle() *title_option*. For instance, in

. scatter mpg weight, by(foreign, total)

by() labeled the graphs "Domestic", "Foreign", and "Total". The subtitle "Total" is what by() uses when the total option is specified. The other two subtitles by() obtained from the by-variable foreign.

by() may be used with numeric or string variables. In this case, foreign is numeric but happens to have a value label associated with it. by() obtained the subtitles "Domestic" and "Foreign" from the value label. If foreign had no value label, the first two graphs would have been subtitled "0" and "1", the numeric values of variable foreign. If foreign had been a string variable, the subtitles would been the string contents of foreign.

If you wish to suppress the subtitle, type

. scatter mpg weight, subtitle("") by(foreign, total)

If you wished to add "*Extra info*" to the subtitle, type

. scatter mpg weight, subtitle("*Extra info*", suffix) by(foreign, total)

Be aware, however, that "*Extra info*" will appear above each of the individual graphs.

Placement of the subtitle()

You can use subtitle()'s suboptions to control the placement of the identifying label. For instance,

. scatter mpg weight,
        subtitle(, ring(0) pos(1) nobexpand) by(foreign, total)

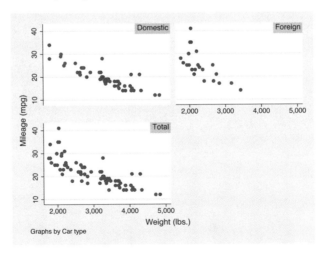

The result will be to move the identifying label inside the individual graphs, displaying it in the northeast corner of each. Type

. scatter mpg weight,
        subtitle(, ring(0) pos(11) nobexpand) by(foreign, total)

and the identifying label will be moved to the northwest corner.

ring(0) moves the subtitle inside the graph's plot region and position() defines the location, indicated as clock positions. nobexpand is rather strange, but just remember to specify it. By default, by() sets subtitles to expand to the size of the box that contains them, which is unusual but makes the default-style subtitles look good with shading.

See [G] *title_options*.

by() uses the overall note()

By default, by() adds an overall note saying "Graphs by ...". When you type

. scatter *yvar xvar*, by(catvar)

results are as if you typed

. scatter *yvar xvar*, by(catvar, note("Graphs by ..."))

If you want to suppress the note, type

. scatter *yvar xvar*, by(catvar, note(""))

If you want to change the overall note to read "My note", type

. scatter *yvar xvar*, by(catvar, note("My note"))

If you want to add your note after the default note, type

. scatter *yvar xvar*, by(catvar, note("My note", suffix))

Use of legends with by()

If you wish to modify or suppress the default legend, you must do that differently when by() is specified. For instance, legend(off)—see [G] *legend_option*—will suppress the legend, yet typing

. line y1 y2 x, by(group) legend(off)

will not have the intended effect. The legend(off) will seemingly be ignored. You must instead type

. line y1 y2 x, by(group, legend(off))

Note that we moved legend(off) inside the by().

Remember that by() repeats the graph command. If you think carefully, you will realize that the legend never was being displayed at the bottom of the individual plots. It is instructive to type

. line y1 y2 x, legend(on) by(group)

This graph will have lots of legends: one underneath each of the plots in addition to the overall legend at the bottom of the graph! by() works exactly as advertised: it repeats the entire graph command for each value of group.

In any case, it is the overall legend() that we want to suppress and that is why we must specify legend(off) inside the by() option; this is the same issue as the one discussed under *Treatment of titles* above.

The issue becomes a little more complicated when, rather than suppressing the legend, we wish to modify the legend's contents or position. In that case, the legend() option to modify contents is specified outside the by() and the legend() option to modify location is specified inside. See *Use of legends with by* in [G] *legend_option*.

By-styles

Option `style(bystyle)` specifies the overall look of by-graphs; see [G] ***bystyle*** for a list of *bystyle* choices. One *bystyle* worth noting is `compact`. Specifying `style(compact)` causes the graph to be displayed in a more compact format. Compare

```
. sysuse lifeexp, clear
(Life expectancy, 1998)
. scatter lexp gnppc, by(region, total)
```

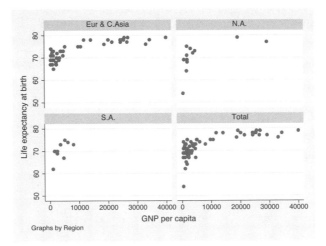

with

```
. scatter lexp gnppc, by(region, total style(compact))
```

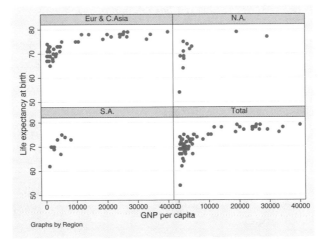

`style(compact)` pushes the graphs together horizontally and vertically, leaving more room for the individual graphs. The disadvantage is that, pushed together, the values on the axes labels sometimes run into each other, as occurred above with the 40,000 of the S.A. graph running into the 0 of the Total graph. That problem could be solved by dividing `gnppc` by 1,000.

Rather than typing out `style(compact)`, you may specify `compact` and you may further abbreviate that as `com`.

Labeling the edges

Consider the graph

    . scatter mpg weight, by(foreign, total)

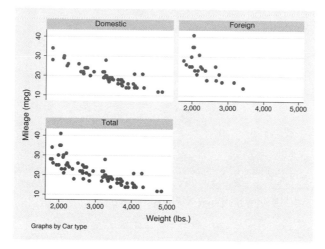

and note that the x axis is labeled in the graph in the (1,2) position. When the last graph of a column does not appear in the last row, its x axis is referred to as an edge. In `style(default)`, the default is to label the edges, but you could type

    . scatter mpg weight, by(foreign, total noedgelabel)

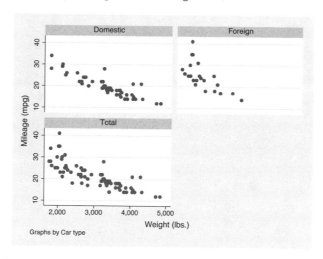

to suppress that. This results in the rows of graphs being closer to each other.

Were you to type

    . scatter mpg weight, by(foreign, total style(compact))

you would discover that the x axis of the (1,2) graph is not labeled. Using `style(compact)`, the default is `noedgelabel`, but you could specify `edgelabel` to override that.

Specifying separate scales for the separate plots

If you type

. scatter *yvar xvar*, by(*catvar*, yrescale)

each graph will be given a separately scaled y axis; if you type

. scatter *yvar xvar*, by(*catvar*, xrescale)

each graph will be given a separately scaled x axis; and if you type

. scatter *yvar xvar*, by(*catvar*, yrescale xrescale)

both scales will be separately set.

History

The twoway scatterplots produced by the by() option are similar to what are known as *casement displays* (see, for instance, Chambers et al. 1983, 141–145). A traditional casement display, however, aligns all the graphs either vertically or horizontally.

References

Chambers, J. M., W. S. Cleveland, B. Kleiner, and P. A. Tukey. 1983. *Graphical Methods for Data Analysis*. Belmont, CA: Wadsworth International Group.

Also See

Complementary: [G] *region_options*, [G] *title_options*

Title

> *bystyle* — Choices for look of by-graphs

Syntax

bystyle may be

bystyle	description
default	determined by scheme
compact	a more compact version of default
stata7	like that provided by Stata 7

Other *bystyles* may be available; type

```
. graph query bystyle
```

to obtain the full list installed on your computer.

Description

bystyles specify the overall look of by-graphs.

Remarks

Remarks are presented under the headings

  What is a by-graph?
  What is a bystyle?

What is a by-graph?

A by-graph is a single graph (image, really) containing an array of separate graphs, each of the same type, and each reflecting a different subset of the data. For instance, a by-graph might contain graphs of mpg versus weight, one for domestic cars and the other for foreign.

By-graphs are produced are produced when you specify the by() option; see [G] *by_option*.

What is a bystyle?

A *bystyle* determines the overall look of the combined graphs, including

1. whether the individual graphs have their own axes and labels or if instead the axes and labels are shared across graphs arrayed in the same row and or in the same column;

2. whether the scales of axes are in common or allowed to be different for each graph; and

3. how close the graphs are placed to each other.

There are options that let you control each of the above attributes—see [G] ***by_option***—but the *bystyle* specifies the starting point.

You need not specify a *bystyle* just because there is something you want to change. You specify a *bystyle* when another style exists that is exactly what you desire or when another style would allow you to specify fewer changes to obtain what you want.

Also See

Complementary: [G] ***by_option***

Title

> *cat_axis_label_options* — Options for specifying look of categorical axis labels

Syntax

The *cat_axis_label_options* are

cat_axis_label_options	description
<u>no</u>labels	suppress axis labels
<u>tic</u>ks	display axis ticks
angle(*anglestyle*)	angle of axis labels
<u>al</u>ternate	offset adjacent labels
tstyle(*tickstyle*)	labels and ticks: overall style
labgap(*relativesize*)	labels: margin between tick and label
labstyle(*textstyle*)	labels: overall style
labsize(*textsizestyle*)	labels: size of text
<u>labc</u>olor(*colorstyle*)	labels: color of text
<u>tl</u>ength(*relativesize*)	ticks: length
tposition(<u>o</u>utside | <u>c</u>rossing | inside)	ticks: position/direction
tlstyle(*linestyle*)	ticks: linestyle of
tlwidth(*linewidthstyle*)	ticks: thickness of line
tlcolor(*colorstyle*)	ticks: color of line
tlpattern(*linepatternstyle*)	ticks: line pattern of line

See [G] *tickstyle*, [G] *relativesize*, [G] *textstyle*, [G] *textsizestyle*, [G] *colorstyle*, [G] *linestyle*, [G] *linewidthstyle*, and [G] *linepatternstyle*.

Description

The *cat_axis_label_options* determine the look of the labels that appear on a categorical x axis produced by graph bar, graph hbar, graph dot, and graph box. These options are specified inside label() of over():

> . graph ..., over(*varname*, ... label(*cat_axis_label_options*) ...)

The most useful *cat_axis_label_options* are angle(), alternate, labcolor(), and labsize().

Options

nolabels suppresses the display of the category labels on the axis. In the case of graph bar and graph hbar, the nolabels option is useful when combined with the blabel() option used to place the labels on the bars themselves; see [G] *blabel_option*.

ticks specifies that ticks are to appear on the categorical x axis. By default, ticks are not presented on categorical axes and it is unlikely that you would want them to be.

angle(*anglestyle*) specifies the angle at which the labels on the axis appear. The default is angle(0), meaning horizontal. With vertical bar charts and other vertically oriented charts, it is sometimes useful to specify angle(90) (vertical text reading bottom to top), angle(-90) (vertical text reading top to bottom), or angle(-45) (angled text reading top-left to bottom-right); see [G] *anglestyle*.

Unix users: if you specify angle(-45), results will appear on your screen as if you specified angle(-90); results will appear correctly when you print.

alternate causes adjacent labels to be offset from one another and is useful when there are many labels or when labels are long. For instance, rather than obtaining an axis labeled,

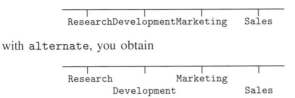

with alternate, you obtain

tstyle(*tickstyle*) specifies the overall look of labels and ticks. In this case, the emphasis is on labels since ticks are usually suppressed on a categorical axis. The options documented below will allow you to change each attribute of the label and tick, but the *tickstyle* specifies the starting point.

You need not specify tstyle() just because there is something you want to change about the look of labels and ticks. You specify tstyle() when another style exists that is exactly what you desire or when another style would allow you to specify fewer changes to obtain what you want.

labgap(), labstyle(), labsize(), and labcolor() specify details about how the labels are presented. Of particular interest are labsize(*textsizestyle*), which specifies the size of the labels, and labcolor(*colorstyle*), which specifies the color of the labels; see [G] *textsizestyle* and [G] *colorstyle* for a list of color choices.

tlength(*relativesize*) specifies the overall length of the ticks.

tposition(outside | crossing | inside) specifies whether the ticks are to extend outside (from the axis out, the usual default), are to extend crossing (crossing the axis line, extending in and out), or are to extend inside (from the axis into the plot region).

tlstyle(), tlwidth(), tlcolor(), tlpattern() specify other details about the look of the ticks. Ticks are just lines. See [G] **lines** for more information.

Remarks

You draw a bar, dot, or box plot of empcost by division:

    . graph ... empcost, over(division)

Seeing the result, you wish to make the text labeling the divisions 20% larger. You type:

    . graph ... empcost, over(division, label(labsize(*1.2)))

Also See

Complementary: [G] **graph bar**, [G] **graph box**, [G] **graph dot**

Title

cat_axis_line_options — Options for specifying look of categorical axis line

Syntax

The *cat_axis_line_options* are

cat_axis_line_options	description
off and on	suppress/force display of axis
fill	allocate space for axis even if off
fextend	extend axis line through plot region and plot region's margin
extend	extend axis line through plot region
noextend	do not extend axis line at all
noline	do not even draw axis line
line	force drawing of axis line
titlegap(*relativesize*)	margin between axis title and tick labels
outergap(*relativesize*)	margin outside of axis title
lstyle(*linestyle*)	overall style of axis line
lcolor(*colorstyle*)	color of axis line
lwidth(*linewidthstyle*)	thickness of axis line
lpattern(*linepatternstyle*)	whether axis solid, dashed, etc.

See [G] *relativesize*, [G] *linestyle*, [G] *colorstyle*, [G] *linewidthstyle* and [G] *linepatternstyle*.

Description

The *cat_axis_line_options* determine the look of the categorical x axis in graph bar, graph hbar, graph dot, and graph box. These options are rarely specified but when specified, are specified inside axis() of over():

. graph ..., over(*varname*, ... axis(*cat_axis_line_options*) ...)

Options

off and on suppress or force the display of the axis.

fill goes with off and is seldom specified. If you turned an axis off but still wanted the space to be allocated for the axis, you could specify fill.

fextend, extend, noextend, line, and noline determine how much of the line representing the axis is to be drawn. They are alternatives.

noline specifies that the line is not to be drawn at all. The axis is there, ticks and labels will appear, but the line that is the axis itself will not be drawn.

93

`line` is the opposite of `noline`, for use if the axis line somehow got turned off.

`noextend` specifies that the axis line should not extend beyond the range of the axis, defined by the first and last categories.

`extend` specifies that the line should be longer than that and extend all the way across the plot region.

`fextend` specifies that the line should be longer than that and extend across the plot region and across the plot region's margins. For a definition of the plot region's margins, see [G] *region_options*. If the plot region has no margins (which would be rare), then `fextend` means the same as `extend`. If the plot region does have margins, `extend` would result in the y and x axes not meeting. With `fextend`, they touch.

`fextend` is the default with most schemes.

`titlegap(`*relativesize*`)` specifies the margin to be inserted between the axis title and the axis' tick labels.

`outergap(`*relativesize*`)` specifies the margin to be inserted outside of the axis title.

`lstyle(`*linestyle*`)`, `lcolor(`*colorstyle*`)`, `lwidth(`*linewidthstyle*`)`, and `lpattern(`*linepatternstyle*`)` determine the overall look of the line that is the axis; see [G] **lines**.

Remarks

The *cat_axis_label_options* are rarely specified.

Also See

Complementary: [G] **graph bar**, [G] **graph box**, [G] **graph dot**

Title

> *clockpos* — Choices for location: direction from central point

Syntax

clockpos is

#	$0 \leq \# \leq 12$, # an integer

Description

clockpos specifies a location or a direction.

clockpos is specified inside options such as the `position()` title option (see [G] *title_options*) or the `mlabposition` marker-label option (see [G] *marker_label_options*):

. graph ..., title(..., position(*clockpos*)) ...

. graph ..., mlabposition(*clockposlist*) ...

In cases where a *clockposlist* is allowed, you may specify a sequence of *clockpos* separated by spaces. Shorthands are allowed to make specifying the list easier; see [G] *stylelists*.

Remarks

clockpos is used to specify a location around a central point:

```
     11   12   1
   10             2
      9    0    3
      8         4
         7   6   5
```

Sometimes the central position is a well-defined object (e.g., in the case of `mlabposition()`, the central point is the marker itself), and sometimes the central position is implied (e.g., in the case of the title-option `position`, the central point is the center of the plot region).

In all cases, note that *clockpos* 0 is allowed: it refers to the center.

Also See

Complementary: [G] *marker_label_options*, [G] *title_options*; [G] *compassdirstyle*

Title

colorstyle — Choices for color

Syntax

colorstyle may be

colorstyle	description
black	erases if black background
gs0	gray scale: 0 = black
gs1	gray scale: very dark gray
gs2	
.	
.	
gs15	gray scale: very light gray
gs16	gray scale: 16 = white
white	erases if white background
blue	
bluishgray	
brown	
cyan	
dimgray	between gs14 and gs15
dkgreen	dark green
dknavy	dark navy blue
dkorange	dark orange
eggshell	
emerald	
forest_green	
gold	
gray	equivalent to gs8
green	
khaki	
lavender	
lime	
ltblue	light blue
ltbluishgray	light blue-gray, used by scheme s2color
ltkhaki	light khaki
magenta	
maroon	
midblue	
midgreen	
mint	
navy	

colorstyle	description
cranberry	
olive	
olive_teal	
orange	
orange_red	
pink	
purple	
red	
sand	
sandb	bright sand
sienna	
stone	
teal	
yellow	
	colors used by *The Economist* magazine:
ebg	background color
ebblue	bright blue
edkblue	dark blue
eltblue	light blue
eltgreen	light green
emidblue	midblue
erose	rose
none	no color; invisible; draws nothing
background or bg	same color as background
foreground or fg	same color as foreground
# # #	RGB value; white = "255 255 255"
color*#	color with adjusted intensity
*#	default color with adjusted intensity

When specifying an RGB value, it is best to enclose the values in quotes; type "128 128 128" and not 128 128 128.

For a color palette showing an individual color, type

    palette color *colorstyle* [, <u>sch</u>eme(*schemename*)]

and for a palette comparing two colors, type

    palette color *colorstyle* *colorstyle* [, <u>sch</u>eme(*schemename*)]

For instance, you might type

    . stata palette color red green

See [G] **palette**.

Note that wherever a *colorstyle* appears, you may specify an RGB value by specifying three numbers in sequence. Each number should be between 0 and 255, and the triplet indicates the amount of red, green, and blue to be mixed. Each of the words above is equivalent to an RGB value.

Other *colorstyles* may be available; type

> . graph query colorstyle

to obtain the full list installed on your computer.

Description

    colorstyle is how you specify the color of a graphical component. *colorstyle* is specified with many different options; all have the form

    ⟨*object*⟩color(*colorstyle*)

For instance, option mcolor() specifies the color of markers and option clcolor() specifies the colors of connecting lines. Anywhere you see *colorstyle*, you may choose from the list above.

    You will sometimes see that a *colorstylelist* is allowed, as in

> . scatter ..., msymbol(*colorstylelist*) ...

    A *colorstylelist* is a sequence of *colorstyles* separated by spaces. Shorthands are allowed to make specifying the list easier; see [G] *stylelists*. When specifying RGB values in *colorstylelists*, remember to enclose the three numbers in quotes.

Remarks

    Remarks are presented under the headings

> *Colors are independent of the background color*
> *White backgrounds and black backgrounds*
> *RGB values*
> *Adjusting intensity*

Colors are independent of the background color

    With the exception of the colors background and foreground, colors do not change because of the background color. Colors background and foreground obviously do change, but otherwise, black means black, red means red, white means white, and so on. White on a black background has high visibility; white on a white background is invisible. Inversely, black on a white background has high visibility; black on a black background is invisible.

    Color foreground always has high visibility.

    Color background is the background color: draw something in this color and you will erase whatever is underneath it.

    Color none is no color at all. Draw something in this color and you will be invisible. Being invisible, it will not hide whatever is underneath it.

White backgrounds and black backgrounds

    The colors do not change because of the background color, but which colors look best depends on the background color.

    Graphs on the screen look best against a black background. With a black background, light colors stand out and dark colors blend into the background.

Graphs on paper are usually presented against a white background. Dark colors stand out and light colors blend into the background.

Because most users need to make printed copies of their graphs, Stata's default is to present graphs on a white background, but you can change that; see [G] **schemes**.

If you want a dark background, it is better to choose a dark-background rather than attempt to darken the background using the *region_option* `graphregion(fcolor())` (see [G] *region_options*); everything else about the graph's scheme will assume a background similar to how it was originally.

`graphregion(fcolor())` (and `graphregion(ifcolor())`, `plotregion(fcolor())`, and `plotregion(ifcolor())`) are best used for adding a little tint to the background.

RGB values

In addition to colors such as `red`, `green`, `blue`, `cyan`, etc., you may mix your own colors by specifying RGB values. An RGB value is a triplet of numbers, each of which specifies on a scale of 0 to 255, the amount of red, green, and blue to be mixed. That is,

```
red    =   255    0    0
green  =     0  255    0
blue   =     0    0  255

cyan   =     0  255  255
purple =   255    0  255
yellow =   255  255    0
white  =   255  255  255
```

The overall scale of the triplet affects intensity, and thus changing 255 to 128 in all of the above would keep the colors the same, but make them dimmer. (Color 128 128 128 is what most people call gray.)

Adjusting intensity

You may specify things such as

`green*.8 red*1 purple*1.2 0 255 255*.8`

to specify a color and modify its intensity (brightness). Multiplying a color by 1 leaves the color unchanged. Multiplying by a number greater than 1 makes the color stand out from the background more; multiplying by a number less than 1 makes the color blend into the background more. For an example using the intensity adjustment, see *Typical use* in [G] **graph twoway kdensity**.

When modifying intensity, the syntax is

color*#

or the color may be omitted,

*#

If the color is omitted, the intensity adjustment is applied to the default color given the context. For instance, you specify `bcolor(*.7)` with `graph twoway bar`—or any other `graph twoway` command that fills an area—to use the default color at 70% intensity. Or you specify `bcolor(*2)` to use the default color at twice its usual intensity.

When you do specify both the color and the adjustment, you must type the color first: `.8*green` will not be understood. Also, do not put a space between the *color* and the *, even when the *color* is an RGB value.

color∗0 makes the color as dim as possible—it is not equivalent to color none. *color*∗255 makes the color as bright as possible, although values much smaller than 255 usually achieve the same result.

Also See

Complementary: [G] **palette**

Related: [G] **schemes**

Title

compassdirstyle — Choices for location

Syntax

compassdirstyle may be

| | ─── synonyms ─── | | | |
compassdirstyle	first	second	third	fourth
north	n	12		top
neast	ne	1	2	
east	e	3		right
seast	se	4	5	
south	s	6		bottom
swest	sw	7	8	
west	w	9		left
nwest	nw	10	11	
center	c	0		

Other *compassdirstyles* may be available; type

    . graph query compassdirstyle

to obtain the full list installed on your computer.

Description

compassdirstyle specifies a direction.

compassdirstyle is specified inside options such as the placement() textbox suboption of title() (see [G] *title_options* and [G] *textbox_options*):

    . graph ..., title(..., placement(*compassdirstyle*)) ...

In some cases, you may see that a *compassdirstylelist* is allowed: A *compassdirstylelist* is a sequence of *compassdirstyles* separated by spaces. Shorthands are allowed to make specifying the list easier; see [G] *stylelists*.

Remarks

Two methods are used for specifying directions—the compass and the clock—some options use the compass and some use the clock. For instance, the textbox option position() uses the compass (see [G] *textbox_options*), but the title option position() uses the clock (see [G] *title_options*). The reason for the difference is that some options need only the eight directions specified by the compass whereas others need more. In any case, synonyms are provided so that you can use the clock notation in all cases.

Also See

Complementary: [G] *textbox_options*

Title

connect_options — Options to connect points with lines

Syntax

The *connect_options* are

connect_options	description
connect(*connectstyle*)	how to connect points
sort [(*varlist*)]	how to sort before connecting
cmissing({ y | n } ...)	whether missing values are ignored
clstyle(*linestyle*)	overall style of line
clpattern(*linepatternstyle*)	whether line solid, dashed, etc.
clwidth(*linewidthstyle*)	thickness of line
clcolor(*colorstyle*)	color of line

See [G] **connectstyle**, [G] **linestyle**, [G] **linepatternstyle**, [G] **linewidthstyle**, and [G] **colorstyle**.

All options are *rightmost*; see [G] **repeated options**. If both sort and sort(*varlist*) are specified, sort is ignored and sort(*varlist*) is honored.

Description

The *connect_options* specify how points on a graph are to be connected.

In certain contexts (e.g., scatter; see [G] **graph twoway scatter**), the clstyle(), clpattern(), clwidth(), and clcolor() options may be specified with a list of elements, with the first element applying to the first variable, the second element to the second variable, and so on. For information on specifying lists, see [G] **stylelists**.

Options

connect(*connectstyle*) specifies whether points are to be connected and, if so, how the line connecting them is to be shaped. The line between each pair of points can connect them directly or in stairstep fashion.

sort and sort(*varlist*) specify how the data should be sorted before the points are connected.

sort specifies that the data should be sorted by the x variable.

sort(*varlist*) specifies that the data are to be sorted by the specified variables.

sort is the option usually specified. Unless you are after a special effect or your data are already sorted, do not forget to specify this option. If you are after a special effect, and if the data are not already sorted, you can specify sort(*varlist*) to specify exactly how the data should be sorted.

Specifying `sort` or `sort(`*varlist*`)` when it is not necessary will slow `graph` down a little. It is usually necessary that you specify `sort` if you specify the `twoway` option `by()`, and especially if you include the suboption `total`.

Options `sort` and `sort(`*varlist*`)` may not be repeated within the same plot.

`cmissing({ y | n } ...)` specifies whether missing values are ignored. The default is `cmissing(y ...)`, meaning that they are ignored. Consider the following data:

	rval	x
1.	.923	1
2.	3.046	2
3.	5.169	3
4.	.	.
5.	9.415	5
6.	11.538	6

Say you graph these data using "`line rval x`" or equivalently "`scatter rval x, c(l)`". Do you want a break in the line between 3 and 5? If so, you code

```
. line rval x, cmissing(n)
```

or equivalently

```
. scatter rval x, c(l) cmissing(n)
```

If you omit the option (or code `cmissing(y)`), the data are treated as if they contained

	rval	x
1.	.923	1
2.	3.046	2
3.	5.169	3
4.	9.415	5
5.	11.538	6

which is to say that a line will be drawn between (3, 5.169) and (5, 9.415).

If you are plotting more than one variable, you may specify a sequence of y/n answers.

`clstyle(`*linestyle*`)`, `clpattern(`*linepatternstyle*`)`, `clwidth(`*linewidthstyle*`)`, and `clcolor(`*colorstyle*`)` determine the look of the line used to connect the points; see [G] **lines**. Note the `clpattern()` option, which will allow you to specify whether the line is solid, dashed, etc.; see [G] *linepatternstyle* for a list of line-pattern choices.

Remarks

An important option among all the above is `connect()`, which determines whether and how the points are connected. The points need not be connected at all (`connect(i)`), which is `scatter`'s default. Or the points might be connected by straight lines (`connect(l)`), which is `line`'s default (and available in `scatter`). `connect(i)` and `connect(l)` are commonly specified, but there are other possibilities such as `connect(J)` (which connects in stairstep fashion and is appropriate for empirical distributions). See [G] *connectstyle* for a full list of your choices.

Equally as important as `connect()` is `sort`. If you do not specify this, the points will be connected in the order in which they are encountered. That can be useful when you are creating special effects, but, in general, you want the points sorted into ascending order of their x variable. That is what `sort` does.

The remaining connect options specify how the line is to look: Is it solid or dashed? Is it red or green? How thick is it? Option `clpattern()` can be of great importance, especially when printing to a monochrome printer. For a general discussion of lines (which occur in lots of contexts other than connecting points), see [G] **lines**.

Also See

Complementary: [G] **lines**; [G] *colorstyle*, [G] *connectstyle*, [G] *linestyle*,
[G] *linepatternstyle*, [G] *linewidthstyle*

Title

connectstyle — Choices for how points are connected

Syntax

connectstyle may be

connectstyle	synonym	description
none	i	do not connect
direct	l	connect with straight lines
ascending	L	direct, but only if $x_{j+1} > x_j$
stairstep	J	flat, then vertical
stepstair		vertical, then flat

Other *connectstyles* may be available; type

    . graph query connectstyle

to obtain the full list installed on your computer.

Description

connectstyle specifies if and how points in a scatter are to be connected; e.g., via straight lines, stairsteps, etc.

connectstyle is specified inside the connect() option which is allowed, for instance, with scatter:

    . scatter ... , connect(*connectstylelist*) ...

In this case, a *connectstylelist* is allowed. A *connectstylelist* is a sequence of *connectstyles* separated by spaces. Shorthands are allowed to make specifying the list easier; see [G] *stylelists*.

Remarks

Points are connected in the order of the data, so be sure that data are in the desired order (which is usually ascending value of x) before specifying the connect(*connectstyle*) option. Commands that provide connect() also provide a sort option, which will sort by the x variable for you.

connect(l) is the most common choice.

connect(J) is the appropriate way to connect the points of empirical cumulative distribution functions (CDFs).

Also See

Complementary: [G] *connect_options*

Title

eps_options — Options for exporting to Encapsulated PostScript

Syntax

The *eps_options* when exporting to Encapsulated PostScript are

eps_options	description
logo(on | off)	whether to include Stata logo
mag(#)	magnification/shrinkage factor; 100 = default
fontface(*fontname*)	font to use
orientation(portrait | landscape)	whether vertical or horizontal

Current default values may be listed by typing

    . graph set eps

and default values may be set by typing

    . graph set eps *name value*

where *name* is the name of an *eps_option*, omitting the parentheses.

Description

These *eps_options* are used with graph export when creating an Encapsulated PostScript file; see [G] **graph export**.

Options

logo(on) and logo(off) specify whether the Stata logo should be included at the bottom of the graph.

mag(#) specifies that the graph is to be drawn smaller or larger than ordinarily. mag(100) is the default, meaning ordinary size. mag(110) would make the graph 10% larger than usual and mag(90) would make the graph 10% smaller than usual. # must be an integer.

fontface(*fontname*) specifies the name of the PostScript font to be used to render text. If *fontname* includes blanks, *fontname* must be enclosed in double quotes.

orientation(portrait) and orientation(landscape) specify whether the graph is to be presented vertically or horizontally.

Remarks

Remarks are presented under the headings

Using the *eps_options*
Setting defaults

Using the eps_options

You have drawn a graph and wish to create an Encapsulated PostScript file for including the file in a document. You wish, however, to change the text from helvetica to roman, which is "Times" in PostScript jargon:

```
. graph ...                                    (draw a graph)
. graph export myfile.eps, fontface(Times)
```

Setting defaults

If you always wanted `graph export` to use Times when exporting to Encapsulated PostScript files, you could type

```
. graph set eps fontface Times
```

At a future date, you could type

```
. graph set eps fontface Helvetica
```

to set it back. You can list the current *eps_option* settings for Encapsulated PostScript by typing

```
. graph set eps
```

Also See

Complementary: [G] **graph export**

Title

gph files — Using gph files

Description

.gph files contain Stata graphs and, in fact, even include the original data from which the graph was drawn. Below, we discuss how to replay graph files and to obtain the data inside them.

Remarks

Remarks are presented under the headings

> *Background*
> *Gph files are machine/operating system independent*
> *Gph files come in three flavors*
> *Advantages of live-format files*
> *Advantages of as-is format files*
> *Retrieving data from live-format files*

Background

.gph files are created either by including the `saving()` option when you draw a graph,

```
. graph ..., ... saving(myfile)
```

or by using the `graph save` command afterwards:

```
. graph ...
. graph save myfile
```

Either way, file `myfile.gph` is created; for details see [G] *saving_option* and [G] **graph save**.

At some later time, in the same session or in a different session, you can redisplay what is in the .gph file by typing

```
. graph use myfile
```

See [G] **graph use** for details.

Gph files are machine/operating system independent

The .gph files created by `saving()` and `graph save` are binary files written in a machine-and-operating-system independent format. You may send .gph files to other users, and they will be able to read them, even if you use, say, a Macintosh and your colleague uses a Windows or Unix computer.

Gph files come in three flavors

There are three flavors of `graph` files:

1. an old-format Stata 7 or earlier .gph file,

2. a modern-format graph in as-is format,

3. a modern-format graph in live format.

You can find out which type a .gph file is by typing

> . graph describe *filename*

See [G] **graph describe**.

Live-format files contain the data and other information necessary to recreate the graph. As-is format files contain a recording of the picture. When you save a graph, unless you specify the asis option, it is saved in live format.

Advantages of live-format files

A live-format file can be edited at a later date and a live-format file can be displayed using different schemes; see [G] **schemes**. In addition, the data used to create the graph can be retrieved from the .gph file.

Advantages of as-is format files

As-is format files are generally smaller than live-format files.

As-is format files cannot be modified; the rendition is fixed and will appear on anyone else's computer just as it last appeared on yours.

Retrieving data from live-format files

First, verify that you have a live-format file, type

> . graph describe *filename*.gph

Then type

> . discard

This will close any open graphs and eliminate anything stored having to do with previously existing graphs. Now display the graph of interest,

> . graph use *filename*

and then type

> . serset dir

From this point on, you are going to have to do a little detective work, but usually it is not much. Sersets are how graph stores the data corresponding to each plot within the graph. You can see [P] **serset**, but unless you are a programmer curious about how things work, that will not be necessary. We will show you below how to load each of the sersets (often there is just one) and to treat it from then on just as if it came from a .dta file.

Let us set up an example. Pretend that previously we have drawn the graph and saved it by typing

> . sysuse lifeexp, clear
> (Life expectancy 1998)

```
. scatter lexp gnppc, by(region)
```

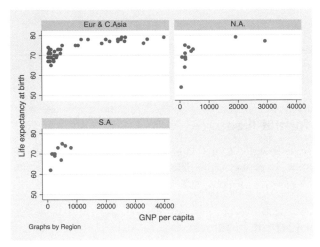

```
. graph save legraph
(file legraph.gph saved)
```

Following the instructions, we now type

```
. graph describe legraph.gph
```

legraph.gph stored on disk

```
       name:  legraph.gph
     format:  live
    created:  20 Nov 2002 13:04:30
   dta file:  C:\stata8\ado\base\lifeexp.dta dated 26 Sep 2002 14:08
    command:  twoway scatter lexp gnppc, by(region)
```

```
. discard
```

```
. graph display legraph
```

```
  0.   44 observations on 2 variables
       lexp gnppc
```

```
  1.   14 observations on 2 variables
       lexp gnppc
```

```
  2.   10 observations on 2 variables
       lexp gnppc
```

We discover that our graph has three sersets. Looking at the graph, that should not surprise us. Although we might think of

```
. scatter lexp gnppc, by(region)
```

as being one plot, it is in fact three if we were to expand it:

```
. scatter lexp gnppc if region==1 ||
  scatter lexp gnppc if region==2 ||
  scatter lexp gnppc if region==3
```

The three sersets numbered 0, 1, and 2 correspond to three pieces of the graph. We can look at the individual sersets. To load a serset, you first set its number and then you type `serset use, clear`:

```
. serset 0
```

```
. serset use, clear
```

If we were now to type describe, we would discover that we have a 44-observation dataset containing two variables: lexp and gnppc. Here is a little bit of the data:

```
. list in 1/5
```

	lexp	gnppc
1.	72	810
2.	74	460
3.	79	26830
4.	71	480
5.	68	2180

These are the data that appeared in the first plot. We could similarly obtain the data for the second plot by typing

```
. serset 1
. serset use, clear
```

If we wanted to put these data back together into one file, we might type

```
. serset 0
. serset use, clear
. generate region=0
. save region0

. serset 1
. serset use, clear
. generate region=1
. save region1

. serset 2
. serset use, clear
. generate region=2
. save region2

. use region0
. append using region1
. append using region2

. erase region0.dta
. erase region1.dta
. erase region2.dta
```

In general, it will not be as much work to retrieve the data because in many graphs, you will find there is only one serset. We chose a complicated .gph file for our demonstration.

Also See

Complementary: [G] **graph display**, [G] **graph save**, [G] *saving_option*,

[G] **graph manipulation**,

[P] **serset**

Title

graph — The graph command

Syntax

graph ...

The commands that draw graphs are

command	description
graph twoway	scatterplots, line plots, etc.
graph matrix	scatterplot matrices
graph bar	bar charts
graph dot	dot charts
graph box	box and whisker plots
graph pie	pie charts
other	more commands to draw statistical graphs

See [G] **graph twoway**, [G] **graph matrix**, [G] **graph bar**,
[G] **graph dot**, [G] **graph box**, [G] **graph pie**, and [G] **graph other**.

The commands that save a previously drawn graph, redisplay previously saved graphs, and combine
graphs are

command	description
graph save	save graph to disk
graph use	redisplay graph stored on disk
graph display	redisplay graph stored in memory
graph combine	combine graphs into one

See [G] **graph save**, [G] **graph use**, [G] **graph display**, and
[G] **graph combine**.

The commands for printing a graph are

command	description
graph print	print currently displayed graph
set printcolor	set how colors are printed
graph export	export .gph file to PostScript, etc.

See [G] **graph print**, [G] **set printcolor**, and [G] **graph export**.

The commands that deal with the graphs currently stored in memory are

command	description
graph display	display graph
graph dir	list names
graph describe	describe contents
graph rename	rename memory graph
graph copy	copy memory graph to new name
graph drop	discard graphs in memory

See [G] **graph manipulation**; [G] **graph display**, [G] **graph dir**, [G] **graph describe**, [G] **graph rename**, [G] **graph copy**, and [G] **graph drop**.

The commands that tell you about available schemes, your default scheme, and allow you to set the default are

command	description
graph query, schemes	list available schemes
query graphics	identify default scheme
set scheme	set default scheme

See [G] **schemes** and see [G] **graph query** and [G] **set scheme**.

The command that tells you about available styles is

command	description
graph query	list available styles

See [G] **graph query**.

The command that allows you to draw graphs without displaying them is

command	description
set graphics	set whether graphs are displayed

See [G] **set graphics**.

Description

graph draws graphs.

Remarks

See [G] **graph intro**.

Also See

Complementary: [G] **graph intro**, [G] **graph other**; [G] **graph export**, [G] **graph print**,
[G] **graph7**

Title

graph7 — The old graph command

Syntax

$\{$ graph7 | gr7 $\}$ [*varlist*] [*weight*] [if *exp*] [in *range*] [, ...]

$\{$ graph7 | gr7 $\}$ using *filename* [*filename* ...] [, ...]

Description

Stata's graph command change markedly as of Stata 8. graph7 and gr7 (they are synonyms) provide access to the old graph command.

Remarks

graph, run under version control, also is the old graph command, and thus old ado-files and commands continue to work although they of course do not pick up the new features available under Stata 8.

Also See

Complementary: [G] **graph**

Title

graph bar — Bar charts

Syntax

graph bar *yvars* [*weight*] [if *exp*] [in *range*] [, *options*]

graph hbar *yvars* [*weight*] [if *exp*] [in *range*] [, *options*]

where *yvars* is

(asis) *varlist*

or is

[(*stat*)] *varname*   [[(*stat*)] ...]

[(*stat*)] *varlist*   [[(*stat*)] ...]

[(*stat*)] *name=varname* [...]   [[(*stat*)] ...]

where *stat* may be any of

mean median p1 p2 ... p99 sum count min max

or

any of the other *stats* defined in [R] **collapse**

mean is the default. p1 means 1st percentile, p2 second, and so on; p50 means the same as median. count means the number of nonmissing values of the specified variable.

and where *options* are

options	description
group_options	groups over which bars are drawn
yvar_options	variables that are the bars
lookofbar_options	how the bars look
legending_options	how *yvars* are labeled
axis_options	how the numerical y axis is labeled
title_and_other_options	titles, added text, aspect ratio, etc.

Each is defined below.

The *group_options*, which define the groups over which the bars are drawn, are

group_options	description
over(*varname* [, *over_subopts*])	categories; option may be repeated
nofill	omit empty categories
missing	keep missing value as category

The *yvar_options*, which define the variables that are the bars, are

yvar_options	description
ascategory	treat *yvars* as first over() group
asyvars	treat first over() group as *yvars*
percentages	show percentages within *yvars*
stack	stack the *yvar* bars
cw	calculate *yvar* statistics omitting missing values of any *yvar*

The *lookofbar_options*, which define how the bars look, are

lookofbar_options	description
outergap([*]#)	gap between edge and first bar and between last bar and edge
bargap(#)	gap between *yvar* bars; default 0
intensity([*]#)	intensity of fill
lintensity([*]#)	intensity of outline
bar(#, barlook_options)	look of #th *yvar* bar

See [G] **barlook_options**.

The *legending_options*, which define how the *yvars* are labeled, are

legending_options	description
legend_option	control of *yvar* legend
nolabel	use *yvar* names, not labels, in legend
yvaroptions(over_subopts)	*over_subopts* for *yvars*; rarely specified
showyvars	label *yvars* on *x* axis; rarely specified
blabel(...)	add labels to bars

See [G] **legend_option** and [G] **blabel_option**.

The *axis_options*, which define how the numerical *y* axis is labeled, are

axis_options	description
yalternate	put numerical *y* axis on right (top)
xalternate	put categorical *x* axis on top (right)
exclude0	do not force *y* axis to include 0
yreverse	reverse *y* axis
axis_scale_options	*y*-axis scaling and look
axis_label_options	*y*-axis labeling
ytitle(...)	*y*-axis titling

See [G] **axis_scale_options**, [G] **axis_label_options**, and [G] **axis_title_options**.

The *title_and_other_options* are

title_and_other_options	description
text(...)	add text on graph; x range $\begin{bmatrix} 0, 100 \end{bmatrix}$
yline(...)	add y lines to graph
std_options	titles, aspect ratio, saving to disk
by(*varlist*, ...)	repeat for subgroups

See [G] *added_text_option*, [G] *added_line_options*, [G] *std_options*, and [G] *by_option*.

The *over_subopts*, used in over(*varname*, *over_subopts*) and, on rare occasion, in yvaroptions(*over_subopts*), are

over_subopts	description
relabel(# "text" ...)	change axis labels
label(*cat_axis_label_options*)	rendition of labels
axis(*cat_axis_line_options*)	rendition of axis line
gap($\begin{bmatrix} * \end{bmatrix}$#)	gap between bars within over() category
sort(*varname*)	put bars in prespecified order
sort(#)	put bars in height order
sort((*stat*)) *varname*	put bars in derived order
descending	reverse default or specified bar order

See [G] *cat_axis_label_options* and [G] *cat_axis_line_options*.

aweights, fweights, and pweights are allowed; see [U] **14.1.6 weight** and see note concerning weights in [R] **collapse**.

Description

graph bar draws bar charts, vertically presented. In a vertical bar chart, the y axis is numerical and the x axis is categorical.

. graph bar (mean) *numeric_var*, over(*cat_var*)

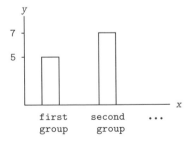

numeric_var must be numeric; statistics of it are shown on the y axis.

cat_var may be numeric or string; it is shown on the categorical x axis.

graph hbar draws horizontal bar charts. In a horizontal bar chart, the numerical axis is still called the y axis and the categorical axis is still called the x axis, but y is presented horizontally and x, vertically.

. graph hbar (mean) *numeric_var*, over(*cat_var*)

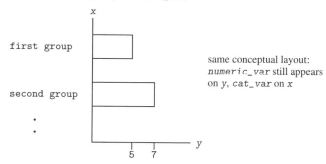

Note that the syntax for vertical and horizontal bar charts is the same; all that is required is changing `bar` to `hbar` or `hbar` to `bar`.

Options

Group options

over(*varname*[, *over_subopts*]) specifies a categorical variable over which the *yvars* are to be repeated. *varname* may be string or numeric. Up to two over() options may be specified when multiple *yvars* are specified, and up to three over()s may be specified when a single *yvar* is specified; options may be specified; see *Examples of syntax* and *Multiple over()s* under *Remarks* below.

nofill specifies that missing subcategories are to be omitted. For instance, consider

. graph bar (mean) y, over(division) over(region)

Say that one of the divisions has no data for one of the regions, either because there are no such observations or because y==. for such observations. In the resulting chart, the bar will be missing:

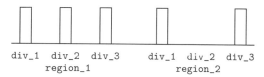

If you specify `nofill`, the missing category will be removed from the chart:

missing specifies that missing values of the over() variables be kept as their own categories, one for ., another for .a, etc. The default is to act as if such observations simply did not appear in the dataset; the observations are ignored. An over() variable is considered to be missing if it is numeric and contains a missing value or if it is string and contains " ".

yvar options

ascategory specifies that the *yvars* be treated as the first over() group; see *Treatment of bars* under *Remarks* below. ascategory is a very useful option.

When you specify ascategory, results are as if you specified a single *yvar* and introduced a new first over() variable. Anyplace you read in the documentation that something is done over the first over() category, or using the first over() category, it will be done over or using *yvars*.

Think like this: you specified

. graph bar y1 y2 y3, ascategory *whatever_other_options*

and results will be as if you typed

. graph bar y, over(*newcategoryvariable*) *whatever_other_options*

with a long rather than wide dataset in memory.

asyvars specifies that the first over() group be treated as *yvars*. See *Treatment of bars* under *Remarks* below. asyvars is seldom specified.

When you specify asyvars, are as if you removed the first over() group and introduced multiple *yvars*. If you previously had k *yvars* and, in your first over() category, G groups, results will be as if you specified $k \times G$ *yvars* and removed the over(). Anyplace you read in the documentation that something is done over the *yvars* or using the *yvars*, it will be done over or using the first over() group.

Think like this: you specified

. graph bar y, over(group) asyvars *whatever_other_options*

and results will be as if you typed

. graph bar *y1 y2 y3* ... , *whatever_other_options*

with a wide rather than a long dataset in memory. Variables *y1*, *y2*, ..., are sometimes called the virtual *yvars*.

percentages specifies that bar heights are to be based on percentages *yvar_i* represents of all the *yvars*. That is,

. graph bar (mean) inc_male inc_female

would produce a chart with bar height reflecting average income.

. graph bar (mean) inc_male inc_female, percentage

would produce a chart with the bar heights being $100 \times$ inc_male$/($inc_male $+$ inc_female$)$ and $100 \times$ inc_female$/($inc_male $+$ inc_female$)$.

If you have a single *yvar* and want percentages calculated over the first over() group, specify the asyvars option. For instance,

. graph bar (mean) wage, over(*i*) over(*j*)

would produce a chart where bar heights reflect mean wages.

. graph bar (mean) wage, over(*i*) over(*j*) asyvars percentages

would produce a chart where bar heights are

$$100 \times \left(\frac{\text{mean}_{ij}}{\sum_i \text{mean}_{ij}} \right)$$

Option `stack` is often combined with option `percentage`.

`stack` specifies that the *yvar* bars are to be stacked.

    . graph bar (mean) inc_male inc_female, over(region) percentage stack

would produce a chart with all bars being the same height, 100%. Each bar would be two bars stacked, percent of inc_male and percent of inc_female, and so the division would show the relative shares of inc_male and inc_female of total income.

If you wish to stack bars over the first `over()` group, specify the `asyvars` option:

    . graph bar (mean) wage, over(sex) over(region) asyvars percentage stack

`cw` specifies casewise deletion. If `cw` is specified, observations for which any of the *yvars* are missing are ignored. The default is to calculate the requested statistics using all the data possible.

lookofbar options

`outergap(*#)` and `outergap(#)` specify the gap between the edge of the graph to the beginning of the first bar and the end of the last bar to the edge of the graph.

> `outergap(*#)` specifies that the default is to be modified. Specifying `outergap(*1.2)` increases the gap by 20% and specifying `outergap(*.8)` reduces the gap by 20%.

> `outergap(#)` specifies the gap in percent of bar-width units. `outergap(50)` specifies that the gap should be half the bar width.

`bargap(#)` specifies the gap to be left between *yvar* bars in percent of bar-width units. The default is `bargap(0)`, meaning bars touch.

> `bargap()` may be specified as positive or negative numbers. `bargap(10)` puts a small gap between the bars (the precise amount being 10% of the width of the bars). `bargap(-30)` overlaps the bars by 30%.

> Note that `bargap()` affects only the *yvar* bars. If you want to change the gap for the first, second, or third `over()` groups, specify the *over_subopt* `gap()` inside the `over()` itself; see *Suboptions for use with over() and yvaroptions()* below.

`intensity(#)` and `intensity(*#)` specify the intensity of the color used to fill the inside of the bar. `intensity(#)` specifies the intensity and `intensity(*#)` specifies the intensity relative to the default.

> By default, the bar is filled with the color of its border, attenuated. Specify `intensity(*#)`, #< 1, to attenuate it more and specify `intensity(*#)`, #> 1, to amplify it.

> Specify `intensity(0)` if you do not want the bar filled at all. Specify `intensity(100)` if you want the bar to have the same intensity as the bar's outline.

`lintensity(#)` and `lintensity(*#)` specify the intensity of the line used to outline the bar. `lintensity(#)` specifies the intensity and `lintensity(*#)` specifies the intensity relative to the default.

> By default, the bar is outlined at the same intensity at which it is filled or at an amplification of that, which depending on your chosen scheme; see [G] **schemes**. If you want the bar outlined in the darkest possible way, specify `intensity(255)`. If you wish simply to amplify the outline, specify `intensity(*#)`, # > 1, and if you wish to attenuate the outline, specify `intensity(*#)`, # < 1.

bar(#, *barlook_options*) specifies the look of the *yvar* bars. bar(1, ...) refers to the bar associated with the first *yvar*, bar(2, ...) refers to the bar associated with the second, and so on. The most useful *barlook_option* is bcolor(*colorstyle*), which sets the color of the bar. For instance, you might specify bar(1, bcolor(green)) to make the bar associated with the first *yvar* green. See [G] *colorstyle* for a list of color choices and see [G] *barlook_options* for information on the other *barlook_options*.

Legending options

legend_option controls the legend. If more than one *yvar* is specified, a legend is produced. Otherwise, no legend is needed because the over() groups are labeled on the categorical *x* axis. See [G] *legend_option* and see *Treatment of bars* under *Remarks* below.

nolabel specifies that, in the automatic construction of the legend, the variable names of the *yvars* be used in preference to "mean of *varname*" or "sum of *varname*", etc.

yvaroptions(*over_subopts*) allows you to specify *over_subopts* for the *yvars*. This is very rarely done.

showyvars specifies that, in addition to building a legend, the identities of the *yvars* are to be shown on the categorical *x* axis. If showyvars is specified, it is typical to also specify legend(off).

blabel() allows you to add labels on top of the bars; see [G] *blabel_option*.

Axis options

yalternate and xalternate switch the side on which the axes appear.

Used with graph bar, yalternate moves the numerical *y* axis from the left to the right; xalternate moves the categorical *x* axis from the bottom to the top.

Used with graph hbar, yalternate moves the numerical *y* axis from the bottom to the top; xalternate moves the categorical *x* axis from the left to the right.

If your scheme by default puts the axes on the opposite sides, then yalternate and xalternate reverse their actions.

exclude0 specifies that the numerical *y* axis need not be scaled so as to include 0.

yreverse specifies that the numerical *y* axis have its scale reversed, so that it runs from maximum to minimum. Note that this option causes bars to extend down rather than up (graph bar) or right-to-left rather than left-to-right (graph hbar).

axis_scale_options specify how the numerical *y* axis is scaled and how it looks; see [G] *axis_scale_options*. There you will also see option xscale() in addition to yscale(). Ignore xscale(); it is irrelevant in the case of bar charts.

axis_label_options specify how the numerical *y* axis is to be labeled. The *axis_label_options* also allow you to add and suppress grid lines; see [G] *axis_label_options*. There you will see that, in addition to options ylabel(), ytick(), ..., ymtick(), options xlabel(), ..., xmtick() are allowed. Ignore the x*() options; they are irrelevant in the case of bar charts.

ytitle() overrides the default title for the numerical *y* axis; see [G] *axis_title_options*. There you will also find option xtitle() documented; it is irrelevant in the case of bar charts.

Title and other options

text() adds text to a specified location on the graph; see [G] **added_text_option**. The basic syntax of text() is

text(#_y #_x "*text*")

text() is documented in terms of twoway graphs. When used with bar charts, the "numeric" x axis is scaled to run from 0 to 100.

yline() adds horizontal (bar) or vertical (hbar) lines at specified y values; see [G] **added_line_options**. The xline() option, also documented there, is irrelevant in the case of bar charts. If your interest is in adding grid lines, see [G] **axis_label_options**.

std_options allow you to add titles, control the aspect ratio, save the graph on disk, and much more; see [G] **std_options**.

by(*varlist*, ...) draws separate plots within a single graph; see [G] **by_option** and see *Use with by()* under *Remarks* below.

Suboptions for use with over() and yvaroptions()

relabel(# "*text*" ...) specifies text to override the default labeling of the categories. Pretend variable sex took on two values and you typed

    . graph bar ..., ... over(sex, relabel(1 "Male" 2 "Female"))

The result would be to relabel the first value of sex to be "Male" and the second value, "Female"; "Male" and "Female" would appear on the categorical x axis to label the bars. This would be the result regardless of whether variable sex were string or numeric and regardless of the codes actually stored in the variable to record sex.

That is, # refers to category number, determined by sorting the unique values of the variable (in this case, sex) and assigning 1 to the first value, 2 to the second, and so on. If you are unsure as to what that ordering would be, the easy way to find out is to type

        over(*varname*, relabel(0 "Male" 1 "Female"))

    . tabulate sex

will change the axis labels to Male and Female. Note that # in relabel(# ...) refers to the values of *varname*. In this example, variable *varname* takes on values 0 and 1. Also note that, on the axis, the category will still be presented Male first. With numeric variables, the default ordering is the ordering implied by the numeric values.

If you also plan on specifying graph bar's or graph hbar's missing option,

    . graph bar ..., ... missing over(sex, relabel(...))

then type

    . tabulate sex, missing

to determine the coding. See [R] **tabulate**.

Relabeling the values does not change the order in which the bars are displayed.

You may create multiple-line labels using quoted strings within quoted strings:

over(*varname*, relabel(1 '" "Male" "patients" "' 2 '" "Female" "patients" "'))

When specifying quoted strings within quoted strings, remember to use compound double quotes '" and "' on the outer level.

relabel() may also be specified inside yvaroptions(). By default, the identity of the *yvars* is revealed in the legend, so specifying yvaroptions(relabel()) changes the legend. Because it is the legend that is changed, use of legend(label()) is preferred; see *Legending Options* above. In any case, specifying

yvaroptions(relabel(1 "Males" 2 "Females"))

changes the text that appears in the legend for the first *yvar* and the second *yvar*. Note that # in relabel(# ...) refers to *yvar* number. In this case, you may not use the nested-quotes to create multiline labels; use the legend(label()) option because it provides multiline capabilities.

label(*cat_axis_label_options*) determines other aspects of the look of the category labels on the *x* axis. With the exception of label(labcolor()) and label(labsize()), these options are seldom specified; see [G] *cat_axis_label_options*.

axis(*cat_axis_line_options*) specifies how the axis line is rendered. This is a rarely specified option. See [G] *cat_axis_line_options*.

gap(#) and gap(*#) specify the gap between the bars in this over() group. gap(#) is specified in percent-of-bar-width units, so gap(67) means two-thirds the width of a bar. gap(*#) allows modifying the default gap. gap(*1.2) would increase the gap by 20% and gap(*.8) would decrease the gap by 20%.

To understand the distinction between over(..., gap()) and option bargap(), consider

    . graph bar revenue profit, bargap(...) over(division, gap(...))

bargap() sets the distance between the revenue and profit bars. over(,gap()) sets the distance between the bars for the first division and the second division, the second division and the third, and so on. Similarly, in

    . graph bar revenue profit, bargap(...)
                            over(division, gap(...))
                            over(year,   gap(...))

over(division, gap()) sets the gap between divisions and over(year, gap()) sets the gap between years.

sort(*varname*), sort(#), and sort((*stat*) *varname*) control how bars are ordered. See *How bars are ordered* and *Reordering the bars* under *Remarks* below.

sort(*varname*) puts the bars in the order of *varname*; see *Putting the bars in a prespecified order* under *Remarks* below.

sort(#) puts the bars in height order. # refers to the *yvar* number on which the ordering should be performed; see *Putting the bars in height order* under *Remarks* below.

sort((*stat*) *varname*) puts the bars in an order based on a calculated statistic; see *Putting the bars in a derived order* under *Remarks* below.

descending specifies that, whatever the order of the bars, default or as specified by sort(), it is to be reversed.

(Continued on next page)

Remarks

Remarks are presented under the headings

Introduction
Examples of syntax
Treatment of bars
Treatment of data
Multiple bars (overlapping the bars)
Controlling the text of the legend
Multiple over()s (repeating the bars)
Nested over()s
Charts with lots of categories
How bars are ordered
Reordering the bars
Putting the bars in a prespecified order
Putting the bars in height order
Putting the bars in a derived order
Reordering the bars, example
Use with by()
Note: Repeated yvar names
History

Introduction

Let us show you some bar charts:

```
. sysuse citytemp
(City Temperature Data)

. graph bar (mean) tempjuly tempjan, over(region)
        bargap(-30)
        legend( label(1 "July") label(2 "January") )
        ytitle("Degrees Fahrenheit")
        title("Average July and January temperatures")
        subtitle("by regions of the United States")
        note("Source:  U.S. Census Bureau, U.S. Dept. of Commerce")
```

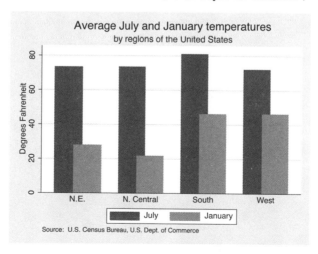

```
. sysuse citytemp, clear
(City Temperature Data)
```

```
. graph hbar (mean) tempjan, over(division) over(region) nofill
       ytitle("Degrees Fahrenheit")
       title("Average January temperature")
       subtitle("by region and division of the United States")
       note("Source:  U.S. Census Bureau, U.S. Dept. of Commerce")
```

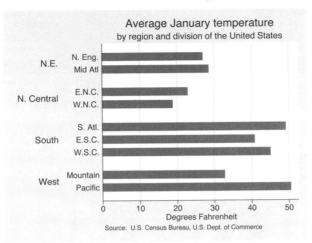

```
. sysuse nlsw88, clear
(NLSW, 1988 extract)

. graph bar (mean) wage, over(smsa) over(married) over(collgrad)
       title("Average Hourly Wage, 1988, Women Aged 34-46")
       subtitle("by College Graduation, Marital Status,
                and SMSA residence")
       note("Source:  1988 data from NLS, U.S. Dept. of Labor,
            Bureau of Labor Statistics")
```

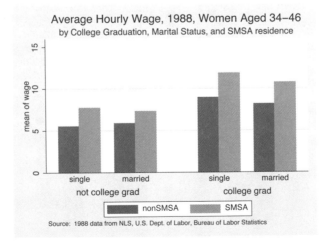

```
. sysuse educ99gdp, clear
(Education and GDP)

. generate total = private + public
```

```
. graph hbar (asis) public private,
    over(country, sort(total) descending) stack
    title( "Spending on tertiary education as % of GDP, 1999",
        span pos(11) )
    subtitle(" ")
    note("Source:  OECD, Education at a Glance 2002", span)
```

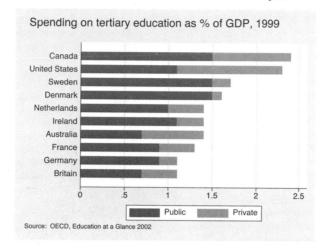

In the sections that follow, we explain how each of the above—and others—works.

Examples of syntax

Below we show you some graph bar commands and tell you what each would do:

graph bar revenue
    One big bar showing average revenue.

graph bar revenue profit
    Two bars, one showing average revenue, and the other showing average profit.

graph bar revenue, over(division)
    #_of_divisions bars showing average revenue for each division.

graph bar revenue profit, over(division)
    $2 \times$ *#_of_divisions* bars showing average revenue and average profit for each division. The grouping would look like this (assuming 3 divisions):

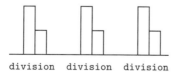

graph bar revenue, over(division) over(year)
    #_of_divisions $\times$ *#_of_years* bars showing average revenue for each division, repeated for each of the years. The grouping would look like this (assuming 3 divisions and 2 years):

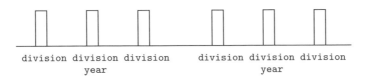

`graph bar revenue, over(year) over(division)`

same as above, but ordered differently. In the previous example, we typed `over(division)` `over(year)`. This time, we reverse it:

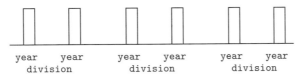

`graph bar revenue profit, over(division) over(year)`

$2 \times \#\_of\_divisions \times \#\_of\_years$ bars showing average revenue and average profit for each division, repeated for each of the years. The grouping would look like this (assuming 3 divisions and 2 years):

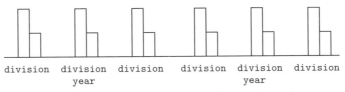

`graph bar (sum) revenue profit, over(division) over(year)`

$2 \times \#\_of\_divisions \times \#\_of\_years$ bars showing the sum of revenue and sum of profit for each division, repeated for each of the years.

`graph bar (median) revenue profit, over(division) over(year)`

$2 \times \#\_of\_divisions \times \#\_of\_years$ bars showing the median of revenue and median of profit for each division, repeated for each of the years.

`graph bar (median) revenue (mean) profit, over(division) over(year)`

$2 \times \#\_of\_divisions \times \#\_of\_years$ bars showing the median of revenue and mean of profit for each division, repeated for each of the years.

Treatment of bars

Assume someone tells you that the average January temperature in the North East of the United States is 27.9 degrees Fahrenheit, 27.1 degrees in the North Central, 46.1 in the South, and 46.2 in the West. You could enter these statistics and draw a bar chart:

```
. input ne nc south west

           ne         nc      south        west
  1. 27.9 21.7 46.1 46.2
  2. end
```

```
. graph bar (asis) ne nc south west
```

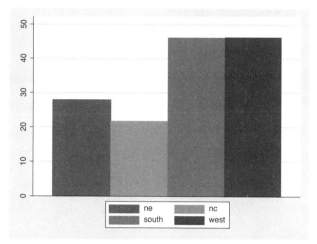

The above is admittedly not a great looking chart, but specifying a few options could fix that. The important thing to see right now is, when we specify multiple *yvars*, (1) the bars touch, (2) the bars are different colors (or at least different shades of gray), and (3) the meaning of the bars is revealed in the legend.

We could enter this data another way,

```
. clear
. input  str8 region  float tempjan
        region  tempjan
  1. N.E. 27.9
  2. "N. Central" 21.7
  3. South 46.1
  4. West 46.2
  5. end
. graph bar (asis) tempjan, over(region)
```

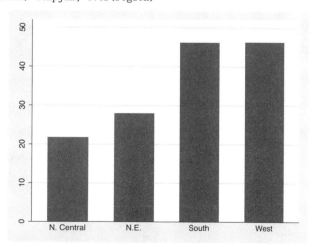

Observe that, when we generate multiple bars via an over() option, (1) the bars do not touch, (2) the bars are all the same color, and (3) the meaning of the bars is revealed by how the categorical x axis is labeled.

These differences in the treatment of the bars in the multiple *yvars* case and the over() case are general properties of graph bar and graph hbar:

	multiple *yvars*	over() groups
bars touch	yes	no
bars different colors	yes	no
bars identified via ...	legend	axis label

Option ascategory will cause multiple *yvars* to be presented as if they were over() groups, and option asyvars will cause over() groups to be presented as if they were *yvars*. Thus,

    . graph bar (asis) tempjan, over(region)

would produce the first chart and

    . graph bar (asis) ne nc south west, ascategory

would produce the second.

Treatment of data

In the previous two examples, we already had the statistics we wanted to plot: 27.9 (North East), 21.7 (North Central), 46.1 (South), and 46.2 (West). We entered the data and we typed

    . graph bar (asis) ne nc south west

or

    . graph bar (asis) tempjan, over(region)

We do not have to know the statistics ahead of time: graph bar and graph hbar can calculate statistics for us. If we had datasets with lots of observations (say, cities of the U.S.), we could type

    . graph bar (mean) ne nc south west

or

    . graph bar (mean) tempjan, over(region)

and obtain the same graphs. All we need do is change (asis) to (mean). In the first example, the data would be organized the wide way,

cityname	ne	nc	south	west
name of city	42	.	.	.
another city	.	28	.	.
...				

and in the second example, the data would be organized the long way:

cityname	region	tempjan
name of city	ne	42
another city	nc	28
...		

We have such a dataset, organized the long way. In `citytemp.dta` we have information on 956 U.S. cities, including the region in which each is located and its average January temperature:

```
. sysuse citytemp, clear
(City Temperature Data)
. list region tempjan if _n < 3 | _n>954
```

	region	tempjan
1.	N.E.	16.6
2.	N.E.	18.2
955.	West	72.6
956.	West	72.6

With this data, we can type

```
. graph bar (mean) tempjan, over(region)
```

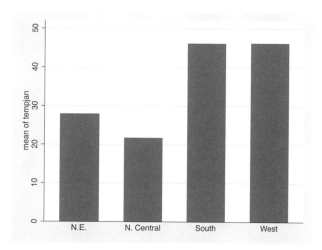

We just produced the same bar chart we previously produced when we entered the statistics 27.9 (North East), 21.7 (North Central), 46.1 (South), and 46.2 (West) and typed

```
. graph bar (asis) tempjan, over(region)
```

When we do not specify (asis) or (mean) (or (median) or (sum) or (p1) or any of the other *stats* allowed), (mean) is assumed. Thus, (...) is often omitted when (mean) is desired, and we could have drawn the previous graph by typing

```
. graph bar tempjan, over(region)
```

Some users even omit typing (...) in the (asis) case because calculating the mean of a single observation results in the number itself. Thus, in the previous section, rather than typing

```
. graph bar (asis) ne nc south west
```

and

```
. graph bar (asis) tempjan, over(region)
```

we could have typed

```
. graph bar ne nc south west
```

and

```
. graph bar tempjan, over(region)
```

Multiple bars (overlapping the bars)

In citytemp.dta, in addition to variable tempjan, there is variable tempjuly, the average July temperature. We can include both averages in a single chart, by region:

```
. sysuse citytemp, clear
(City Temperature Data)
. graph bar (mean) tempjuly tempjan, over(region)
```

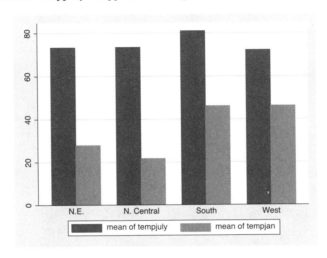

We can improve the look of the chart by

1. Including the *legend_option* legend(label()) to change the text of the legend; see [G] *legend_option*.

2. Including the *axis_title_option* ytitle() to add a title saying "Degrees Fahrenheit"; see [G] *axis_title_options*.

3. Including the *title_options* title(), subtitle(), and note() to say what the graph is about and from where the data came; see [G] *title_options*.

Doing all of that produces,

```
. graph bar (mean) tempjuly tempjan, over(region)
        legend( label(1 "July") label(2 "January") )
        ytitle("Degrees Fahrenheit")
        title("Average July and January temperatures")
        subtitle("by regions of the United States")
        note("Source:  U.S. Census Bureau, U.S. Dept. of Commerce")
```

(*Continued on next page*)

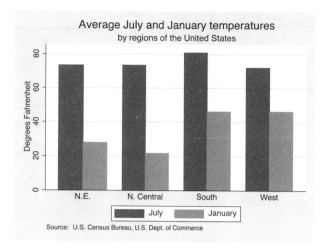

We can make one more improvement to this chart by overlapping the bars. Below, we add the option `bargap(-30)`:

```
. graph bar (mean) tempjuly tempjan, over(region)
        bargap(-30)                                              ←  new
        legend( label(1 "July") label(2 "January") )
        ytitle("Degrees Fahrenheit")
        title("Average July and January temperatures")
        subtitle("by regions of the United States")
        note("Source:  U.S. Census Bureau, U.S. Dept. of Commerce")
```

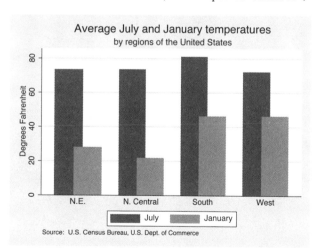

`bargap(#)` specifies the distance between the *yvar* bars (i.e., between the bars for `tempjuly` and `tempjan`); # is in percent-of-bar-width units, so `barwidth(-30)` means that the bars overlap by 30%. `bargap()` may be positive or negative; its default is 0.

Controlling the text of the legend

In the above example, we changed the text of the legend by specifying the legend option:

`legend( label(1 "July") label(2 "January") )`

We could just as well have changed the text of the legend by typing

`yvaroptions( relabel(1 "July" 2 "January") )`

Which you use makes no difference, but we prefer `legend(label())` to `yvaroptions(relabel())` because `legend(label())` is how you modify the contents of a legend in a twoway graph; so why do bar charts differently?

Multiple over()s (repeating the bars)

Option `over(`*varname*`)` repeats the *yvar* bars for each unique value of *varname*. Using `citytemp.dta`, were we to type

  . graph bar (mean) tempjuly tempjan

we would obtain two (fat) bars. When we type

  . graph bar (mean) tempjuly tempjan, over(region)

we obtain two (thinner) bars for each of the four regions. (We typed exactly this command in *Multiple bars* above.)

You may repeat the `over()` option. You may specify `over()` twice when you specify two or more *yvars* and up to three times when you specify just one *yvar*.

In dataset `nlsw88.dta`, we have information on 2,246 women:

```
. sysuse nlsw88, clear
(NLSW, 1988 extract)
. graph bar (mean) wage, over(smsa) over(married) over(collgrad)
        title("Average Hourly Wage, 1988, Women Aged 34-46")
        subtitle("by College Graduation, Marital Status,
                and SMSA residence")
        note("Source:  1988 data from NLS, U.S. Dept. of Labor,
                Bureau of Labor Statistics")
```

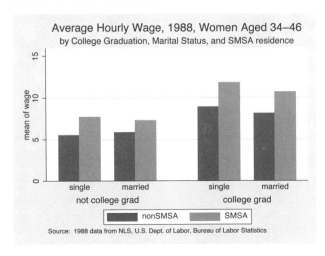

If you strip away the *title_options*, the above command reads

```
. graph bar (mean) wage, over(smsa) over(married) over(collgrad)
```

Note that in this three-over() case, the first over() is treated like multiple *yvars*: the bars touch, the bars are assigned different colors, and the meaning of the bars is revealed in the legend. When you specify three over() groups, the first is treated in the same way as multiple *yvars*. This means that if we wanted to separate the bars, we could specify option bargap(#), #>0, and if we wanted them to overlap, we could specify bargap(#), #<0.

Nested over()s

Sometimes you have multiple over() groups with one group explicitly nested within the other. In the citytemp.dta dataset, we have variables region and division, and division is nested within region. The Census Bureau divides the U.S. into four regions and into nine divisions, which work like this

Region	Division
1. North East	1. New England
	2. Mid Atlantic
2. North Central	3. East North Central
	4. West North Central
3. South	5. South Atlantic
	6. East South Central
	7. West South Central
4. West	8. Mountain
	9. Pacific

Were we to type

```
. graph bar (mean) tempjuly tempjan, over(division) over(region)
```

we would obtain a chart with space allocated for 9*4 = 36 groups, of which only 9 would be used:

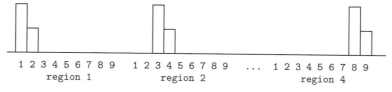

```
1 2 3 4 5 6 7 8 9    1 2 3 4 5 6 7 8 9    ...    1 2 3 4 5 6 7 8 9
     region 1             region 2                    region 4
```

The nofill option prevents the chart from including the unused categories:

```
. sysuse citytemp, clear
(City Temperature Data)

. graph bar tempjuly tempjan, over(division) over(region) nofill
        bargap(-30)
        ytitle("Degrees Fahrenheit")
        legend( label(1 "July") label(2 "January") )
        title("Average July and January temperatures")
        subtitle("by region and division of the United States")
        note("Source:  U.S. Census Bureau, U.S. Dept. of Commerce")
```

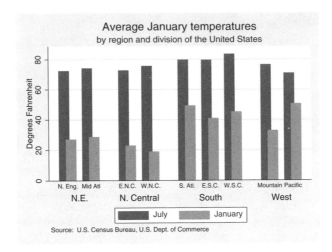

The above chart, if we omit one of the temperatures, also looks good horizontally:

```
. graph hbar (mean) tempjan, over(division) over(region) nofill
        ytitle("Degrees Fahrenheit")
        title("Average January temperature")
        subtitle("by region and division of the United States")
        note("Source:  U.S. Census Bureau, U.S. Dept. of Commerce")
```

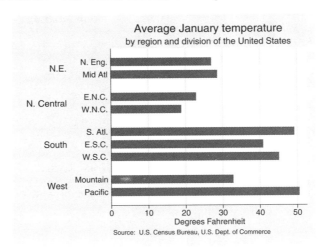

Charts with lots of categories

Using `nlsw88.dta`, we want to draw the chart

```
. sysuse nlsw88
(NLSW, 1988 extract)
. graph bar wage, over(industry) over(collgrad)
```

Variable `industry` records industry of employment in 12 categories and variable `collgrad` records whether the woman is a college graduate. Thus, we will have 24 bars. We draw the above and quickly discover that the long labels associated with industry result in considerable amounts of overprinting along the horizontal x axis.

Horizontal bar charts work better than vertical bar charts when labels are long. We change our command to read

```
. graph hbar wage, over(ind) over(collgrad)
```

That works better, but now we have overprinting problems of a different sort: the letters of one line are touching the letters of the next.

Graphs are by default 4 × 5: 4 inches tall by 5 inches wide. In this case, we need to make the chart taller, and that is the job of the region_option ysize(). Below we make a chart that is 5 inches tall:

```
. sysuse nlsw88, clear
(NLSW, 1988 extract

. graph hbar wage, over(ind, sort(1)) over(collgrad)
        title("Average hourly wage, 1988, women aged 34-46", span)
        subtitle(" ")
        note("Source:  1988 data from NLS, U.S. Dept. of Labor,
              Bureau of Labor Statistics", span)
        ysize(5)
```

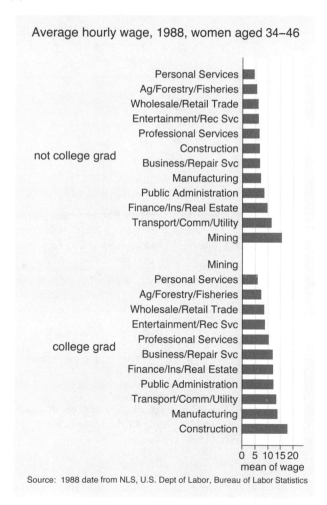

The important option in the above is `ysize(5)` which made the graph taller than usual; see [G] *region_options*. Concerning the other options:

`over(ind, sort(1)) over(collgrad)`
> `sort(1)` we specified so that the bars would be sorted on mean wage. The 1 says to sort on the first *yvar*; see *Reordering the bars* below.

`title("Average hourly wage, 1988, women aged 34-46", span)`
> `span` we specified so that the title, rather than be centered over the plot region, would be centered over the entire graph. In this case, the plot region (the part of the graph where the real chart appears, ignoring the labels) is very narrow, and centering over that was not going to work. See [G] *region_options* for a description of the graph region and plot region, and see [G] *title_options* and [G] *textbox_options* for a description of `span`.

`subtitle(" ")`
> We specified this because the title looked too close to the graph without it. We could have done things properly and specified a `margin()` suboption within the `title()`, but we often find it easier to include a blank subtitle. Note that we typed `subtitle(" ")` and not `subtitle("")`. We had to include the blank or otherwise the subtitle would not have appeared.

`note("Source: 1988 data from NLS, ...", span)`
> `span` we specified so that the note would be left justified in the graph rather than just in the plot region.

How bars are ordered

The default is to place the bars in the order of the *yvars* and to order each `over(`*varname*`)` groups according to the values of *varname*. Let us consider some examples:

`graph bar (sum) revenue profit`
> Bars appear in the order specified, revenue and profit.

`graph bar (sum) revenue, over(division)`
> Bars are ordered according to the values of variable `division`.

> If `division` is a numeric variable, the lowest division number comes first, followed by the next lowest, and so on. This is true even if variable `division` has a value label. Say division 1 has been labeled "Sales" and division 2, labeled "Development". The bars will be in the order Sales followed by Development.

> If `division` is a string variable, the bars will be ordered by the sort order of the values of `division` (which is to say, alphabetically, but with capital letters placed before lowercase letters). If variable `division` contains the values "Sales" and "Development", the bars will be in the order Development followed by Sales.

`graph bar (sum) revenue profit, over(division)`
> Bars appear in the order specified, `revenue` and `profit`, and are repeated for each division, which will be ordered as explained above.

`graph bar (sum) revenue, over(division) over(year)`
> Bars appear ordered by the values of division, as previously explained, and then that is repeated for each of the years. The years are ordered according to the values of the variable `year`, following the same rules as applied to the variable `division`.

`graph bar (sum) revenue profit, over(division) over(year)`
> Bars appear in the order specified, profit and revenue, repeated for division ordered on the values of variable `division`, repeated for year ordered on the values of variable `year`.

Reordering the bars

There are three ways you may wish to reorder the bars.

1. You want to control the order in which the elements of each `over()` group appear. Your divisions might be named Development, Marketing, Research, and Sales, alphabetically speaking, but you want them to appear in the more logical order Research, Development, Marketing, and Sales.

2. You wish to order the bars according to their heights. You wish to draw the graph

   ```
   . graph bar (sum) empcost, over(division)
   ```

   and you want the divisions ordered by total employee cost.

3. You wish to order on some other derived value.

We will consider each of these desires separately.

Putting the bars in a prespecified order

You have drawn the graph

```
. graph (sum) bar empcost, over(division)
```

Variable `division` is a string containing "Development", "Marketing", "Research", and "Sales". You wish to draw the chart placing the divisions in the order Research, Development, Marketing, and Sales.

To do that, you create a new numeric variable that orders division as you would like:

```
. generate order = 1 if division=="Research"
. replace  order = 2 if division=="Development"
. replace  order = 3 if division=="Marketing"
. replace  order = 4 if division=="Sales"
```

You may name the variable and create it however you wish, just be sure that there is a one-to-one correspondence between the new variable and the `over()` group's values. You then specify the `over()`'s `sort(`*varname*`)` option:

```
. graph bar (sum) empcost, over( division, sort(order) )
```

If you want to reverse the order, you may specify the `descending` suboption:

```
. graph bar (sum) empcost, over(division, sort(order) descending)
```

Putting the bars in height order

You have drawn the graph

```
. graph bar (sum) empcost, over(division)
```

and now wish to put the bars in height order, shortest first. You type

```
. graph bar (sum) empcost, over( division, sort(1) )
```

If you wanted the tallest first, you type

```
. graph bar empcost, over(division, sort(1) descending)
```

The 1 in sort(1) refers to the first (and in this case, only) *yvar*. If you had multiple *yvars*, you might type

```
. graph bar (sum) empcost othcost, over( division, sort(1) )
```

and you would have a chart showing employee cost and other cost, sorted on employee cost. If you typed

```
. graph bar (sum) empcost othcost, over( division, sort(2) )
```

the graph would be sorted on other cost.

You can use sort(#) on the second over() group as well:

```
. graph bar (sum) empcost, over( division, sort(1) )
                           over( country,  sort(1) )
```

Country will be ordered on the sum of the heights of the bars.

Putting the bars in a derived order

You have employee cost broken into two categories: empcost_direct and empcost_indirect. Variable emp_cost is the sum of the two. You wish to make a chart showing the two costs, stacked, over division, and you want the bars ordered on the total height of the stacked bars. You type

```
. graph bar (sum) empcost_direct empcost_indirect,
             stack
             over( division, sort((sum) empcost) )
             descending
```

Reordering the bars, example

We have a dataset showing the spending on tertiary education as a percent of GDP from the 2002 edition of *Education at a Glance: OECD Indicators 2002*,

```
. sysuse educ99gdp, clear
(Education and GDP)
. list
```

	country	public	private
1.	Australia	.7	.7
2.	Britain	.7	.4
3.	Canada	1.5	.9
4.	Denmark	1.5	.1
5.	France	.9	.4
6.	Germany	.9	.2
7.	Ireland	1.1	.3
8.	Netherlands	1	.4
9.	Sweden	1.5	.2
10.	United States	1.1	1.2

We wish to graph total spending on education and simultaneously show the distribution of that total between public and private expenditures. We want the bar sorted on total expenditures:

```
. generate total = private + public
```

```
. graph hbar (asis) public private,
        over(country, sort(total) descending) stack
        title( "Spending on tertiary education as % of GDP, 1999",
            span pos(11) )
        subtitle(" ")
        note("Source:  OECD, Education at a Glance 2002", span)
```

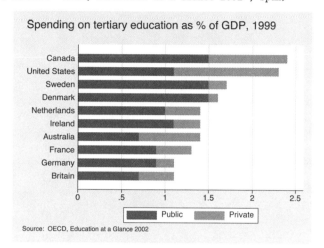

Alternatively, perhaps we wish to disguise the total expenditures and focus the graph exclusively on the share of spending that is public and private:

```
. generate frac = private/(private + public)
. graph hbar (asis) public private,
        over(country, sort(frac) descending) stack percent
        title("Public and private spending on tertiary education, 1999",
            span pos(11) )
        subtitle(" ")
        note("Source:  OECD, Education at a Glance 2002", span)
```

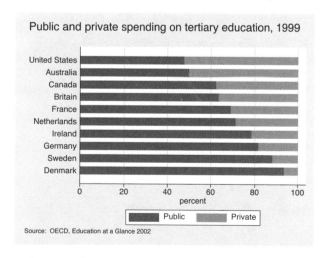

The only differences between the two `graph hbar` commands are

1. We added option `percentage` to change the *yvars* public and private from spending amounts to percent each is of the total.

2. We changed the order of the bars.

3. We changed the title.

Use with by()

`graph bar` and `graph hbar` may be used with `by()`, but in general, you want to use `over()` in preference to `by()`. Bar charts are explicitly categorical and do an excellent job of presenting summary statistics for multiple groups in a single chart.

A good use of `by()`, however, is when you are ordering the bars and you wish to emphasize that the ordering is different for different groups. For instance,

```
. sysuse nlsw88, clear
(NLSW, 1988 extract)
. graph hbar wage, over(occ, sort(1)) by(union)
```

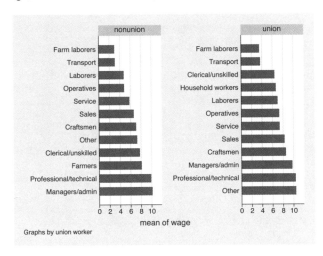

In the above graph we have ordered the bars by height (hourly wage), and note that the orderings are different for union and nonunion workers.

Note: Repeated yvar names

Consider a command in which the same *yvar* is used more than once:

```
. graph bar (p25) wage (p75) wage, over(status)
```

The above will generate an error message. To prevent that, you need to type

```
. graph bar (p25) w25=wage (p75) w75=wage, over(status)
```

The `w25` and `w75`—which may be names of your choosing—provide new names by which the calculated results (p25) `wage` and (p75) `wage` are to be known. When you do not specify the new name, results are as if you typed

```
. graph bar (p25) wage=wage (p75) wage=wage, over(status)
```

and therein lies the problem. `graph bar` attempts to store two different calculated results under the same name.

In the example above, it is not necessary that you specify new names for both calculated results; you could type

```
. graph bar (p25) w25=wage (p75)    wage, over(status)
```

or

```
. graph bar (p25)    wage (p75) w75=wage, over(status)
```

Understand that even when you specify a name, the results are not left behind in our dataset; you are merely helping `graph bar` out. If you want to obtain the statistics, see [R] **collapse**. If you want to put the statistics in a table, see [R] **table**.

History

The first published bar chart appeared in William Playfair's *Commercial and Political Atlas* (1786). See Tufte (1983, 32–33) or Beniger and Robyn (1978) for additional historical information.

References

Beniger, J. R. and D. L. Robyn. 1978. Quantitative graphics in statistics: a brief history. *The American Statistician* 32: 1–11.

Tufte, E. R. 1983. *The Visual Display of Quantitative Information.* Cheshire, CT: Graphics Press.

Playfair, W. 1786. *Commercial and Political Atlas: Representing, by Copper-Plate Charts, the Progress of the Commerce, Revenues, Expenditure, and Debts of England, during the Whole of the Eighteenth Century.* London: Corry.

Also See

Complementary:	[G] **graph dot**;
	[R] **collapse**, [R] **table**

Title

graph box — Box plots

Syntax

graph box *yvars* [*weight*] [if *exp*] [in *range*] [, *options*]

graph hbox *yvars* [*weight*] [if *exp*] [in *range*] [, *options*]

where *yvars* is a *varlist* and where *options* are

options	description
group_options	groups over which boxes are drawn
yvar_options	variables that are the boxes
boxlook_options	how the boxes look
legending_options	how variables are labeled
axis_options	how numerical y axis is labeled
title_and_other_options	titles, added text, aspect ratio, etc.

Each is defined below.

The *group_options*, which define the groups over which the boxes are drawn, are

group_options	description
over(*varname* [, *over_subopts*])	categories; option may be repeated
nofill	omit empty categories
missing	keep missing value as category

The *yvar_options*, which define the variables that are the boxes, are

yvar_options	description
ascategory	treat *yvars* as first over() group
asyvars	treat first over() group as *yvars*
cw	calculate variable statistics omitting missing values of any variable

(Continued on next page)

144

The *boxlook_options*, which define how the boxes look, are

boxlook_options	description
<u>noout</u>sides	do not plot outside values
box(#, *barlook_options*)	look of #th box
<u>inten</u>sity([*]#)	intensity of fill
<u>linten</u>sity([*]#)	intensity of outline
<u>medt</u>ype(line|cline|marker)	how median indicated in box
<u>medl</u>ine(*line_options*)	look of line if medtype(cline)
<u>medm</u>arker(*marker_options*)	look of marker if medtype(marker)
<u>cw</u>hiskers	use custom whiskers
<u>lines</u>(*line_options*)	look of custom whiskers
<u>al</u>size(#)	width of adjacent line, default 67
<u>cap</u>size(#)	height of cap on adjacent line, default 0
<u>marker</u>(# *marker_options*)	look of #th marker for outside values
outergap([*]#)	gap between edge and first box and between last box and edge
boxgap(#)	gap between boxes; default 33

See [G] ***barlook_options*** *(sic)*, [G] ***line_options***, and [G] ***marker_options***.

The *legending_options*, which define how the *yvar* boxes are labeled, are

legending_options	description
legend_option	control of *yvar* legend
<u>nolabel</u>	use *yvar* names, not labels, in legend
<u>yvar</u>options(*over_subopts*)	*over_subopts* for *yvars*; rarely specified
showyvars	label *yvars* on *x* axis; rarely specified

See [G] ***legend_option***.

The *axis_options*, which define how the numerical *y* axis is labeled, are

axis_options	description
<u>yal</u>ternate	put numerical *y* axis on right (top)
<u>xal</u>ternate	put categorical *x* axis on top (right)
<u>yre</u>verse	reverse *y* axis
axis_scale_options	*y*-axis scaling and look
axis_label_options	*y*-axis labeling
ytitle(...)	*y*-axis titling

See [G] ***axis_scale_options***, [G] ***axis_label_options***, and [G] ***axis_title_options***.

The *title_and_other_options* are

title_and_other_options	description
text(...)	add text on graph; x range $[0, 100]$
yline(...)	add y lines to graph
std_options	titles, aspect ratio, saving to disk
by(*varlist*, ...)	repeat for subgroups

See [G] *added_text_option*, [G] *added_line_options*, [G] *std_options*, and [G] *by_option*.

The *over_subopts*, used in over(*varname*, *over_subopts*) and, on rare occasion, in yvaroptions(*over_subopts*), are

over_subopts	description
total	add total group
relabel(# "*text*" ...)	change axis labels
label(*cat_axis_label_options*)	rendition of labels
axis(*cat_axis_line_options*)	rendition of axis line
gap([*]#)	gap between boxes within over() category
sort(*varname*)	put boxes in prespecified order
sort(#)	put boxes in median order
descending	reverse default or specified box order

See [G] *cat_axis_label_options* and [G] *cat_axis_line_options*.

aweights, fweights, and pweights are allowed; see [U] **14.1.6 weight** and see note concerning weights in [R] **collapse**.

Description

graph box draws box plots, vertically presented. In a vertical box plot, the y axis is numerical and the x axis is categorical.

. graph box *y1 y2*, over(*cat_var*)

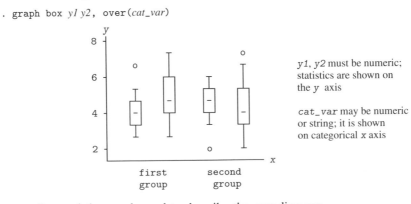

y1, *y2* must be numeric; statistics are shown on the y axis

cat_var may be numeric or string; it is shown on categorical x axis

The encoding and the words used to describe the encoding are

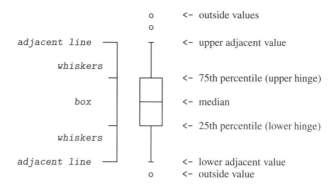

graph hbox draws horizontal box plots. In a horizontal box plot, the numerical axis is still called the y axis and the categorical axis is still called the x axis, but y is presented horizontally and x, vertically.

. graph hbox *y1 y2*, over(*cat_var*)

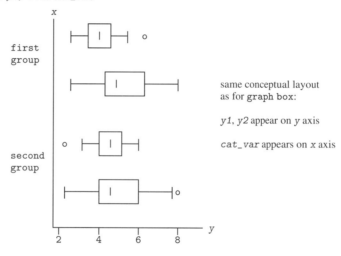

Options

Group options

over(*varname*[, *over_subopts*]) specifies a categorical variable over which the *yvars* are to be repeated. *varname* may be string or numeric. Up to two over() options may be specified when multiple *yvars* are specified, and up to three over()s may be specified when a single *yvar* is specified; see *Examples of syntax* under *Remarks* below.

nofill specifies that missing subcategories are to be omitted. See the description of the nofill option in [G] **graph bar**.

missing specifies that missing values of the over() variables be kept as their own categories, one for ., another for .a, etc. The default is to act as if such observations simply did not appear in the dataset; the observations are ignored. An over() variable is considered to be missing if it is numeric and contains a missing value or if it is string and contains "".

yvar options

ascategory specifies that the *yvars* be treated as the first over() group. The important effect of this is to move the captioning of the variables from the legend to the categorical *x* axis. See the description of ascategory in [G] **graph bar**.

asyvars specifies that the first over() group be treated as *yvars*. The important effect of this is to move the captioning of the first over group from the categorical *x* axis to the legend. See the description of asyvars in [G] **graph bar**.

cw specifies casewise deletion. If cw is specified, observations for which any of the *yvars* are missing are ignored. The default is to calculate statistics for each box using all the data possible.

boxlook options

nooutsides specifies that the outside values are not be plotted, nor are they to be used in setting the scale of the *y* axis.

box(*#*, *barlook_options*) specifies the look of the *yvar* boxes. box(1, ...) refers to the box associated with the first *yvar*, box(2, ...) refers to the box associated with the second, and so on.

Note that you specify *barlook_options*. Those options are borrowed from graph bar in the case of boxes. The most useful *barlook_option* is bcolor(*colorstyle*), which sets the color of the box. For instance, you might specify box(1, bcolor(green)) to make the box associated with the first *yvar* green. See [G] *colorstyle* for a list of color choices and see [G] *barlook_options* for information on the other *barlook_options*.

intensity(*#*) and intensity(**#*) specify the intensity of the color used to fill the inside of the box. intensity(*#*) specifies the intensity and intensity(**#*) specifies the intensity relative to the default.

By default, the box is filled with the color of its border, attenuated. Specify intensity(**#*), *# < 1*, to attenuate it more and specify intensity(**#*), *# > 1*, to amplify it.

Specify intensity(0) if you do not want the box filled at all. If you are using a scheme that draws the median line in the background color such as s2mono, also specify option medtype(line) to change the median line to be in the color of the outline of the box.

lintensity(*#*) and lintensity(**#*) specify the intensity of the line used to outline the box. lintensity(*#*) specifies the intensity and lintensity(**#*) specifies the intensity relative to the default.

By default, the box is outlined at the same intensity at which it is filled or at an amplification of that, which depending on your chosen scheme; see [G] **schemes**. If you want the box outlined in the darkest possible way, specify intensity(255). If you wish simply to amplify the outline, specify intensity(**#*), *# > 1*, and if you wish to attenuate the outline, specify intensity(**#*), *# < 1*.

medtype(), medline(), and medmarker() specify how the median is to be indicated in the box.

medtype(line) is the default. A line is drawn across the box at the median. In this case, options medline() and medmarker() are irrelevant.

medtype(cline) specifies a custom line be drawn across the box at the median. The default custom line is usually of a different color. You can, however, specify option medline(*line_options*) to control exactly how the line is to look; see [G] *line_options*.

medtype(marker) specifies a marker be placed in the box at the median. In this case, you may also specify option medmarker(*marker_options*) to specify the look of the marker; see [G] *marker_options*.

cwhiskers, lines(*line_options*), alsize(#), and capsize(#) specify the look of the whiskers.

cwhiskers specifies that custom whiskers are desired. The default custom whiskers are usually dimmer, but you may specify option lines(*line_options*) to specify how the custom whiskers are to look; see [G] *line_options*.

alsize(#) and capsize(#) specify the width of the adjacent line and the height of the cap on the adjacent line. You may specify these options whether or not you specify cwhiskers. alsize() and capsize() are specified in percent-of-box-width units; the defaults are alsize(67) and capsize(0). Thus, the adjacent lines extend two-thirds the width of a box and, by default, have no caps. Caps refer to whether the whiskers look like

If you want caps, try capsize(5).

marker(# *marker_options*) specifies the marker to be used to display the outside values. See [G] *marker_options*.

outergap(*#) and outergap(#) specify the gap between the edge of the graph to the beginning of the first box and the end of the last box to the edge of the graph.

outergap(*#) specifies that the default is to be modified. Specifying outergap(*1.2) increases the gap by 20% and specifying outergap(*.8) reduces the gap by 20%.

outergap(#) specifies the gap in percent-of-box-width units. outergap(50) specifies that the gap should be half the box width.

boxgap(#) specifies the gap to be left between *yvar* boxes in percent-of-box-width units. The default is boxgap(33).

Note that boxgap() affects only the *yvar* boxes. If you want to change the gap for the first, second, or third over() groups, specify the *over_subopt* gap() inside the over() itself; see *Suboptions for use with over() and yvaroptions()* below.

Legending options

legend_option allows you to control the legend. If more than one *yvar* is specified, a legend is produced. Otherwise, no legend is needed because the over() groups are labeled on the categorical x axis. See [G] *legend_option* and see *Treatment of multiple yvars versus treatment of over() groups* under *Remarks* below.

nolabel specifies that, in the automatic construction of the legend, the variable names of the *yvars* be used in preference to their labels.

yvaroptions(*over_subopts*) allows you to specify *over_subopts* for the *yvars*. This is very rarely done.

showyvars specifies that, in addition to building a legend, the identities of the *yvars* are to be shown on the categorical x axis. If showyvars is specified, it is typical to also specify legend(off).

Axis options

yalternate and xalternate switch the side on which the axes appear.

Used with graph box, yalternate moves the numerical y axis from the left to the right; xalternate moves the categorical x axis from the bottom to the top.

Used with graph hbox, yalternate moves the numerical y axis from the bottom to the top; xalternate moves the categorical x axis from the left to the right.

If your scheme by default puts the axes on the opposite sides, then yalternate and xalternate reverse their actions.

yreverse specifies that the numerical y axis have its scale reversed, so that it runs from maximum to minimum.

axis_scale_options specify how the numerical y axis is scaled and how it looks; see [G] ***axis_scale_options***. There you will also see option xscale() in addition to yscale(). Ignore xscale(); it is irrelevant in the case of box plots.

axis_label_options specify how the numerical y axis is to be labeled. The *axis_label_options* also allow you to add and suppress grid lines; see [G] ***axis_label_options***. There you will see that, in addition to options ylabel(), ytick(), ..., ymtick(), options xlabel(), ..., xmtick() are allowed. Ignore the x*() options; they are irrelevant in the case of box plots.

ytitle() overrides the default title for the numerical y axis; see [G] ***axis_title_options***. There you will also find option xtitle() documented; it is irrelevant in the case of box plots.

Title and other options

text() adds text to a specified location on the graph; see [G] ***added_text_option***. The basic syntax of text() is

text(*#_y #_x "text"*)

text() is documented in terms of twoway graphs. When used with box plots, the "numeric" x axis is scaled to run from 0 to 100.

yline() adds horizontal (box) or vertical (hbox) lines at specified y values; see [G] ***added_line_options***. The xline() option, also documented there, is irrelevant in the case of box plots. If your interest is in adding grid lines, see [G] ***axis_label_options***.

std_options allow you to add titles, control the aspect ratio, save the graph on disk, and much more; see [G] ***std_options***.

by(*varlist*, ...) draws separate plots within a single graph; see [G] ***by_option*** and see *Use with by()* under *Remarks* below.

Suboptions for use with over() and yvaroptions()

total specifies that, in addition to the unique values of over(*varname*), a group be added reflecting all the observations. When multiple over()s are specified, total may only be specified in one of them.

relabel(*# "text"* ...) specifies text to override the default labeling of the categories. See the description of the relabel() option in [G] **graph bar** for more information on this very useful option.

label(*cat_axis_label_options*) determines other aspects of the look of the category labels on the *x* axis. With the exception of label(labcolor()) and label(labsize()), these options are seldom specified; see [G] *cat_axis_label_options*.

axis(*cat_axis_line_options*) specifies how the axis line is rendered. This is a rarely specified option. See [G] *cat_axis_line_options*.

gap(*#*) and gap(**#*) specify the gap between the boxes in this over() group. gap(*#*) is specified in percent-of-box-width units, so gap(67) means two-thirds the width of a box. gap(**#*) allows modifying the default gap. gap(**1.2*) would increase the gap by 20% and gap(**.8*) would decrease the gap by 20%.

To understand the distinction between over(..., gap()) and option boxgap(), consider

        . graph box before after, boxgap(...) over(sex, gap(...))

boxgap() sets the distance between the before and after boxes. over(,gap()) sets the distance between the boxes for males and females. Similarly, in

        . graph box before after, boxgap(...)
                            over(sex,   gap(...))
                            over(agegrp, gap(...))

over(sex, gap()) sets the gap between males and females and over(agegrp, gap()) sets the gap between age groups.

sort(*varname*) and sort(*#*) control how the boxes are ordered. See *How boxes are ordered* and *Reordering the boxes* under *Remarks* below.

sort(*varname*) puts the boxes in the order of *varname*; see *Putting the boxes in a prespecified order* under *Remarks* below.

sort(*#*) puts the boxes in order of their medians. *#* refers to the *yvar* number on which the ordering should be performed; see *Putting the boxes in median order* under *Remarks* below.

descending specifies that, whatever the order of the boxes, default or as specified by sort(), it is to be reversed.

Remarks

Remarks are presented under the headings

> *Introduction*
> *Examples of syntax*
> *Treatment of multiple yvars versus treatment of over() groups*
> *How boxes are ordered*
> *Reordering the boxes*
> *Putting the boxes in a prespecified order*
> *Putting the boxes in median order*
> *Use with by()*
> *History*

Also see [G] **graph bar**. Most of what is said there applies equally well to box charts.

Introduction

graph box draws box plots, vertically presented:

```
. sysuse bplong, clear
(fictional blood-pressure data)

. graph box bp, over(when) over(sex)
        ytitle("Systolic blood pressure")
        title("Response to Treatment, by Sex")
        subtitle("(120 Preoperative Patients)" " ")
        note("Source:  Fictional Drug Trial, Stata Corporation, 2003")
```

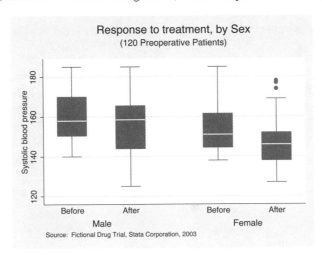

graph hbox draws box plots, horizontally presented:

```
. sysuse nlsw88, clear
(NLSW, 1988 extract)

. graph hbox wage, over(ind, sort(1)) nooutside
        ytitle("")
        title("Hourly wage, 1988, woman aged 34-46", span)
        subtitle(" ")
        note("Source:  1988 data from NLS, U.S. Dept of Labor,
                       Bureau of Labor Statistics", span)
```

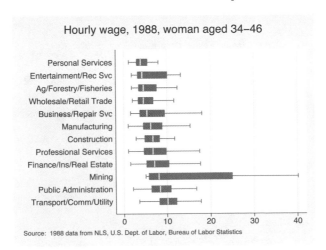

Examples of syntax

Below we show you some `graph box` commands and tell you what each would do:

`graph box bp`
One big box showing statistics on blood pressure.

`graph box bp_before bp_after`
Two boxes, one showing average blood pressure before, and the other, after.

`graph box bp, over(agegrp)`
#_of_agegrp boxes showing blood pressure for each age group.

`graph box bp_before bp_after, over(agegrp)`
2×*#_of_agegrp* boxes showing blood pressure, before and after, for each age group. The grouping would look like this (assuming 3 age groups):

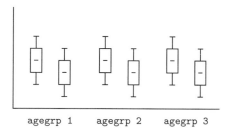

`graph box bp, over(agegrp) over(sex)`
#_of_agegrps × *#_of_sexes* boxes showing blood pressure for each age group, repeated for each sex. The grouping would look like this:

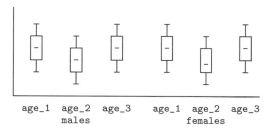

`graph box bp, over(sex) over(agegrp)`
same as above, but ordered differently. In the previous example we typed `over(agegrp)` `over(sex)`. This time, we reverse it:

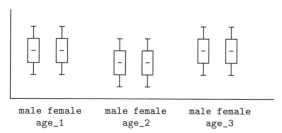

```
graph box bp_before bp_after, over(agegrp) over(sex)
```
$2 \times \#\_of\_agegrps \times \#\_of\_sexes$ boxes showing blood pressure, before and after, for each age group, repeated for each sex. The grouping would look like this:

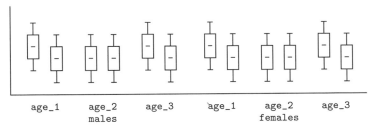

Treatment of multiple yvars versus treatment of over() groups

Consider two datasets containing the same data, organized differently. The datasets contain blood pressure, before and after an intervention. In the first dataset, the data are organized the wide way; each patient is an observation. A little bit of the data is

patient	sex	agegrp	bp_before	bp_after
1	Male	30–45	143	153
2	Male	30–45	163	170
3	Male	30–45	153	168

In the second dataset, the data are organized the long way; each patient is a pair of observations. The corresponding observations in the second dataset are

patient	sex	agegrp	when	bp
1	Male	30–45	Before	143
1	Male	30–45	After	153
2	Male	30–45	Before	163
2	Male	30–45	After	170
3	Male	30–45	Before	153
3	Male	30–45	After	168

Using the first dataset, we might type

```
. sysuse bpwide, clear
(fictional blood-press data)
```

```
. graph box bp_before bp_after, over(sex)
```

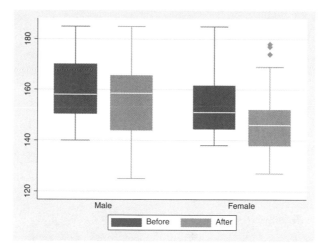

Using the second dataset, we could type

```
. sysuse bplong, clear
. graph box bp, over(when) over(sex)
```

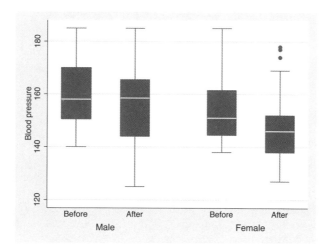

The two graphs are virtually identical. They differ in that

	multiple *yvars*	over() groups
boxes different colors	yes	no
boxes identified via ...	legend	axis label

Option `ascategory` will cause multiple *yvars* to be presented as if they were the first `over()` group, and option `asyvars` will cause the first `over()` group to be presented as if they were multiple *yvars*. Thus,

```
. graph box bp, over(when) over(sex) asyvars
```

would produce the first chart and

```
. graph box bp_before bp_after, over(sex) ascategory
```

would produce the second.

How boxes are ordered

The default is to place the boxes in the order of the *yvars* and to order each over(*varname*) group according to the values of *varname*. Let us consider some examples:

graph box bp_before bp_after
Boxes appear in the order specified, bp_before and bp_after.

graph box bp, over(when)
Boxes are ordered according to the values of variable when.

If variable when is a numeric, the lowest when number comes first, followed by the next lowest, and so on. This is true even if variable when has a value label. Say when = 1 has been labeled "Before" and when = 2, labeled "After". The boxes will be in the order Before followed by After.

If variable when is a string, the boxes will be ordered by the sort order of the values of the variable (which is to say, alphabetically, but with capital letters placed before lowercase letters). If variable when contains "Before" and "After", the boxes will be in the order After followed by Before.

graph box bp_before bp_after, over(sex)
Boxes appear in the order specified, bp_before and bp_after, and are repeated for each sex, which will be ordered as explained above.

graph box bp_before bp_after, over(sex) over(agegrp)
Boxes appear in the order specified, bp_before and bp_after, repeated for sex ordered on the values of variable sex, repeated for agegrp ordered on the values of variable agegrp.

Reordering the boxes

There are two ways you may wish to reorder the boxes.

1. You want to control the order in which the elements of each over() group appear. String variable when might contain "After" and "Before", but you want the boxes to appear in the order Before and After.

2. You wish to order the boxes according to their median values. You wish to draw the graph

```
. graph box wage, over(industry)
```

and you want the industries ordered by wage.

We will consider each of these desires separately.

Putting the boxes in a prespecified order

You have drawn the graph

```
. graph box bp, over(when) over(sex)
```

Variable when is a string containing "Before" and "After". You wish the boxes to be in that order.

To do that, you create a new numeric variable that orders the group as you would like:

```
. generate order = 1 if when=="Before"
. replace  order = 2 if when=="After"
```

You may name the variable and create it however you wish, just be sure that there is a one-to-one correspondence between the new variable and the over() group's values. You then specify over()'s sort(*varname*) option:

```
. graph box bp, over(when, sort(order)) over(sex)
```

If you want to reverse the order, you may specify the descending suboption:

```
. graph box bp, over(when, sort(order) descending) over(sex)
```

Putting the boxes in median order

You have drawn the graph

```
. graph hbox wage, over(industry)
```

and now wish to put the boxes in median order, lowest first. You type

```
. graph hbox wage, over( industry, sort(1) )
```

If you wanted the largest first, you type

```
. graph hbox wage, over(industry, sort(1) descending)
```

The 1 in sort(1) refers to the first (and in this case, only) *yvar*. If you had multiple *yvars*, you might type

```
. graph hbox wage benefits, over( industry, sort(1) )
```

and you would have a chart showing wage and benefits sorted on wage. If you typed

```
. graph hbox wage benefits, over( industry, sort(2) )
```

the graph would be sorted on benefits.

Use with by()

graph box and graph hbox may be used with by(), but in general, you want to use over() in preference to by(). Box charts are explicitly categorical and do an excellent job of presenting summary statistics for multiple groups in a single chart.

A good use of by(), however, is when the graph would otherwise be very long. Consider the graph

```
. sysuse nlsw88, clear
(NLSW, 1988 extract)
. graph hbox wage, over(ind) over(union)
```

In the above graph there are 12 industry categories and 2 union categories, resulting in 24 separate boxes. The graph, presented at normal size, would be virtually unreadable. One way around that problem would be to make the graph longer than usual,

```
. graph hbox wage, over(ind) over(union) ysize(7)
```

See *Charts with lots of categories* in [G] **graph bar** for more information about that solution. The other solution would be to introduce union as a by() category rather than an over() category:

```
. graph hbox wage, over(ind) by(union)
```

Below we do precisely that, adding some extra options to produce a good-looking chart:

```
. graph hbox wage, oer(ind, sort(1)) nooutside
       ylabel(, grid)  ytitle("")
       by(
              union,
              title("Hourly wage, 1988, women aged 34-46", span)
              subtitle(" ")
              note("Source:  1988 data from NLS, U.S. Dept. of Labor,
                     Bureau of Labor Statistics", span)
       )
```

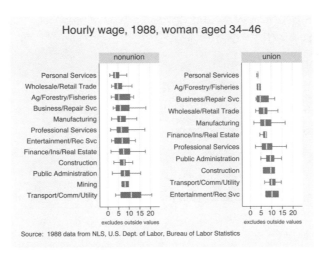

The title options were specified inside the by() so that they would not be applied to each graph separately; see [G] *by_option*.

History

Box plots have been used in geography and climatology, under the name "dispersion diagrams", since at least 1933; see Crowe (1933). His Figure 1 shows all the data points, medians, quartiles, and octiles by month for monthly rainfalls for Glasgow, 1868–1917. His Figure 2, a map of Europe with several climatic stations, shows monthly medians, quartiles, and octiles.

Methods and Formulas

For a description of box plots, see Cleveland (1993, 25–27).

Summary statistics are obtained from summarize; see [R] **summarize**.

The upper and lower adjacent values are as defined by Tukey (1977):

Let x represent a variable for which adjacent values are being calculated. Define $x_{(i)}$ as the ith ordered value of x and define $x_{[25]}$ and $x_{[75]}$ as the 25th and 75th percentiles.

Define U as $x_{[75]} + \frac{3}{2}(x_{[75]} - x_{[25]})$. The upper adjacent value is defined as x_i such that $x_{(i)} \leq U$ and $x_{(i+1)} > U$.

Define L as $x_{[25]} - \frac{3}{2}(x_{[75]} - x_{[25]})$. The lower adjacent value is defined as x_i such that $x_{(i)} \geq L$ and $x_{(i+1)} < L$.

References

Chambers, J. M., W. S. Cleveland, B. Kleiner, and P. A. Tukey. 1983. *Graphical Methods for Data Analysis*. Belmont, CA: Wadsworth International Group.

Cleveland, W. S. 1993. *Visualizing Data*. Summit, NJ: Hobart Press.

——. 1994. *The Elements of Graphing Data*. Summit, NJ: Hobart Press.

Crowe, P. R. 1933. The analysis of rainfall probability. A graphical method and its application to European data. *Scottish Geographical Magazine* 49: 73–91.

Nash, J. C. 1996. gr19: Misleading or confusing boxplots. *Stata Technical Bulletin* 29: 14–17. Reprinted in *Stata Technical Bulletin Reprints*, vol. 5. pp. 60–64.

Tukey, J. W. 1977. *Exploratory Data Analysis*. Reading, MA: Addison–Wesley Publishing Company.

Also See

Complementary:	[G] **graph bar**;
	[R] **lv**, [R] **summarize**

Title

graph combine — Combine multiple graphs into one

Syntax

graph combine *name* [*name* ...] [, *combine_options*]

where *name* is

name	description
simplename	name of graph in memory
name.gph	name of graph stored on disk
"*name*"	name of graph stored on disk

and where *combine_options* are

combine_options	description
colfirst	display down columns
rows(#) | cols(#)	display in # rows or # columns
holes(*numlist*)	positions to leave blank
iscale([*]#)	size of text and markers
imargin(*marginstyle*)	margins for individual graphs
ycommon	make y axes have common scale
xcommon	make x axes have common scale
title_options	titles to appear on combined graph
region_options	outlining, shading, aspect ratio
commonscheme	put graphs on common scheme
scheme(*schemename*)	overall look
nodraw	suppress display of combined graph
name(*name*, ...)	specify name for combined graph
saving(*filename*, ...)	save combined graph in file

See [U] **14.1.8 numlist**, [G] *marginstyle*, [G] *title_options*, [G] *region_options*, [G] *nodraw_option*, [G] *name_option*, and [G] *saving_option*.

Description

graph combine arrays separately drawn graphs into one.

Options

colfirst, rows(#), cols(#), and holes(*numlist*) specify how the resulting graphs are arrayed. These are the same options described in [G] **by_option**.

iscale(#) and iscale(*#) specify a size adjustment (multiplier) to be used to scale the text and markers used in the individual graphs.

By default, iscale() gets smaller and smaller the larger is G, the number of graphs being combined. The default is parameterized as a multiplier $f(G)$—$0 < f(G) < 1$, $f'(G) < 0$—that is used to multiply msize(), $\{$ y | x $\}$label(,labsize()), etc., in the individual graphs.

If you specify iscale(#), the number you specify is substituted for $f(G)$. iscale(1) means text and markers should appear at the same size as they were originally. iscale(.5) displays text and markers at half that size. We recommend that you specify a number between 0 and 1, but you are free to specify numbers larger than 1.

If you specify iscale(*#), the number you specify is multiplied by $f(G)$ and that product is used to scale the text and markers. iscale(*1) is the default. iscale(*1.2) means text and markers should appear at 20% larger than graph combine would ordinarily choose. iscale(*.8) would make them 20% smaller.

imargin(*marginstyle*) specifies margins to be put around the individual graphs. See [G] **marginstyle**.

ycommon and xcommon specify that the individual graphs previously drawn by graph twoway, and for which the by() option was not specified, are to be put on common y or x axes scales. See *Combining twoway graphs* under *Remarks* below.

title_options allow you to specify titles, subtitles, notes, and captions to be placed on the combined graph; see [G] **title_options**.

region_options allow you to control the aspect ratio, size, etc. of the combined graph; see [G] **region_options**. Important among these options are ysize(#) and xsize(#), which specify the overall size of the resulting graph. It is sometimes desirable to make the combined graph wider or longer than usual.

commonscheme and scheme(*schemename*) are for use when combining graphs that use different schemes. By default, each subgraph will be drawn according to its own scheme.

commonscheme specifies that all subgraphs be drawn using the same scheme and, by default, that scheme will be your default scheme; see [G] **schemes**.

scheme(*schemename*) specifies that the *schemename* be used instead.

nodraw causes the combined graph to be constructed but not displayed; see [G] **nodraw_option**.

name(*name*[, replace]) specifies the name of the resulting combined graph. name(Graph, replace) is the default. See [G] **name_option**.

saving(*filename*[, asis replace]) specifies the combined graph should be saved as *filename*. If *filename* is specified without an extension, .gph is assumed. asis specifies that the graph be saved in as-is format. replace specifies that, if the file already exists, it is okay to replace it. See [G] **saving_option**.

(Continued on next page)

Remarks

Remarks are presented under the headings

Typical use
Typical use with memory graphs
Combining twoway graphs
Advanced use
Controlling the aspect ratio of subgraphs

Typical use

You have previously drawn

```
. sysuse uslifeexp, clear
(U.S. life expectancy, 1900-1999)
. line le_male    year, saving(male)
. line le_female year, saving(female)
```

You now wish to combine these two graphs:

```
. gr combine male.gph female.gph
```

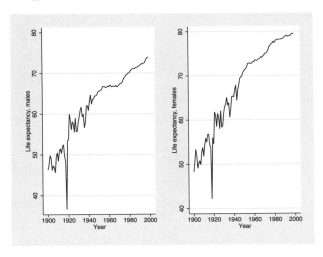

This graph would look better combined into a single column and if we specified `iscale(1)` to prevent the font from shrinking:

(Continued on next page)

```
. gr combine male.gph female.gph, col(1) iscale(1)
```

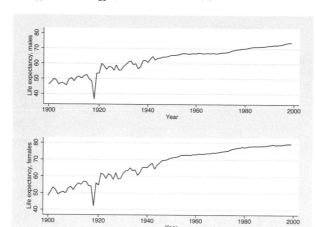

Typical use with memory graphs

Note that in both the above examples, we explicitly typed the .gph suffix on the ends of the filenames:

```
. gr combine male.gph female.gph
. gr combine male.gph female.gph, col(1) iscale(1)
```

We must do that or we must enclose the filenames in quotes:

```
. gr combine "male" "female"
. gr combine "male" "female", col(1) iscale(1)
```

If we did neither, graph combine would assume that the graphs were stored in memory and would then have issued the error that the graphs could not be found. Had we wanted to do these examples using memory graphs rather than disk files, we would have substituted name() for saving on the individual graphs

```
. sysuse uslifeexp, clear
(U.S. life expectancy, 1990-1999)
. line le_male   year, name(male)
. line le_female year, name(female)
```

and then we would type the names without quotes on the graph combine commands:

```
. gr combine male female
. gr combine male female, col(1) iscale(1)
```

Combining twoway graphs

In the first example of *Typical use*, note that the y axes of the two graphs did not align: One had a minimum of 40, while the other was approximately 37. Option ycommon will put all twoway graphs on a common y scale.

```
. sysuse uslifeexp, clear
(U.S. life expectancy, 1990-1999)

. line le_male   year, saving(male)

. line le_female year, saving(female)

. gr combine male.gph female.gph, ycommon
```

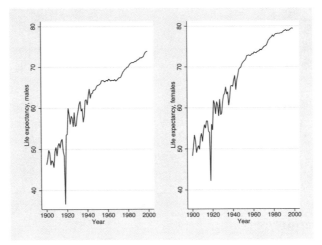

Advanced use

```
. sysuse lifeexp, clear
(Life expectancy, 1998)

. generate loggnp = log10(gnppc)

. label var loggnp "Log base 10 of GNP per capita"

. scatter lexp loggnp,
        ysca(alt) xsca(alt)
        xlabel(, grid gmax)        saving(yx)

. twoway histogram lexp, fraction
        xsca(alt reverse) horiz   saving(hy)

. twoway histogram loggnp, fraction
        ysca(alt reverse)
        ylabel(,nogrid)
        xlabel(,grid gmax)        saving(hx)

. graph combine hy.gph yx.gph hx.gph,
        hole(3)
        imargin(0 0 0 0) graphregion(margin(l=22 r=22))
        title("Life expectancy at birth vs. GNP per capita")
        note("Source:  1998 data from The World Bank Group")
```

(Continued on next page)

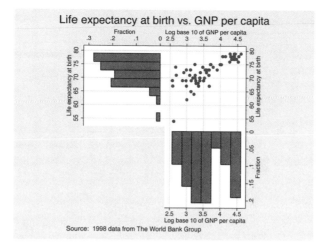

Note our specification of

imargin(0 0 0 0) graphregion(margin(l=22 r=22))

on the graph combine statement. imargin(), we specified to push the graphs together; we eliminated the margins around them. graphregion(margin()), we specified to make the graphs more square—to control the aspect ratio.

Controlling the aspect ratio of subgraphs

The above graph can be converted to look like this:

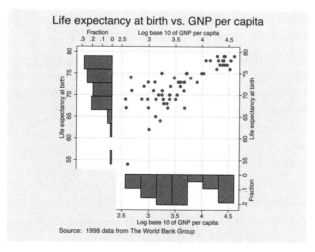

To do the above, we added fysize(25) to the drawing of the histogram for the x axis,

```
. twoway histogram loggnp, fraction
       ysca(alt reverse)
       xlabel(,grid gmax)       saving(hx)
       fysize(25)                                        ← new
```

and we added fxsize(25) to the drawing of the histogram for the y axis,

```
. twoway histogram lexp, fraction
        xsca(alt reverse) horiz
        ylabel(,grid)            saving(hy)
        fxsize(25)                              ← new
```

and that was all we did. The graph combine command remained unchanged.

The *forced_size_options* fysize() and fxsize() are allowed with any graph; their syntax being

forced_size_options	description
fysize(*relativesize*)	use only percent of height available
fxsize(*relativesize*)	use only percent of width available

See [G] *relativesize*.

There are three ways to control the aspect ratio of a graph:

1. Specify the *region_options* ysize(#) and xsize(#); # is specified in inches.

2. Specify the *region_option* graphregion(margin(*marginstyle*)).

3. Specify the *forced_size_options* fysize(*relativesize*) and fxsize(*relativesize*).

Now let us distinguish between

a. controlling the aspect ratio of the overall graph, and

b. controlling the aspect ratio of individual graphs in a combined graph.

Concerning problem (a), methods (1) and (2) are best. We used method (2) when we constructed the overall combined graph above—we specified graphregion(margin(l=22 r=22)). Methods 1 and 2 are discussed under *Controlling the aspect ratio* in [G] *region_options*.

Concerning problem (b), method (1) will not work and methods (2) and (3) do different things.

Method (1) controls the physical size at which the graph appears and so it indirectly controls the aspect ratio. graph combine, however, discards this physical-size information.

Method (2) is one way of controlling the aspect-ratio of subgraphs. graph combine honors margins assuming that you do not specify graph combine's imargin() option, which overrides the original margin information. In any case, if we want the subgraph long and narrow, or short and wide, we need only specify the appropriate graphregion(margin()) at the time we draw the subgraph. When we combine the resulting graph with other graphs, it will look exactly as we want it. The long-and-narrow or short-and-wide graph will appear in the array adjacent to all the other graphs. The way this works, each graph is allocated an equal-sized area in the array, and the oddly shaped graph is drawn into it.

Method (3) is the only way you can obtain unequally sized areas. In the case of the combined graph above, we specified graph combine's imargin() option and that alone precluded our use of method (2), but most importantly, we did not want an array of four equally sized areas:

1	2
histogram	scatter
3	4
	histogram

We wanted

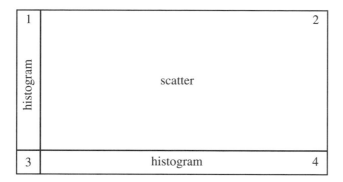

The *forced_size_options* allowed us to achieve that. You specify the *forced_size_options* fysize() and fxsize() with the commands that draw the subgraphs, not with graph combine. Inside the parentheses, you specify the percent of the graph region to be used. Although you could use fysize() and fxsize() to control aspect ratio in ordinary cases, there is no reason to do that. Use fysize() and fxsize() to control aspect when you are going to use graph combine and you want unequally sized areas or when you will be specifying graph combine's imargin() option.

Also See

Complementary: [G] **graph use**; [G] **graph save**, [G] *saving_option*; [G] **gph files**

Title

---
graph copy — Copy graph in memory

Syntax

graph copy [*oldname*] *newname* [, replace]

If *oldname* is not specified, the name of the current graph is assumed.

Description

graph copy makes a copy of a graph stored in memory under a new name.

Options

replace specifies that it is okay to replace *newname* if it already exists.

Remarks

See [G] **graph manipulation** for an introduction to the graph manipulation commands.

graph copy is rarely used. Perhaps you have a graph displayed in the Graph window (known as the current graph) and you wish to experiment with changing its aspect ratio or scheme using the graph display command. Before starting your experiments, you make a copy of the original:

```
. graph copy backup
. graph display ...
```

Also See

Complementary: [G] **graph manipulation**; [G] **graph rename**

Title

graph describe — Describe contents of graph in memory or on disk

Syntax

graph <u>des</u>cribe [*name*]

where *name* may be

name	description
simplename	name of graph in memory
filename.gph	name of graph on disk
"*filename*"	name of graph on disk

If *name* is not specified, the graph currently displayed in the Graph window is described.

Description

graph describe describes the contents of a graph in memory or a graph stored on disk.

Remarks

See [G] **graph manipulation** for an introduction to the graph manipulation commands.

graph describe describes the contents of a graph, which may be stored in memory or on disk. Without arguments, the graph stored in memory named Graph is described:

```
. sysuse auto
(1978 Automobile Data)
. scatter mpg weight
. graph describe
Graph stored in memory
        name:  Graph
      format:  live
     created:  20 Nov 2002 14:35:51
      scheme:  default
        size:  4 x 5.5
    dta file:  auto.dta dated 3 Oct 2002 13:53
     command:  twoway scatter mpg weight
```

In the above, note that the size is reported as *ysize* × *xsize*, not the other way around.

When you type a name ending in .gph, the disk file is described:

```
. graph save myfile
. graph describe myfile.gph
```
myfile.gph stored on disk
```
   name:  myfile.gph
 format:  live
created:  9 Nov 2002 14:26:12
 scheme:  default
   size:  4 x 5.5
dta file:  auto.dta dated 3 Oct 2002 13:53
command:  twoway scatter mpg weight
```

If the file is saved in asis format—see [G] **gph files**— only the name and format are listed:

```
. graph save picture, asis
. graph describe picture.gph
```
picture.gph stored on disk
```
  name:  picture.gph
format:  old
```

Saved Results

Returned in r() is

>Macro
>
>| | |
>|---|---|
>| r(fn) | *filename* or *filename*.gph |
>| r(ft) | "old", "asis", or "live" |

and, if r(ft)=="live",

>Macro
>
>| | |
>|---|---|
>| r(command) | command |
>| r(command_date) | date on which command was run |
>| r(command_time) | time at which command was run |
>| r(scheme) | scheme name |
>| r(ysize) | ysize() value |
>| r(xsize) | xsize() value |
>| r(dtafile) | .dta file in memory at command_time |
>| r(dtafile_date) | .dta file date |

Note that any of r(command), ..., r(dtafile_date) may be undefined, so refer to contents using macro quoting.

Also See

Complementary: [G] **graph manipulation**; [G] **graph dir**

Title

> **graph dir** — List names of graphs in memory and on disk

Syntax

> graph dir $\big[$ *pattern* $\big]$ $\big[$, *dir_options* $\big]$

where *dir_options* are

dir_options	description
memory	list only graphs stored in memory
gph	list only graphs stored on disk
detail	produce detailed listing

and where *pattern* is allowed by Stata's match() function: * means 0 or more characters go here and ? means exactly one character goes here; see [R] **functions**.

Description

> graph dir lists the names of graphs stored in memory and stored on disk in the current directory.

Options

> memory and gph restrict what is listed; memory lists only the names of graphs stored in memory and gph lists only the names of graphs stored on disk.

> detail specifies that, in addition to the names, the commands that created the graphs be listed.

Remarks

> See [G] **graph manipulation** for an introduction to the graph manipulation commands.

> graph dir without options lists in column format the names of the graphs stored in memory and those stored on disk in the current directory.

```
. graph dir
        Graph       figure1.gph     large.gph     s7.gph
        dot.gph     figure2.gph     old.gph       yx_lines.gph
```

Graphs in memory are listed first followed by graphs stored on disk. In the example above, we have only one graph in memory: Graph.

You may specify a pattern to restrict the files listed:

```
. mydir fig*
        figure1.gph   figure2.gph
```

The detail option lists the names and the commands that drew the graphs:

```
. mydir fig*, detail
  name          command

  figure1.gph  matrix  h-tempjul, msy(p) name(myview)
  figure2.gph  twoway scatter mpg weight, saving(figure2)
```

Saved Results

graph dir returns in macro r(list) the names of the graphs.

Also See

Complementary: [G] **graph manipulation**; [G] **graph describe**

Title

> **graph display** — Redisplay graph in memory

Syntax

$$\underline{\text{gr}}\text{aph } \underline{\text{di}}\text{splay } \left[\textit{name} \right] \left[, \; \textit{display\_options} \right]$$

where *display_options* are

display_options	description
ysize(*#*)	change height of graph (in inches)
xsize(*#*)	change width of graph (in inches)
margins(*marginstyle*)	change outer margins
scheme(*schemename*)	change overall look

See [G] ***marginstyle*** and [G] ***scheme_option***.

If *name* is not specified, the name of the current graph—the graph displayed in the Graph window—is assumed.

Description

graph display redisplays a graph stored in memory.

Options

ysize(*#*) and xsize(*#*) specify in inches the height and width of the entire graph (also known as the *available area*). The defaults are the original height and width of the graph. These two options can be used to change the aspect ratio; see *Changing the size and aspect ratio* under *Remarks* below.

margins(*marginstyle*) specifies the outer margins: the margins between the outer graph region and the inner graph region as shown in the diagram in [G] ***region_options***. See *Changing the margins* under *Remarks* below and see [G] ***marginstyle***.

scheme(*schemename*) specifies the overall look of the graph. The default is the original scheme with which the graph was drawn. See *Changing the scheme* under *Remarks* below and see [G] ***scheme_option***.

Remarks

See [G] **graph manipulation** for an introduction to the graph manipulation commands.

Remarks are presented under the headings

> *Changing the size and aspect ratio*
> *Changing the margins and aspect ratio*
> *Changing the scheme*

Changing the size and aspect ratio

Under *Controlling the aspect ratio* in [G] *region_options*, we compared

```
. sysuse auto, clear
(1978 Automobile Data)
. scatter mpg weight
```

with

```
. scatter mpg weight, ysize(5)
```

You do not need to reconstruct the graph merely to change the `ysize()` or `xsize()`. We could start with some graph

```
. scatter mpg weight
```

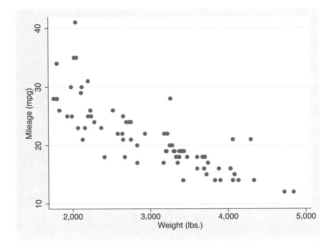

and then we could redisplay it with different `ysize()` and/or `xsize()` values:

(Continued on next page)

```
. graph display, ysize(5)
```

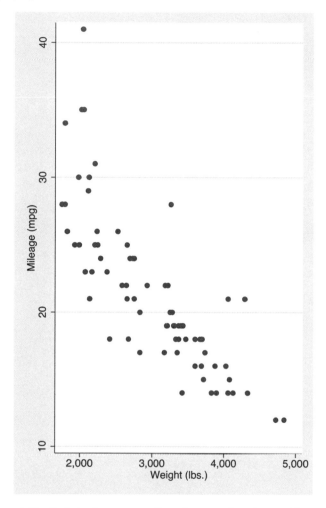

In this way you can quickly find the best ysize() and xsize() values. This works particularly well when the graph you have drawn required lots of options:

```
. sysuse uslifeexp, clear
(U.S. life expectancy, 1900-1999)
```

(Continued on next page)

```
. line le_wm year, yaxis(1 2) xaxis(1 2)
|| line le_bm year
|| line diff  year
|| lfit diff  year
||,
    ylabel(0(5)20, axis(2) gmin angle(horizontal))
    ylabel(0 20(10)80,    gmax angle(horizontal))
    ytitle("", axis(2))
    xlabel(1918, axis(2)) xtitle("", axis(2))
    ytitle("Life expectancy at birth (years)")
    title("White and black life expectancy")
    subtitle("USA, 1900-1999")
    note("Source: National Vital Statistics, Vol 50, No. 6"
 "(1918 dip caused by 1918 Influenza Pandemic)")
    legend(label(1 "White males") label(2 "Black males"))
```

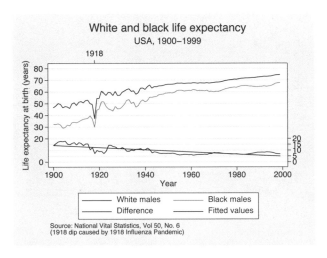

(Continued on next page)

```
. graph display, ysize(5.25)
```

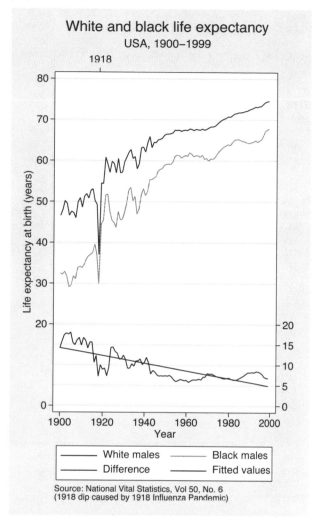

In addition, note that you can change sizes of graphs you have previously drawn and stored on disk:

```
. graph use ...
. graph display, ysize(...) xsize(...)
```

You may not remember what `ysize()` and `xsize()` values were used (the defaults are `ysize(4)` and `xsize(5.5)`). In that case, use `graph describe` to describe the file; it reports the `ysize()` and `xsize()` values; see [G] **graph describe**.

Changing the margins and aspect ratio

You can change the size of a graph or change its margins to control the aspect ratio; this is discussed in *Controlling the aspect ratio* in [G] *region_options*. There is shown the example

```
scatter mpg weight, by(foreign, total graphregion(margin(l+10 r+10)))
```

This too can be done in two steps:

```
. scatter mpg weight, by(foreign, total)
. graph display, margins(l+10 r+10)
```

graph display's `margin()` option corresponds to `graphregion(margin())` used at the time you construct graphs.

Changing the scheme

Schemes determine the overall look of a graph such as where axes, titles, and legends are placed, the color of the background, etc.; see [G] **schemes**. Changing the scheme after a graph has been constructed sometimes works well and sometimes works poorly.

Here is an example in which it works well:

```
. sysuse uslifeexp2, clear
(U.S. life expectancy, 1900-1940)

. line le year, sort
        title("Line plot")
        subtitle("Life expectancy at birth, U.S.")
        note("1")
        caption("Source: National Vital Statistics Report,
        Vol. 50 No. 6")
```

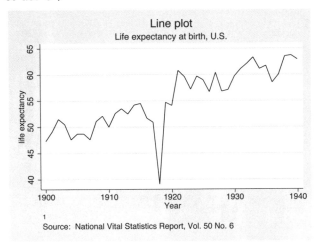

```
. graph display, scheme(economist)
```

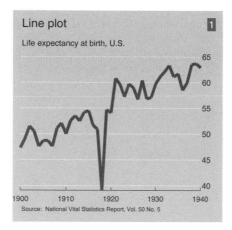

The above example works well because no options were specified to move from their default location of things such as axes, titles, and legends, and no options were specified overriding default colors. The issue is simple: if you draw a graph and say, "move the title from its default location to over here", it may turn out that over here is a terrible place for the title once you change schemes. Or if you override a color and make it magenta, it may turn out that magenta clashes terribly.

The above does not mean that the graph command need be simple. The example shown in *Changing the size and aspect ratio* above,

```
. line le_wm year, yaxis(1 2) xaxis(1 2)
  || line le_bm year
  || line diff  year
  || lfit diff  year
  ||,
     ylabel(0(5)20, axis(2) gmin angle(horizontal))
     ylabel(0 20(10)80,    gmax angle(horizontal))
     ytitle("", axis(2))
     xlabel(1918, axis(2)) xtitle("", axis(2))
     ytitle("Life expectancy at birth (years)")
     title("White and black life expectancy")
     subtitle("USA, 1900-1999")
     note("Source: National Vital Statistics, Vol 50, No. 6"
  "(1918 dip caused by 1918 Influenza Pandemic)")
     legend(label(1 "White males") label(2 "Black males"))
```

moves across schemes just fine, the only potential problem being our specification of an-gle(horizontal) for the labeling of the two y axes. That might not look good with some schemes.

If you are concerned about moving between schemes, when you specify options, specify style options in preference to options that directly control the outcome. For example, to make two sets of points have the same color, specify the mstyle() option rather than changing the color of one set to match the color you currently see of the other set.

There is an additional issue when moving between styles that have different background colors. Styles are said to have naturally white or naturally black background colors; see [G] **schemes**. When you move from one type of scheme to another, if the colors were not changed, colors that previously stood out would blend into the background and vice versa. To prevent this, graph display changes all the colors to be in accordance with the scheme, except that graph display does not change colors you specify by name (e.g., you specify `mcolor(magenta)` or `mcolor("255 0 255")` to change the color of a symbol).

Our recommendation is that you do not use `graph display` to change graphs from having naturally black to naturally white backgrounds. As long as you print in monochrome, `print` does an excellent job translating black to white backgrounds, so there is no need to change styles for that reason. If you are printing in color, then we recommend you change your default scheme to a naturally white scheme; see [G] **set scheme**.

Also See

Complementary: [G] **graph manipulation**

Title

> **graph dot** — Dot charts

Syntax

$$\underline{\text{gr}}\text{aph dot } \textit{yvars } \left[\textit{weight}\right] \left[\text{if } \textit{exp}\right] \left[\text{in } \textit{range}\right] \left[, \textit{ options }\right]$$

where *yvars* is

        (asis) *varlist*

or is

$\left[\,(\textit{stat})\,\right]$ *varname*	$\left[\,\left[\,(\textit{stat})\,\right]\,\ldots\,\right]$
$\left[\,(\textit{stat})\,\right]$ *varlist*	$\left[\,\left[\,(\textit{stat})\,\right]\,\ldots\,\right]$
$\left[\,(\textit{stat})\,\right]$ *name=varname* $\left[\ldots\right]$	$\left[\,\left[\,(\textit{stat})\,\right]\,\ldots\,\right]$

where *stat* may be any of

        mean median p1 p2 ...p99 sum count min max

or

        any of the other *stats* defined in [R] **collapse**

mean is the default. p1 means 1st percentile, p2 second, and so on; p50 means the same as median. count means the number of nonmissing values of the specified variable.

and where *options* are

options	description
group_options	groups over which lines of dots are drawn
yvar_options	variables that are the dots
linelook_options	how the lines of dots look
legending_options	how *yvars* are labeled
axis_options	how numerical y axis is labeled
title_and_other_options	titles, added text, aspect ratio, etc.

Each is defined below.

The *group_options*, which define the groups over which the lines of dots (lines) are drawn are

group_options	description
<u>over</u>(*varname*$\left[\,, \textit{over\_subopts}\,\right]$)	categories; option may be repeated
nofill	omit empty categories
<u>missing</u>	keep missing value as category

The *yvar_options*, which define the variables that are the lines, are

yvar_options	description
<u>a</u>scategory	treat *yvars* as first over() group
<u>asy</u>vars	treat first over() group as *yvars*
<u>percent</u>ages	show percentages within *yvars*
cw	calculate *yvar* statistics omitting missing values of any *yvar*

The *linelook_options*, which define how the lines look, are

linelook_options	description
<u>outer</u>gap([*]#)	gap between top and first line and between last line and bottom
<u>linegap</u>(#)	gap between *yvar* lines; default 0
<u>m</u>arker(#, *marker_options*)	marker used for #th *yvar* line
<u>linet</u>ype(dot | line | rectangle)	type of line
<u>ndots</u>(#)	# of dots if linetype(dot); default 100
<u>dots</u>(*marker_options*)	look if linetype(dot)
<u>lines</u>(*line_options*)	look if linetype(line)
<u>rectan</u>gles(*area_options*)	look if linetype(rectangle)
rwidth(*relativesize*)	rectangle width if linetype(rectangle)
[<u>no</u>]<u>extend</u>line	whether line extends through plot region margins; extendline is usual default
<u>lowext</u>ension(*relativesize*)	extend line through axis (advanced)
<u>highext</u>ension(*relativesize*)	extend line through axis (advanced)

See [G] *marker_options*, [G] *line_options*, [G] *area_options*, and [G] *relativesize*.

The *legending_options*, which define how the *yvars* are labeled, are

legending_options	description
legend_option	control of *yvar* legend
<u>nolabel</u>	use *yvar* names, not labels, in legend
<u>yvar</u>options(*over_subopts*)	*over_subopts* for *yvars*; rarely specified
<u>showy</u>vars	label *yvars* on *x* axis; rarely specified

See [G] *legend_option*.

The *axis_options*, which define how the numerical y axis is labeled, are

axis_options	description
yalternate	put numerical y axis on right (top)
xalternate	put categorical x axis on top (right)
exclude0	do not force y axis to include 0
yreverse	reverse y axis
axis_scale_options	y-axis scaling and look
axis_label_options	y-axis labeling
ytitle(...)	y-axis titling

See [G] ***axis_scale_options***, [G] ***axis_label_options***, and [G] ***axis_title_options***.

The *title_and_other_options* are

title_and_other_options	description
text(...)	add text on graph; x range $\left[0, 100\right]$
yline(...)	add y lines to graph
std_options	titles, aspect ratio, saving to disk
by(*varlist*, ...)	repeat for subgroups

See [G] ***added_text_option***, [G] ***added_line_options***, [G] ***std_options***, and [G] ***by_option***.

The *over_subopts*, used in over(*varname*, *over_subopts*) and, on rare occasion, in yvaroptions(*over_subopts*), are

over_subopts	description
relabel(# "*text*" ...)	change axis labels
label(*cat_axis_label_options*)	rendition of labels
axis(*cat_axis_line_options*)	rendition of axis line
gap($\left[*\right]$#)	gap between lines within over() category
sort(*varname*)	put lines in prespecified order
sort(#)	put lines in height order
sort((*stat*) *varname*)	put lines in derived order
descending	reverse default or specified line order

See [G] ***cat_axis_label_options*** and [G] ***cat_axis_line_options***.

aweights, fweights, and pweights are allowed; see [U] **14.1.6 weight** and see note concerning weights in [R] **collapse**.

Description

graph dot draws dot charts, horizontally presented. In a dot chart, the categorical axis is presented vertically and the numerical axis, horizontally. Even so, the numerical axis is called the y axis and the categorical axis is still called the x axis:

. graph dot (mean) *numeric_var*, over(*cat_var*)

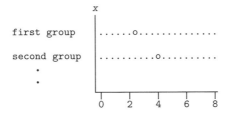

The syntax for dot charts is identical to that for bar charts; see [G] **graph bar**.

We use the following words to describe a dot chart:

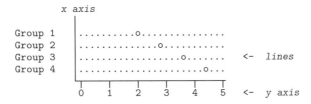

The above dot chart contains four *lines*. The words used to describe a line are

Options

Group options

over(*varname* [, *over_subopts*]) specifies a categorical variable over which the *yvars* are to be repeated. *varname* may be string or numeric. Up to two over() options may be specified when multiple *yvars* are specified, and up to three over()s may be specified when a single *yvar* is specified; options may be specified; see *Appendix: Examples of syntax* below.

nofill specifies that missing subcategories are to be omitted. For instance, consider

. graph dot (mean) y, over(division) over(region)

Say that one of the divisions has no data for one of the regions, either because there are no such observations or because y==. for such observations. In the resulting chart, the marker will be missing:

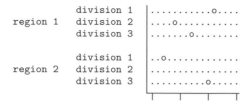

If you specify `nofill`, the missing category will be removed from the chart:

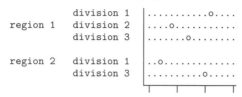

`missing` specifies that missing values of the `over()` variables be kept as their own categories, one for `.`, another for `.a`, etc. The default is to act as if such observations simply did not appear in the dataset; the observations are ignored. An `over()` variable is considered to be missing if it is numeric and contains a missing value or if it is string and contains " ".

Yvar options

`ascategory` specifies that the *yvars* be treated as the first `over()` group.

When you specify `ascategory`, results are as if you specified a single *yvar* and introduced a new first `over()` variable. Anyplace you read in the documentation that something is done over the first `over()` category, or using the first `over()` category, it will be done over or using *yvars*.

Think like this: you specified

    . graph dot y1 y2 y3, ascategory *whatever_other_options*

and results will be as if you typed

    . graph dot y, over(*newcategoryvariable*) *whatever_other_options*

with a long rather than wide dataset in memory.

`asyvars` specifies that the first `over()` group be treated as *yvars*.

When you specify `asyvars`, results are as if you removed the first `over()` group and introduced multiple *yvars*. Note that we said in most ways, not all ways, but let's ignore that for a moment. If you previously had *k* *yvars* and, in your first `over()` category, *G* groups, results will be as if you specified *k*G* yvars and removed the `over()`. Anyplace you read in the documentation that something is done over the *yvars* or using the *yvars*, it will be done over or using the first `over()` group.

Think like this: you specified

    . graph dot y, over(group) asyvars *whatever_other_options*

and results will be as if you typed

    . graph dot *y1 y2 y3* ... , *whatever_other_options*

with a wide rather than long dataset in memory. Variables *y1*, *y2*, ..., are sometimes called the virtual *yvars*.

percentages specifies that marker positions be based on percentages *yvar_i* represents of all the *yvars*. That is,

> . graph dot (mean) inc_male inc_female

would produce a chart with the markers reflecting average income.

> . graph dot (mean) inc_male inc_female, percentage

would produce a chart with the markers being located at $100 \times \text{inc\_male}/(\text{inc\_male} + \text{inc\_female})$ and $100 \times \text{inc\_female}/(\text{inc\_male} + \text{inc\_female})$.

If you have a single *yvar* and want percentages calculated over the first over() group, specify the asyvars option. For instance,

> . graph dot (mean) wage, over(*i*) over(*j*)

would produce a chart where marker positions reflect mean wages.

> . graph dot (mean) wage, over(*i*) over(*j*) asyvars percentages

would produce a chart where marker positions are $100 \times (\text{mean}_{ij}/(\text{Sum}_i \text{mean}_{ij}))$

cw specifies casewise deletion. If cw is specified, observations for which any of the *yvars* are missing are ignored. The default is to calculate each statistic using all the data possible.

Linelook options

outergap(*#) and outergap(#) specify the gap between the top of the graph and beginning of the first line, and the last line and the bottom of the graph.

outergap(*#) specifies that the default is to be modified. Specifying outergap(*1.2) increases the gap by 20% and specifying outergap(*.8) reduces the gap by 20%.

outergap(#) specifies the gap in "percent-of-bar-width units" (*sic*). graph dot is related to graph bar. Just remember that outergap(50) specifies a sizable but not excessive gap.

linegap(#) specifies the gap to be left between *yvar* lines. The default is linegap(0), meaning that multiple *yvars* appear on the same line. For instance, typing

> . graph dot y1 y2, over(group)

results in

```
        group 1  |..x....o........
        group 2  |........x..o....
        group 3  |.......x.....o..
                  └──┴──┴──┴──┘
```

In the above, o represents the symbol for y1 and x the symbol for y2. If you want to have separate lines for the separate *yvars*, specify linegap(20):

> . graph dot y1 y2, over(group) linegap(20)

```
        group 1  |.......o........
                 |..x............
                 |
        group 2  |...........o....
                 |.......x........
                 |
        group 3  |............o..
                 |.......x........
                  └──┴──┴──┴──┘
```

Specify a number smaller or larger than 20 to reduce or increase the distance between the y1 and y2 lines.

Alternatively, and generally preferred, is to specify option `ascategory`. That will result in

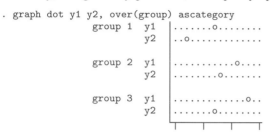

Note that `linegap()` affects only the *yvar* lines. If you want to change the gap for the first, second, or third `over()` groups, specify the *over_subopt* `gap()` inside the `over()` itself.

`marker(#, marker_options)` specifies the shape, size, color, etc., of the marker to be used to mark the value of the #th *yvar* variable. `marker(1, ...)` refers to the marker associated with the first *yvar*, `marker(2, ...)` refers to the marker associated with the second, and so on. A particularly useful *marker_option* is `mcolor(colorstyle)`, which sets the color of the marker. For instance, you might specify `marker(1, mcolor(green))` to make the marker associated with the first *yvar* green. See [G] *colorstyle* for a list of color choices and see [G] *marker_options* for information on the other *marker_options*.

`linetype(dot)`, `linetype(line)`, and `linetype(rectangle)` specify the style of the line.

`linetype(dot)` is the usual default. In this style, dots are used to fill the line around the marker:

········o········

`linetype(line)` specifies that a solid line be used to fill the line around the marker:

————o————

`linetype(rectangle)` specifies that a long "rectangle" (which looks more like two parallel lines) be used to fill the area around the marker:

========o=======

`ndots(#)` and `dots(marker_options)` are relevant only in the `linetype(dots)` case.

`ndots(#)` specifies the number of dots to be used to fill the line. The default is `ndots(100)`.

`dots(marker_options)` specifies the marker symbol, color, and size to be used as the dot symbol. The default is to use `dots(msymbol(p))`. See [G] *marker_options*.

`lines(line_options)` is relevant only if `linetype(line)` is specified. It specifies the look of the line to be used; see [G] *line_options*.

`rectangles(area_options)` and `rwidth(relativesize)` are relevant only if `linetype(rectangle)` is specified.

`rectangles(area_options)` specifies the look of the parallel lines (rectangle); see [G] *area_options*.

`rwidth(relativesize)` specifies the width (height) of the rectangle (the distance between the parallel lines). The default is usually `rwidth(.45)`; see [G] *relativesize*.

noextendline and extendline are relevant in all cases. They specify whether the line (be it dots, a line, or a rectangle) is to extend through the plot region margin and so touch the axes. The usual default is extendline, so noextendline is the option. See [G] *region_options* for a definition of the plot region.

lowextension(*relativesize*) and highextension(*relativesize*) are advanced options. They specify the amount by which the line (be it dots, line or a rectangle) is extended through the axes. The usual defaults are lowextension(0) and highextension(0). See [G] *relativesize*.

Legending options

legend_option allows you to control the legend. If more than one *yvar* is specified, a legend is produced. Otherwise, no legend is needed because the over() groups are labeled on the categorical x axis. See [G] *legend_option*.

nolabel specifies that, in the automatic construction of the legend, the variable names of the *yvars* be used in preference to "mean of *varname*" or "sum of *varname*", etc.

yvaroptions(*over_subopts*) allows you to specify *over_subopts* for the *yvars*. This is very rarely done.

showyvars specifies that, in addition to building a legend, the identities of the *yvars* are to be shown on the categorical x axis. If showyvars is specified, it is typical to also specify legend(off).

Axis options

yalternate and xalternate switch the side on which the axes appear. yalternate moves the numerical y axis from the bottom to the top; xalternate moves the categorical x axis from the left to the right. If your scheme by default puts the axes on the opposite sides, then yalternate and xalternate reverse their actions.

exclude0 specifies that the numerical y axis need not be scaled so as to include 0.

yreverse specifies that the numerical y axis have its scale reversed, so that it runs from maximum to minimum.

axis_scale_options specify how the numerical y axis is scaled and how it looks; see [G] *axis_scale_options*. There you will also see option xscale() in addition to yscale(). Ignore xscale(); it is irrelevant in the case of dot plots.

axis_label_options specify how the numerical y axis is to be labeled. The *axis_label_options* also allow you to add and suppress grid lines; see [G] *axis_label_options*. There you will see that, in addition to options ylabel(), ytick(), ymlabel(), and ymtick(), options xlabel(), ..., xmtick() are allowed. Ignore the x*() options; they are irrelevant in the case of dot charts.

ytitle() overrides the default title for the numerical y axis; see [G] *axis_title_options*. There you will also find option xtitle() documented; it is irrelevant in the case of dot charts.

Title and other options

text() adds text to a specified location on the graph; see [G] *added_text_option*. The basic syntax of text() is

text(#$_y$#$_x$ "*text*")

text() is documented in terms of twoway graphs. When used with dot charts, the "numeric" x axis is scaled to run from 0 to 100.

yline() adds vertical lines at specified y values; see [G] ***added_line_options***. The xline() option, also documented there, is irrelevant in the case of dot charts. If your interest is in adding grid lines, see [G] ***axis_label_options***.

std_options allow you to add titles, control the aspect ratio, save the graph on disk, and much more; see [G] ***std_options***.

by(*varlist*, ...) draws separate plots within a single graph; see [G] ***by_option***.

Suboptions for use with over() and yvaroptions()

relabel(*#* "*text*" ...) specifies text to override the default labeling of the categories. See the description of the relabel() option under *Suboptions for use with over() and yvaroptions()* in [G] **graph bar**.

label(*cat_axis_label_options*) determines other aspects of the look of the category labels on the x axis. With the exception of label(labcolor()) and label(labsize()), these options are seldom specified; see [G] ***cat_axis_label_options***.

axis(*cat_axis_line_options*) specifies how the axis line is rendered. This is a rarely specified option. See [G] ***cat_axis_line_options***.

gap(*#*) and gap(**#*) specify the gap between the lines in this over() group. gap(*#*) is specified in "percent-of-bar-width units" *(sic)*. Just remember gap(50) is a considerable, but not excessive width. gap(**#*) allows modifying the default gap. gap(*1.2) would increase the gap by 20% and gap(*.8) would decrease the gap by 20%.

sort(*varname*), sort(*#*), and sort((*stat*) *varname*) control how the lines are ordered. See *How bars are ordered* and *Reordering the bars* in [G] **graph bar**.

sort(*varname*) puts the lines in the order of *varname*.

sort(*#*) puts the markers in distance order. *#* refers to the *yvar* number on which the ordering should be performed.

sort((*stat*) *varname*) puts the lines in an order based on a calculated statistic.

descending specifies that, whatever the order of the lines, default or as specified by sort(), it is to be reversed.

Remarks

Remarks are presented under the headings

> *Relationship between dot plots and horizontal bar charts*
> *Examples*
> *Appendix: Examples of syntax*

Relationship between dot plots and horizontal bar charts

Despite appearances, graph hbar and graph dot are in fact the same command, meaning that concepts and options are the same:

. graph hbar y, over(group)

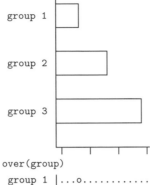

. graph dot y, over(group)

```
group 1 |...o............
group 2 |........o.......
group 3 |.............o.
```

There is only one substantive difference between the two commands: Given multiple *yvars*, graph hbar draws multiple bars:

. graph hbar y1 y2, over(group)

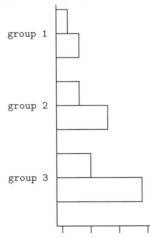

graph dot draws multiple markers on single lines:

. graph dot y1 y2, over(group)

```
group 1 |.x.o............
group 2 |...x....o.......
group 3 |.....x.........o.
```

The way around this problem (if it is a problem) is to specify option ascategory or to specify option linegap(*#*). Specifying ascategory is usually best.

Read about graph hbar in [G] **graph bar**.

Examples

Since `graph dot` and `graph hbar` are so related, the following examples should require little by way of explanation:

```
. sysuse nlsw88, clear
(NLSW, 1988 extract)

. graph dot wage, over(occ, sort(1))
        ytitle("")
        title("Average hourly wage, 1988, women aged 34-46", span)
        subtitle(" ")
        note("Source:  1988 data from NLS, U.S. Dept. of Labor,
              Bureau of Labor Statistics", span)
```

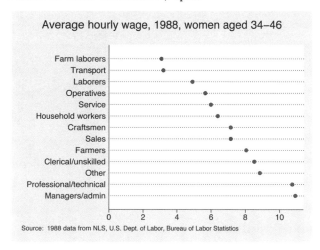

```
. graph dot (p10) p10=wage (p90) p90=wage,
        over(occ, sort(2))
        legend(label(1 "10th percentile") label(2 "90th percentile"))
        title("10th and 90th percentiles of hourly wage", span)
        subtitle("Women aged 34-46, 1988" " ", span)
        note("Source:  1988 data from NLS, U.S. Dept. of Labor,
              Bureau of Labor Statistics", span)
```

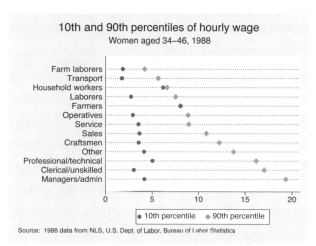

```
. graph dot (mean) wage,
      over(occ, sort(1))
      by(collgrad,
          title("Average hourly wage, 1988, women aged 34-46", span)
          subtitle(" ")
          note("Source:  1988 data from NLS, U.S. Dept. of Labor,
              Bureau of Labor Statistics", span)
      )
```

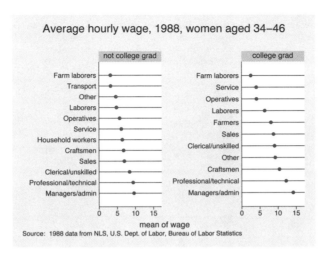

Appendix: Examples of syntax

Let us consider some graph dot commands and what they do:

graph dot revenue
: One line showing average revenue.

graph dot revenue profit
: One line with two markers, one showing average revenue and the other, average profit.

graph dot revenue, over(division)
: #_of_divisions lines, each with one marker showing average revenue for each division.

graph dot revenue profit, over(division)
: #_of_divisions lines, each with two markers, one showing average revenue and the other, average profit for each division.

graph dot revenue, over(division) over(year)
: #_of_divisions×#_of_years lines, each with one marker showing average revenue for each division, repeated for each of the years. The grouping would look like this (assuming 3 divisions and 2 years):

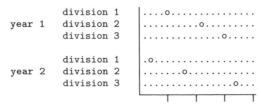

```
graph dot revenue, over(year) over(division)
```
same as above, but ordered differently. In the previous example, we typed over(division)
over(year). This time, we reverse it:

```
division 1    year 1    |....o...............
              year 2    |.o..................

division 2    year 1    |..........o.........
              year 2    |.......o............

division 3    year 1    |..............o.....
              year 2    |...............o...
                        └────┬────┬────┬────┬
```

```
graph dot revenue profit, over(division) over(year)
```
#_of_divisions × *#_of_years* lines each with two markers, one showing average revenue and the
other, average profit for each division, repeated for each of the years.

```
graph dot (sum) revenue profit, over(division) over(year)
```
#_of_divisions × *#_of_years* lines each with two markers, the first showing the sum of revenue
and the second, sum of profit for each division, repeated for each of the years.

```
graph dot (median) revenue profit, over(division) over(year)
```
#_of_divisions × *#_of_years* lines each with two markers, showing the median of revenue and
median of profit for each division, repeated for each of the years.

```
graph dot (median) revenue (mean) profit, over(division) over(year)
```
#_of_divisions × *#_of_years* lines each with two markers, showing the median of revenue and
mean of profit for each division, repeated for each of the years.

Also See

Complementary:	[G] **graph bar**;
	[R] **collapse**

Title

> **graph drop** — Drop graphs from memory

Syntax

> graph drop *name* [*name* ...]
>
> graph drop _all

Description

> graph drop *name* drops (discards) the specified graph(s) from memory.
>
> graph drop _all drops all graphs from memory.

Remarks

See [G] **graph manipulation** for an introduction to the graph manipulation commands.

Remarks are presented under the headings

> *Typical use*
> *Relationship between graph drop _all and discard*
> *Erasing graphs on disk*

Typical use

Graphs contain the data they display so, when datasets are large, graphs have the potential to consume considerable amounts of memory. graph drop frees that memory. Remember that Graph is the name of a graph when you do not specify otherwise.

```
. graph twoway scatter faminc educ, ms(p)
. ...
. graph drop Graph
```

One often uses graphs in memory to prepare the pieces for graph combine:

```
. graph ..., ... name(p1)
. graph ..., ... name(p2)
. graph ...   , ... name(p3)
. graph combine p1 p2 p3, ... saving(result, replace)
. graph drop _all
```

Relationship between graph drop _all and discard

The discard command performs graph drop _all and more:

1. discard eliminates prior estimation results and automatically loaded programs, and thereby frees even more memory.

2. discard closes any open dialog boxes and thereby frees even more memory.

We nearly always type discard in preference to graph drop _all if only because discard has fewer characters. The exception to that is when we have fitted a model and still plan on redisplaying prior results, performing tests on that model, or referring to _b[], _se[], etc.

See [P] **discard** for a description of the discard command.

Erasing graphs on disk

graph drop is not used to erase .gph files; instead use Stata's standard erase command:

. erase matfile.gph

Also See

Complementary:	[G] **graph manipulation**
Related:	[R] **erase**,
	[P] **discard**

Title

> **graph export** — Export current graph

Syntax

> graph export *newfilename.suffix* [, *export_options*]

where *export_options* are

export_options	description
as(*fileformat*)	desired format of output
replace	*newfilename* may already exist
override_options	override defaults in conversion

If as() is not specified, the output format is determined by the suffix of *newfilename.suffix*:

suffix	implied option	output format
.ps	as(ps)	PostScript
.eps	as(eps)	Encapsulated PostScript
.wmf	as(wmf)	Windows Metafile
.emf	as(emf)	Windows Enhanced Metafile
.pict	as(pict)	Macintosh PICT format
other		must specify as()

ps and eps are available with all versions of Stata; wmf and emf are available only with Stata for Windows; and pict is available only with Stata for Macintosh.

The *override_options* are

override_options	description
ps_options	when exporting to ps
eps_options	when exporting to eps
mf_options	when exporting to wmf or emf
pict_options	when exporting to pict

See [G] *ps_options*, [G] *eps_options*, [G] *mf_options*, and [G] *pict_options*.

Description

graph export exports to a file the graph currently displayed in the Graph window.

Options

as(*fileformat*) specifies the file format to which the graph is to be exported. This option is rarely specified since, by default, graph export determines the format based on the suffix of the file being created.

replace specifies that it is okay to replace *filename.suffix* if it already exists.

override_options modify how the graph is converted. See [G] ***ps_options***, [G] ***eps_options***, [G] ***mf_options***, and [G] ***pict_options***.

Remarks

Graphs are exported by displaying them on the screen and then typing

. graph export *filename.suffix*

Remarks are presented under the headings

Exporting the graph currently displayed in the Graph window
Exporting a graph stored on disk
Exporting a graph stored in memory

If your interest is simply in printing a graph, see [G] **graph print**.

Exporting the graph currently displayed in the Graph window

There are two ways to export the graph currently displayed in the Graph Window:

1. Pull down **File** and choose **Save Graph...** and choose the appropriate **Save as type**.

2. Type "graph export *filename.suffix*" in the Command window.

Both are equivalent. The advantage of graph export is that you can include it in do-files:

. graph ...   (draw a graph)
. graph export *filename.suffix* (and export it)

By default, graph export determines the output type by the *suffix*. If we wanted to create an Encapsulated PostScript file, we might type

. graph export figure57.eps

Exporting a graph stored on disk

To export a graph stored on disk, type

. graph use *gph_filename*
. graph export *output_filename.suffix*

Do not specify graph use's nodraw option; see [G] **graph use**.

Note for Stata-for-Unix(console) users: Follow the instructions just given even though you have no Graph window and cannot see what has just been "displayed": Use the graph and then export it.

Exporting a graph stored in memory

To export a graph stored in memory but not currently displayed, type

. graph display *name*

. graph export *filename*.*suffix*

Do not specify graph display's nodraw option; see [G] **graph display**.

Note for Stata-for-Unix(console) users: Follow the instructions just given even though you have no Graph window and cannot see what has just been "displayed": Display the graph and then export it.

Also See

Complementary: [G] *ps_options*, [G] *eps_options*, [G] *mf_options*, [G] *pict_options*;

[G] **graph display**, [G] **graph use**; [G] **graph print**

Title

> **graph manipulation** — Graph manipulation commands

Syntax

The graph manipulation commands are

command	description
graph dir	list names of graphs
graph describe	describe contents of graph
graph drop	discard graph stored in memory
graph rename	rename graph stored in memory
graph copy	copy graph stored in memory
graph use	load graph on disk into memory and display it
graph display	redisplay graph stored in memory
graph combine	combine one or more graphs into one graph

See [G] **graph dir**, [G] **graph describe**, [G] **graph drop**, [G] **graph rename**, [G] **graph copy**, [G] **graph use**, [G] **graph display**, and [G] **graph combine**.

Description

The graph manipulation commands manipulate graphs stored in memory or stored on disk.

Remarks

Remarks are presented under the headings

> *Overview of graphs in memory and graphs on disk*
> *Summary of graph manipulation commands*

Overview of graphs in memory and graphs on disk

Graphs are stored in memory and on disk. When you draw a graph, such as by typing

    . graph twoway scatter mpg weight

the resulting graph is stored in memory and, in particular, it is stored under the name Graph. Were you next to type

    . graph matrix mpg weight dispel

this new graph replaces the existing graph named Graph.

Graph is the default name used to record graphs in memory and, when you draw graphs, they replace what was previously recorded in Graph.

You can specify the name() option—see [G] *name_option*—to record graphs under different names:

    . graph twoway scatter mpg weight, name(scat)

Now we have two graphs in memory: Graph containing a scatterplot matrix and scat containing a graph of mpg versus weight.

Graphs in memory are forgotten when you exit Stata and they are forgotten at other times, too, such as when you type clear or discard; see [R] **drop** and [P] **discard**.

Graphs can be stored on disk, where they will reside permanently until you erase them. They are saved in files known as .gph files—files whose names end in .gph; see [G] **gph files**.

You can save on disk the graph currently showing in the Graph window by typing

```
. graph save mygraph.gph
```

The result is to create a new file mygraph.gph; see [G] **graph save**. Or—see [G] *saving_option*— you can save on disk graphs when you originally draw them:

```
. graph twoway scatter mpg weight, saving(mygraph.gph)
```

Either way, graphs saved on disk can be reloaded:

```
. graph use mygraph.gph
```

loads mygraph.gph into memory under the name—you guessed it—Graph. Of course, you could load it under a different name:

```
. graph use mygraph.gph, name(memcp)
```

Having brought this graph back into memory, things are just as if we had drawn the graph for the first time. Anything we could do back then—such as combine the graph with other graphs or change its aspect ratio—we can do now. And, of course, after making any changes, we can save the result on disk, either replacing file mygraph.gph or saving it under a new name.

There is only one final, and very minor, wrinkle: Graphs on disk can be saved in either of two formats, known as live and asis. live is preferred and the default, and what we just said above applies only to live-format files. asis files are more like pictures—all you can do is admire them and make copies. To save a file in asis format, you type

```
. graph save ..., asis
```

or

```
. graph ..., ... saving(..., asis)
```

asis format is discussed in [G] **gph files**.

Oh yes, and there is a third format called old. old is like asis except that it refers to graphs made by older versions of Stata prior to Stata 8. That is discussed in [G] **gph files**, too.

Summary of graph manipulation commands

The graph manipulation commands help you manage your graphs, whether stored in memory or on disk. The commands are

graph dir
Lists the names under which graphs are stored, both in memory and on disk; see [G] **graph dir**.

graph describe
Provides details about a graph, whether stored in memory or on disk; see [G] **graph describe**.

graph drop
Eliminates from memory graphs stored there; see [G] **graph drop**.

`graph rename`
Changes the name of a graph stored in memory; see [G] **graph rename**.

`graph copy`
Makes a copy of a graph stored in memory; see [G] **graph copy**.

`graph use`
Copies a graph on disk into memory and displays it; see [G] **graph use**.

`graph display`
Redisplays a graph stored in memory; see [G] **graph display**.

`graph combine`
Combines into one graph graphs stored in memory or on disk; see [G] **graph combine**.

Also See

Complementary: [G] **graph dir**, [G] **graph describe**, [G] **graph drop**, [G] **graph rename**,
[G] **graph copy**, [G] **graph use**, [G] **graph display**, [G] **graph combine**;
[G] **gph files**, [G] **graph save**, [G] *name_option*, [G] *saving_option*,
[R] **drop**,
[P] **discard**

Title

graph matrix — Matrix graphs

Syntax

<u>gr</u>aph matrix *varlist* [*weight*] [if *exp*] [in *range*] [, *matrix_options*]

where *matrix_options* are

matrix_options	description
half	draw lower triangle only
marker_options	look of markers
marker_label_options	include labels on markers
jitter(*relativesize*)	perturb location of markers
<u>dia</u>gonal(*stringlist*, ...)	override text on diagonal
scale(#)	overall size of symbols, labels, ...
iscale([*]#)	size of symbols, labels, within plots
<u>max</u>es(*axis_scale_options* *axis_label_options*)	labels, ticks, grids, log scales, ...
axis_label_options	axis-by-axis control
by(*varlist*, ...)	repeat for subgroups
std_options	title, aspect ratio, saving to disk

Note: All options allowed by graph twoway scatter are also allowed, but they are ignored. half, diagonal(), scale(), and iscale() are *unique*; jitter() is *rightmost* and maxes() is *merged-implicit*; see [G] **repeated options**.

See [G] *marker_options*, [G] *marker_label_options*, [G] *relativesize*, [G] *scale_option*, [G] *axis_scale_options*, [G] *axis_label_options*, [G] *by_option*, and [G] *std_options*.

stringlist, ..., the argument allowed by diagonal(), is defined

$$[\{ . \mid "string" \}] [\{ . \mid "string" \} ...] [, \textit{textbox\_options}]$$

aweights, fweights, and pweights are allowed; see [U] **14.1.6 weight**. Weights affect the size of the markers. See *Weighted markers* in [G] **graph twoway scatter**.

Description

graph matrix draws scatterplot matrices.

Options

half specifies that only the lower triangle of the scatterplot matrix be drawn.

marker_options specify the look of the markers used to designate the location of the points. The important *marker_options* are msymbol(), mcolor(), and msize().

The default symbol used is msymbol(O)—solid circles. You specify msymbol(Oh) if you want hollow circles (a recommended alternative). If you have lots of observations, we recommend specifying msymbol(p); see *Marker symbols and the number of observations* under *Remarks* below. See [G] *symbolstyle* for a list of marker symbol choices.

The default mcolor() is dictated by the scheme; see [G] **schemes**. See [G] *colorstyle* for a list of color choices.

Be careful specifying the msize() option. In graph matrix, the size of the markers varies with the number of variables specified; see option iscale() below. If you specify msize(), that will override the automatic scaling.

See [G] *marker_options* for more information on markers.

marker_label_options allow placing identifying labels on the points. To obtain this, you specify the *marker_label_option* mlabel(*varname*); see [G] *marker_label_options*. These options are of little use in the case of scatterplot matrices because they make the graph seem too busy.

jitter(*relativesize*) adds spherical random noise to the data before plotting. This is useful when plotting data which otherwise would result in points plotted on top of each other. See *Jittered markers* in [G] **graph twoway scatter** for an explanation of jittering.

diagonal([*stringlist*][, *textbox_options*]) specifies text and its style to be displayed along the diagonal. This text serves to label the graphs (axes). By default, what appears along the diagonals are the variable labels of the variables of *varlist* or, if a variable has no variable label, its name. Typing

        . graph matrix mpg weight displ, diag(. "Weight of car")

would change the text appearing in the cell corresponding to variable weight. Note that we specified period (.) to leave the text in the first cell unchanged, and we did not bother to type a third string or a period; so we left the third element unchanged, too.

You may specify *textbox_options* following *stringlist* (which may itself be omitted) and a comma. These options will modify the style in which the text is presented but are of little use in this case. We recommend that you do not specify diagonal(,size()) to override the default sizing of the text. By default, the size of text varies with the number of variables specified; see option iscale() below. If you specify diagonal(,size()), that will override the automatic size scaling. See [G] *textbox_options* for more information on textboxes.

scale(#) specifies a multiplier that affects the size of all text and markers in a graph. scale(1) is the default and scale(1.2) would make all text and markers 20% larger.
See [G] *scale_option*.

iscale(#) and iscale(*#) specify an adjustment (multiplier) to be used to scale the markers, the text appearing along the diagonals, and the labels and ticks appearing on the axes.

By default, iscale() gets smaller and smaller the larger is n, the number of variables specified in *varlist*. The default is parameterized as a multiplier $f(n)$—$0 < f(n) < 1$, $f'(n) < 0$—that is used as a multiplier for msize(), diagonal(,size()), maxes(labsize()), and maxes(tlength()).

If you specify iscale(#), the number you specify is substituted for $f(n)$. We recommend that you specify a number between 0 and 1, but you are free to specify numbers larger than 1.

If you specify iscale(*#), the number you specify is multiplied by f(n) and that product is used to scale text. In this case, you should specify #>0; #>1 merely means you want the text to be bigger than graph matrix would otherwise choose.

maxes(*axis_scale_options axis_label_options*) affect the scaling and look of the axes. This is a case where you specify options within options.

Consider the *axis_scale_options* { y|x }scale(log), which produces logarithmic scales. Type maxes(yscale(log) xscale(log)) to draw the scatterplot matrix using log scales. Remember to specify both xscale(log) and yscale(log) unless you really want just the y axis or just the x axis logged.

Or consider the *axis_label_options* { y|x }label(,grid), which adds grid lines. Specify maxes(ylabel(,grid)) to add grid lines across, maxes(xlabel(,grid)) to add grid lines vertically, and both options to add grid lines in both directions. Concerning the both case, you can specify the maxes() option twice—maxes(ylabel(,grid)) maxes(xlabel(,grid))— or once combined—maxes(ylabel(,grid) xlabel(,grid))—it makes no difference because maxes() is *merged-implicit*; see [G] **repeated options**.

See [G] *axis_scale_options* and [G] *axis_label_options* for the suboptions that may appear inside maxes(). In reading those entries, ignore the axis(#) suboption; graph matrix will ignore it if you specify it.

axis_label_options allow you to assert axis-by-axis control over the labeling. Do not confuse this with maxes(*axis_label_options*). maxes(*axis_label_options*) specify options that affect all the axes. *axis_label_options* specified outside the maxes() option specify options that affect just one of the axes. *axis_label_options* can be repeated for each axis.

When you specify *axis_label_options* outside of maxes(), you must specify the axis-label suboption axis(#). For instance, you might type

    . graph mpg weight displ, yaxis(0(5)40, axis(1))

The effect of that would be to label the specified values on the first y axis (the one appearing on the far right). The axes are numbered as follows:

	x axis(2)		x axis(4)		
	v1/v2	v1/v3	v1/v4	v1/v5	y axis(1)
y axis(2) v2/v1		v2/v3	v2/v4	v2/v5	
	v3/v1	v3/v2		v3/v4	v3/v5 y axis(3)
y axis(4) v4/v1	v4/v2	v4/v3		v4/v5	
	v5/v1	v5/v2	v5/v3	v5/v4	y axis(5)
	x axis(1)		x axis(3)		x axis(5)

and if half is specified, the numbering scheme is

y axis(2)	v2/v1			
y axis(3)	v3/v1	v3/v2		
y axis(4)	v4/v1	v4/v2	v4/v3	
y axis(5)	v5/v1	v5/v2	v5/v3	v5/v4

x axis(1)	x axis(2)	x axis(3)	x axis(4)

See [G] *axis_label_options*; remember to specify the axis(#) suboption, and do not specify the *matrix_option* maxes().

by(*varlist*, ...) allows drawing multiple graphs for each subgroup of the data. See *Use with by()* under *Remarks* below and see [G] *by_option*.

std_options allow you to specify titles (see *Adding titles* under *Remarks* below and see [G] *title_options*), allow you to control the aspect ratio and background shading (see [G] *region_options*), allow you to control the overall look of the graph (see [G] *scheme_option*), and allow you to save the graph to disk (see [G] *saving_option*).

See [G] *std_options* for an overview of the standard options.

Remarks

Remarks are presented under the headings

> *Typical use*
> *Marker symbols and the number of observations*
> *Controlling the axes labeling*
> *Adding grid lines*
> *Adding titles*
> *Use with by()*
> *History*

Typical use

graph matrix provides an excellent alternative to correlation matrices (see [R] **correlate**) as a quick way to examine the relationships among variables:

```
. sysuse lifeexp, clear
(Life expectancy, 1998)
```

(Continued on next page)

. graph matrix popgrowth-safewater

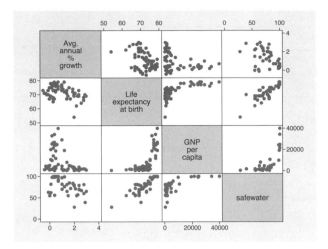

Seeing the above, we are tempted to transform gnppc into log units:

. generate lgnppc = ln(gnppc)
. graph matrix popgr lexp lgnp safe

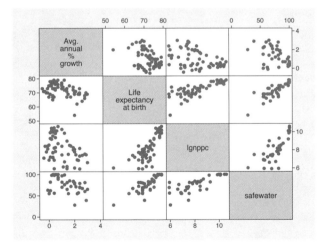

Some people prefer showing just half the matrix, moving the "dependent" variable to the end of the list:

```
. gr matrix popgr lgnp safe lexp, half
```

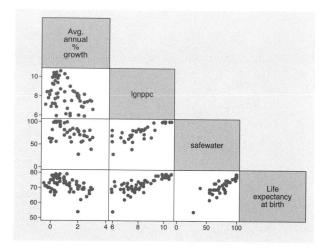

Marker symbols and the number of observations

The msymbol() option—abbreviation ms()—allows you to control the marker symbol used; see [G] *marker_options*. Hollow symbols sometimes work better as the number of observations increases:

```
. sysuse auto, clear
(1978 Automobile Data)
. gr mat mpg price weight length, ms(Oh)
```

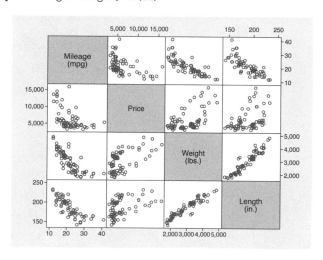

Points work best when you have lots of data:

```
. sysuse citytemp, clear
(City Temperature Data)

. gr mat heatdd-tempjuly, ms(p)
```

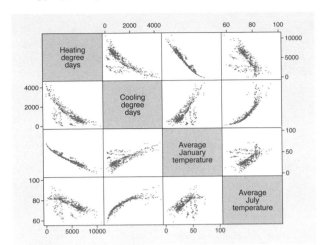

Controlling the axes labeling

By default, approximately 3 values are labeled and ticked on the y and x axes. When graphing only a few variables, increasing this often works well:

```
. sysuse citytemp, clear
(City Temperature Data)

. gr mat heatdd-tempjuly, ms(p) maxes(ylab(#4) xlab(#4))
```

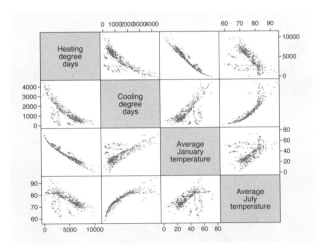

Specifying #4 does not guarantee four labels; it specifies that approximately four are to be used; see [G] *axis_label_options*. Also see *axis_label_options* under *Options* above for instructions on controlling the axes individually.

Adding grid lines

To add horizontal grid lines, specify maxes(ylab(,grid)) and to add vertical grid lines, specify maxes(xlab(,grid)). Below we do both and we specify that four values be labeled:

```
. sysuse lifeexp, clear
(Life expectancy, 1998)

. generate lgnppc = ln(gnppc)

. graph matrix popgr lexp lgnp safe, maxes(ylab(#4, grid) xlab(#4, grid))
```

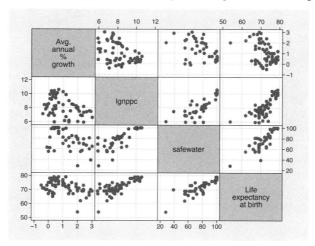

Adding titles

The standard title options may be used with graph matrix:

```
. sysuse lifeexp, clear
(Life expectancy, 1998)

. generate lgnppc = ln(gnppc)

. label var lgnppc "ln GNP per capita"

. graph matrix popgr lexp lgnp safe, maxes(ylab(#4, grid) xlab(#4, grid))
                subtitle("Summary of 1998 life-expectancy data")
                note("Source:  The World Bank Group")
```

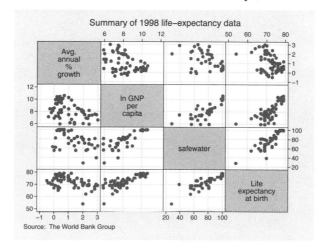

Use with by()

graph matrix may be used with by():

. sysuse auto, clear
(1978 Automobile Data)

. gr matrix mpg weight displ, by(foreign)

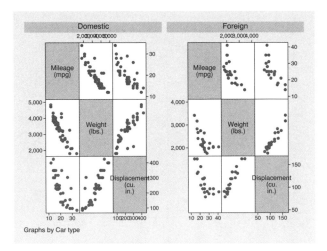

See [G] *by_option*.

History

The origin of the scatterplot matrix is unknown, although early written discussions may be found in Hartigan (1975), Tukey and Tukey (1981), and Chambers et al. (1983). The scatterplot matrix has also been called the *draftman's display* and *pairwise scatterplot*. Regardless of the name used, we believe the first "canned" implementation was by Becker and Chambers in a system called S—see Becker and Chambers (1984)—although S predates 1984. We also believe that Stata provided the second implementation, in 1985.

References

Becker, R. A. and J. M. Chambers. 1984. *S: An Interactive Environment for Data Analysis and Graphics*. Belmont, CA: Wadsworth Advanced Book Program.

Chambers, J. M., W. S. Cleveland, B. Kleiner, and P. A. Tukey. 1983. *Graphical Methods for Data Analysis*. Belmont, CA: Wadsworth International Group.

Hartigan, J. A. 1975. Printer graphics for clustering. *Journal of Statistical Computing and Simulation* 4: 187–213.

Tukey, P. A. and J. W. Tukey. 1981. Preparation; prechosen sequences of views. In *Interpreting Multivariate Data*, ed. V. Barnett, 189–213. Chichester, U. K.: John Wiley & Sons.

Also See

Complementary: [G] **graph**; [G] **graph twoway scatter**

Title

graph other — Other graphics commands

Syntax

Distributional diagnostic plots:

command	description
histogram	histograms
symplot	symmetry plots
quantile	quantiles plots
qnorm	quantile/normal plots
pnorm	normal probability plots, standardized
qchi	chi-squared quantile plots
pchi	chi-squared probability plots
qqplot	quantiles/quantiles plots
gladder	ladder-of-powers plots
qladder	ladder-of-powers quantiles
spikeplot	rootograms

See [R] **histogram**, [R] **diagnostic plots**, [R] **ladder**, and [R] **spikeplot**.

Smoothing and densities:

command	description
kdensity	kernel density estimation, univariate
lowess	lowess smoothing

See [R] **kdensity** and [R] **lowess**.

Regression diagnostics:

command	description
avplot	added-variable (leverage) plots
cprplot	component-plus-residual plots
lvr2plot	L-R (leverage versus squared residual) plots
rvfplot	residual-versus-fitted plots
rvpplot	residual-versus-predicted plots

See [R] **regression diagnostics**.

Time series:

command	description
ac	correlograms
pac	partial correlograms
pergram	periodograms
cumsp	spectral distribution plots, cumulative
xcorr	cross-correlograms for bivariate time series
wntestb	Bartlett's periodogram-based white noise test

See [TS] **corrgram**, [TS] **pergram**, [TS] **cumsp**, [TS] **xcorr**, and [TS] **wntestb**.

Vector autoregression (VAR):

command	description
varfcast graph	forecasts
varirf graph	impulse response functions (IRF) and forecast error variance decompositions (FEVD)
varirf ograph	overlaid IRFs and FEVDs
varirf cgraph	combined IRFS and FEVDs

See [TS] **varfcast graph**, [TS] **varirf graph**, [TS] **varirf ograph**, and [TS] **varirf cgraph**.

Survival analysis:

command	description
sts graph	survivor and cumulative-hazard functions
strate	failure rates and cumulative hazard comparisons
ltable	life tables
stci	means & percentiles of survival time, with CIs
stphtest and stphplot	verify proportional hazards assumption
stcoxkm	verify proportional hazards assumption after stcox
stcurve	survivor & hazard functions after streg

See [ST] **sts**, [ST] **strate**, [ST] **ltable**, [ST] **stci**, [ST] **stcox**, [ST] **stphplot**, and [ST] **streg**.

ROC analysis:

command	description
roctab	ROC curve
rocplot	parametric ROC curve
roccomp	multiple ROC curves, compared
lroc	ROC curve after logistic
lsens	sensitivity & specificity vs. probability cutoff

See [R] **roc** and [R] **logistic**.

Quality-control charts:

command	description
cusum	cusum plots
cchart	c charts
pchart	p charts
rchart	R charts
xchart	$\overline{X}$ charts
shewhart	$\overline{X}$ charts, vertically aligned
serrbar	standard error bar charts

See [R] **cusum**, [R] **qc**, and [R] **serrbar**.

Other statistical graphs:

command	description
cltree	dendrograms for hierarchical cluster analysis
greigen	eigenvalues after factor analysis
tabodds	odds of failure versus categories
pkexamine	summarize pharmacokinetic data
dotplot	means or medians by group

See [CL] **cluster dendrogram**, [R] **factor**, [ST] **epitab**, [R] **pkexamine**, and [R] **dotplot**.

Description

In addition to graph, there are many other commands that draw graphs. They are listed above.

Remarks

The other graph commands are implemented in terms of graph. graph provides the following capabilities:

command	description
graph bar	bar charts
graph pie	pie charts
graph dot	dot charts
graph matrix	scatterplot matrices
graph twoway	twoway (y-x) graphs, including:
graph twoway scatter	scatterplots
graph twoway line	line plots
graph twoway function	function plots
graph twoway histogram	histograms
graph twoway *	more

See [G] **graph bar**, [G] **graph pie**, [G] **graph dot**, [G] **graph matrix**, and [G] **graph twoway**.

Also See

Related: [G] **graph intro**

Title

graph pie — Pie charts

Syntax

graph pie *yvars* [*weight*] [if *exp*] [in *range*] [, *options*]

where *yvars* is

yvars	description
nothing	if option over() is specified
varname	if option over() is specified
varlist	otherwise

and where *options* are

options	description
over(*varname*)	slices are distinct values of *varname*
missing	do not ignore missing values of *varname*
cw	casewise treatment of missing values
noclockwise	counterclockwise pie chart
angle0(#)	angle of first slice; default angle(90)
sort	put slices in size order
sort(*varname*)	put slices in *varname* order
descending	reverse default or specified order
pie(...)	look of slice, including explosion
plabel(...)	labels to appear on the slice
ptext(...)	text to appear on the pie
intensity([*]#)	color intensity of slices
line(*line_options*)	outline of slices
legend(...)	legend explaining slices
std_options	titles, saving to disk
by(*varlist*, ...)	repeat for subgroups

See [G] *line_options*, [G] *legend_option*, [G] *std_options*, and [G] *by_option*.

The syntax of the pie() option is

   p̲ie({*numlist* | _all} [, *pie_subopts*])

pie_subopts	description
explode	explode slice by *relativesize* = 3.8
explode(*relativesize*)	explode slice by *relativesize*
c̲olor(*colorstyle*)	color of slice

See [G] *relativesize* and [G] *colorstyle*.

The syntax of the plabel() option is

   p̲label({*#* | _all} {sum | p̲ercent | name | "*text*"} [, *plabel_subopts*])

plabel_subopts	description
fo̲rmat(%*fmt*)	display format for sum or percent
gap(*relativesize*)	additional radial distance
textbox_options	look of label

See [G] *relativesize* and [G] *textbox_options*.

The syntax for the ptext() option is

   p̲text(*#*$_a$*#*$_r$ "*text*" ["*text*" ...] [*#*$_a$*#*$_r$ ...] [, *ptext_subopts*])

ptext_subopts	description
textbox_options	look of added text

See [G] *textbox_options*.

aweights, fweights, and pweights are allowed; see [U] **14.1.6 weight**.

Description

graph pie draws pie charts.

graph pie has three modes of operation. The first corresponds to the specification of two or more *yvars*:

   . graph pie div1_revenue div2_revenue div3_revenue

Three pie slices are drawn, the first corresponding to the sum of variable div1_revenue, the second to the sum of div2_revenue, and the third, to the sum of div3_revenue.

The second mode of operation corresponds to the specification of a single *yvar* and the over() option:

   . graph pie revenue, over(division)

Pie slices are drawn for each value of variable `division`; the first slice corresponds to the sum of revenue for the first division, the second to the sum of revenue for the second division, and so on.

The third mode of operation corresponds to the specification of `over()` with no *yvars*:

> . graph pie, over(popgroup)

Pie slices are drawn for each value of variable popgroup; the slices correspond to the number of observations in each group.

Options

over(*varname*) specifies a categorical variable which is to correspond to the pie slices. *varname* may be string or numeric.

missing is for use with over(); it specifies that missing values of *varname* should not be ignored. Instead, separate slices are to be formed for *varname*==., *varname*==.a, ..., or *varname*=="".

cw specifies casewise deletion and is for use when over() is not specified. cw specifies that, in calculating the sums, observations be ignored for which any of the variables in *varlist* contain missing values. The default is to calculate sums for each variable using all nonmissing observations.

noclockwise and angle0(#) specify how the slices are oriented on the pie. The default is to start at 12 o'clock (known as angle(90)) and to proceed clockwise.

noclockwise causes slices to be placed counterclockwise.

angle0(#) specifies the angle at which the first slice is to appear. Angles are recorded in degrees and measured in the usual mathematical way: counterclockwise from the horizontal.

sort, sort(*varname*), and descending specify how the slices are to be ordered. The default is to put the slices in the order specified; see *How slices are ordered* under *Remarks* below.

sort specifies that the smallest slice be put first, followed by the next largest, etc. See *Ordering slices by size* under *Remarks* below.

sort(*varname*) specifies that the slices be put in (ascending) order of *varname*. See *Reordering the slices* under *Remarks* below.

descending, which may be specified whether or not sort or sort(*varname*) is specified, reverses the order.

pie({*numlist* | _all}, *pie_subopts*) specifies the look of a slice or of a set of slices. This option allows you to "explode" (offset) one or more slices of the pie and to control the color of the slices. Examples include

> . graph pie ..., ... pie(2, explode)
> . graph pie ..., ... pie(2, explode color(red))
> . graph pie ..., ... pie(2, explode color(red)) pie(5, explode)

numlist specifies the slices; see [U] **14.1.8 numlist**. The slices (after any sorting) are referred to as slice 1, slice 2, etc. pie(1 ...) would change the look of the first slice. pie(2 ...) would change the look of the second slice. pie(1 2 3 ...) would change the look of the first through third slices, as would pie(1/3 ...). The pie() option may be specified more than once to specify a different look for different slices. Alternatively, you may specify pie(_all ...) to specify a common characteristic for all slices.

The *pie_subopts* are explode, explode(*relativesize*), and color(*colorstyle*).

explode and explode(*relativesize*) specify that the slice is to be offset. Specifying explode is equivalent to specifying explode(3.8). explode(*relativesize*) specifies by how much (measured radially) the slice is to be offset; see [G] ***relativesize***.

color(*colorstyle*) sets the color of the slice. See [G] ***colorstyle*** for a list of color choices.

plabel({#| _all} {sum|percent|name|"*text*"}, *plabel_subopts*) specifies labels to appear on the slice. Slices may be labeled with their sum, their percent of overall sum, their identity, or with text you specify. The default is that no labels appear. Think of the syntax of plabel() as

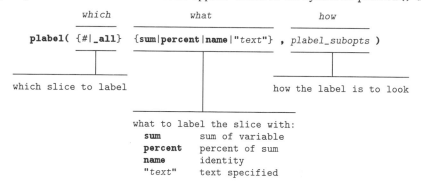

Thus, you might type

   . graph pie ..., ... plabel(_all sum)

   . graph pie ..., ... plabel(_all percent)

   . graph pie ..., ... plabel(1 "New appropriation")

The plabel() option may appear more than once, so you might also type

   . graph pie ..., ... plabel(1 "New appropriation") plabel(2 "old")

If you choose to label the slices with their identities, it is common also to suppress the legend:

   . graph pie ..., ... plabel(_all name) legend(off)

The *plabel_subopts* are format(*%fmt*), gap(*relativesize*), and *textbox_options*.

format(*%fmt*) specifies the display format to be used to format the number when sum or percent is chosen; see [R] **format**.

gap(*relativesize*) specifies a radial distance from the origin by which the usual location of the label is to be adjusted. gap(0) is the default. gap(#), $\# < 0$, moves the text inward. gap(#), $\# > 0$, moves the text outward. See [G] ***relativesize***.

textbox_options specify the size, color, etc. of the text; see [G] ***textbox_options***.

ptext($\#_a\#_r$ "*text*" ["*text*" ...] [$\#_a\#_r$ ...], *ptext_subopts*) specifies additional text to appear on the pie. The position of the text is specified by the polar coordinates $\#_a$ and $\#_r$. $\#_a$ specifies the angle in degrees, and $\#_r$ specifies the distance from the origin in relative-size units; see [G] ***relativesize***.

intensity(#) and intensity(*#) specify the intensity of the color used to fill the slices. intensity(#) specifies the intensity and intensity(*#) specifies the intensity relative to the default.

Specify intensity(*#), # < 1, to attenuate the interior color and specify intensity(*#), # > 1, to amplify it.

Specify intensity(0) if you do not want the slice filled at all.

line(*line_options*) specifies the look of the line used to outline the slices. See [G] ***line_options*** but ignore option lpattern(); it is not allowed in the case of pie charts.

legend() allows you to control the legend. See [G] ***legend_option***.

std_options allow you to add titles, save the graph on disk, and more; see [G] ***std_options***.

by(*varlist*, ...) draws separate pies within a single graph; see [G] ***by_option*** and see *Use with by()* under *Remarks* below.

Remarks

Remarks are presented under the headings

> *Typical use*
> *Data are summed*
> *Data may be long rather than wide*
> *How slices are ordered*
> *Ordering slices by size*
> *Reordering the slices*
> *Use with by()*
> *History*

Typical use

You have been told that the expenditures for XYZ Corp. are $12 million in sales, $14 million in marketing, $2 million in research, and $8 million in development:

```
. input sales marketing research development

    sales  marketing   research  develop~t
1. 12 14 2 8
2. end
. label var sales "Sales"
. label var market "Marketing"
. label var research "Research"
. label var develop  "Development"
. graph pie sales marketing research development,
        plabel(_all name, size(*1.5) color(white))      (Note 1)
        legend(off)                                      (Note 2)
        plotregion(lstyle(none))                         (Note 3)
        title("Expenditures, XYZ Corp.")
        subtitle("2002")
        note("Source:  2002 Financial Report (fictional data)")
```

(Continued on next page)

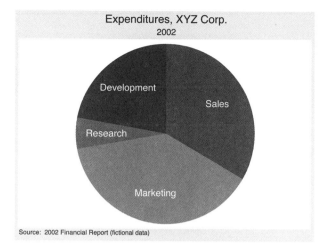

Notes:

1. We specified plabel(all name) to put the division names on the slices. We specified the plabel()'s textbox-option size(*1.5) to make the text 50% larger than usual. We specified plabel()'s textbox-option color(white) to make the text white. See [G] *textbox_options*.

2. We specified the legend-option legend(off) to keep the division names from being repeated in a key at the bottom of the graph; see [G] *legend_option*.

3. We specified the region-option plotregion(lstyle(none)) to prevent a border from being drawn around the plot area; see [G] *region_options*.

Data are summed

Rather than having the above summary data, you have

```
. list, abbrev(11)
```

	qtr	sales	marketing	research	development
1.	1	3	4.5	.3	1
2.	2	4	3	.5	2
3.	2	4	3	.5	2
4.	4	2	2.5	.6	3

Note that the sums of these data are the same as the totals in the previous section. The same graph pie command

```
. graph pie sales marketing research development, ...
```

will result in the same chart.

Data may be long rather than wide

Rather than having the quarterly data in wide form, you have it in the long form:

```
. list, sepby(qtr)
```

	qtr	division	cost
1.	1	Development	1
2.	1	Marketing	4.5
3.	1	Research	.3
4.	1	Sales	3
5.	2	Development	2
6.	2	Marketing	3
7.	2	Research	.5
8.	2	Sales	4
9.	3	Development	2
10.	3	Marketing	4
11.	3	Research	.6
12.	3	Sales	3
13.	4	Development	3
14.	4	Marketing	2.5
15.	4	Research	.6
16.	4	Sales	2

In this case, rather than typing

```
. graph pie sales marketing research development, ...
```

you type

```
. graph pie cost, over(division) ...
```

For example,

```
. graph pie cost, over(division),
        plabel(_all name, size(*1.5) color(white))
        legend(off)
        plotregion(lstyle(none))
        title("Expenditures, XYZ Corp.")
        subtitle("2002")
        note("Source:  2002 Financial Report (fictional data)")
```

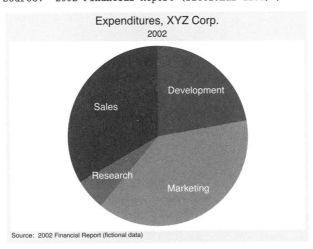

This is the same pie chart as drawn previously except for the order in which the divisions are presented.

How slices are ordered

When you type

```
. graph pie sales marketing research development, ...
```

the slices are presented in the order you specify. When you type

```
. graph pie cost, over(division) ...
```

the slices are presented in the order implied by variable division. If division is numeric, slices are presented in ascending order of division. If division is string, slices are presented in alphabetic order (except that all capital letters occur before lower case letters).

Ordering slices by size

Regardless of whether you type

```
. graph pie sales marketing research development, ...
```

or

```
. graph pie cost, over(division) ...
```

if you add the sort option, slices will be presented in the order of the size, smallest first:

```
. graph pie sales marketing research development, sort ...
. graph pie cost, over(division) sort ...
```

If you also specify the descending option, the largest slice will be presented first:

```
. graph pie sales marketing research development, sort descending ...
. graph pie cost, over(division) sort descending ...
```

Reordering the slices

If you wish to force a particular order, then if you type

```
. graph pie sales marketing research development, ...
```

specify the variables in the desired order. If you type

```
. graph pie cost, over(division) ...
```

then create a numeric variable that has a one-to-one correspondence with the order in which you wish the divisions to appear. For instance, you might type

```
. generate order    = 1 if division=="Sales"
. replace order = 2 if division=="Marketing"
. replace order = 3 if division=="Research"
. replace order = 4 if division=="Development"
```

Then type

```
. graph pie cost, over(division) sort(order) ...
```

Use with by()

You have two years of data on XYZ Corp.:

```
. list, abbrev(11)
```

	year	sales	marketing	research	development
1.	2002	12	14	2	8
2.	2003	15	17.5	8.5	10

```
. graph pie sales marketing research development,
        plabel(_all name, size(*1.5) color(white))
        by(year,
                legend(off)
                title("Expenditures, XYZ Corp.")
                note("Source:  2002 Financial Report (fictional data)")
        )
```

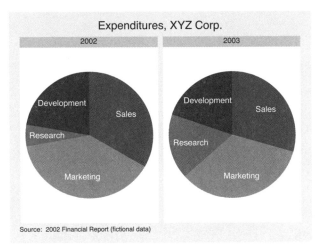

History

The first pie chart is credited to William Playfair (1801). See Beniger and Robyn (1978), Funkhouser (1937, 283–285), or Tufte (1983, 44–45) for further historical details.

References

Beniger, J. R. and D. L. Robyn. 1978. Quantitative graphics in statistics: a brief history. *The American Statistician* 32: 1–11.

Funkhouser, H. G. 1937. Historical development of the graphical representation of statistical data. *Osiris* 3: 269–404.

Tufte, E. R. 1983. *The Visual Display of Quantitative Information.* Cheshire, CT: Graphics Press.

Playfair, W. 1801. *Statistical Breviary; Shewing, on a Principle Entirely New, the Resources of Every State and Kingdom in Europe.* London: Wallis.

Also See

Complementary: [G] **graph**, [G] **graph bar**

Title

> **graph print** — Print current graph

Syntax

> graph print [, *pr_options*]

where *pr_options* are defined in [G] ***pr_options***.

Description

> graph print prints the graph currently displayed in the Graph window.
>
> Stata for Unix users must do some setup before using graph print for the first time; see *Appendix: Setting up Stata for Unix to print graphs* below.

Options

> *override_options* modify how the graph is printed. See [G] ***pr_options***.

Remarks

> Graphs are printed by displaying them on the screen and then typing
>
> . graph print
>
> Remarks are presented under the headings
>
> > *Printing the graph currently displayed in the Graph window*
> > *Printing a graph stored on disk*
> > *Printing a graph stored in memory*
> > *Appendix: Setting up Stata for Unix to print graphs*
>
> Also see [G] **set printcolor**. By default, if the graph being printed has a black background, it is printed in monochrome.
>
> In addition to printing graphs, Stata can export graphs in PostScript, Encapsulated PostScript (EPS), Windows Metafile (WMF), Windows Enhanced Metafile (EMF), and Macintosh PICT formats; see [G] **graph export**.

Printing the graph currently displayed in the Graph window

> There are three ways to print the graph currently displayed in the Graph window:
>
> 1. Pull down **File** and choose **Print Graph...**.
>
> 2. Right click in the Graph window and choose **Print Graph...**.
>
> 3. Type "graph print" in the Command window.
>
> All are equivalent. The advantage of graph print is that you may include it in do-files:
>
> . graph ...   (draw a graph)
> . graph print   (and print it)

Printing a graph stored on disk

To print a graph stored on disk, type

. graph use *filename*
. graph print

Do not specify graph use's nodraw option; see [G] **graph use**.

Note for Stata-for-Unix(console) users: Follow the instructions just given even though you have no Graph window and cannot see what has just been "displayed": Use the graph and then print it.

Printing a graph stored in memory

To print a graph stored in memory but not currently displayed, type

. graph display *name*
. graph print

Do not specify graph display's nodraw option; see [G] **graph display**.

Note for Stata-for-Unix(console) users: Follow the instructions just given even though you have no Graph window and cannot see what has just been "displayed": Display the graph and then print it.

Appendix: Setting up Stata for Unix to print graphs

Before you can print graphs, you must tell Stata the command you ordinarily use to print PostScript files. Out of the box, Stata assumes that command is

$ lpr < *filename*

That command may be correct for you. If, on the other hand, you usually type something like

$ lpr -Plexmark *filename*

you need to tell Stata that by typing

. printer define prn ps "lpr -Plexmark @"

Type an @ where you ordinarily would type the filename. If you want the command to be "lpr -Plexmark < @", type

. printer define prn ps "lpr -Plexmark < @"

Stata assumes the printer you specify understands PostScript format.

Also See

Complementary: [G] *pr_options*, [G] **set printcolor**; [G] **graph display**, [G] **graph use**;
[G] **graph export**

Title

> **graph query** — List available schemes and styles

Syntax

graph query, schemes

graph query

graph query *stylename*

Description

graph query, schemes lists the available schemes.

graph query without options lists the available styles.

graph query *stylename* lists the styles available within *stylename*.

Remarks

This manual may not—probably is not—complete. Schemes and styles can be added by StataCorp via updates (see [R] **update**), by other users and traded over the Internet (see [R] **net** and [R] **ssc**), and by you.

Schemes define how graphs look (see [G] **schemes**), and styles define the features that are available to you (see, for instance, [G] *symbolstyle* or [G] *linestyle*).

To find out which schemes are installed on your computer, type

. graph query, schemes

See [G] **schemes** for information on schemes and how to use them.

To find out which styles are installed on your computer, type

. graph query

Many styles will be listed. How you use those styles is described in this manual. For instance, one of the styles that will be listed is *symbolstyle*. See [G] *symbolstyle* for more information on symbol styles. To find out which symbol styles are available to you, type

. graph query symbolstyle

All styles end in "*style*", and you may omit typing that part:

. graph query symbol

Also See

Complementary:	[G] **schemes**; [G] **palette**; [G] *addedlinestyle*, [G] *alignmentstyle*,
	[G] *anglestyle*, [G] *areastyle*, [G] *bystyle*, [G] *colorstyle*,
	[G] *compassdirstyle*, [G] *connectstyle*, [G] *gridstyle*, [G] *justificationstyle*,
	[G] *legendstyle*, [G] *linepatternstyle*, [G] *linestyle*, [G] *linewidthstyle*,
	[G] *marginstyle*, [G] *markerlabelstyle*, [G] *markersizestyle*,
	[G] *markerstyle*, [G] *orientationstyle*, [G] *pstyle*, [G] *symbolstyle*,
	[G] *textboxstyle*, [G] *textsizestyle*, [G] *textstyle*,
	[G] *tickstyle*

Title

graph rename — Rename graph in memory

Syntax

graph rename [*oldname*] *newname* [, replace]

If *oldname* is not specified, the name of the current graph is assumed.

Description

graph rename changes the name of a graph stored in memory.

Options

replace specifies that it is okay to replace *newname* if it already exists.

Remarks

See [G] **graph manipulation** for an introduction to the graph manipulation commands.

graph rename is most commonly used to rename the current graph—the graph currently displayed in the Graph window—when creating the pieces for graph combine:

```
. graph ..., ...
. graph rename p1
. graph ..., ...
. graph rename p2
. graph combine p1 p2, ...
```

Also See

Complementary: [G] **graph manipulation**; [G] **graph copy**

Title

graph save — Save graph to disk

Syntax

<u>gr</u>aph save [*graphname*] *filename* [, asis replace]

Description

graph save saves the specified graph to disk. If *graphname* is not specified, the graph currently displayed is saved.

If *filename* is specified without an extension, .gph will be assumed.

Options

asis specifies that the graph is to be frozen and saved as is. The alternative—and the default if asis is not specified—is live format. In live format, the graph can be edited in future sessions, and, in addition, the overall look of the graph continues to be controlled by the chosen scheme (see [G] **schemes**).

Pretend that you type

```
. scatter yvar xvar, ...
. graph save mygraph
```

which will create file mygraph.gph. Now pretend that you send the file to a colleague. The way the graph will appear on your colleague's computer might be different from how it appeared on yours. Perhaps you display titles on the top and your colleague has set his scheme to display titles on the bottom. Or perhaps your colleague prefers y axes on the right rather than on the left. Understand that it will still be the same graph, but it might have a different look.

Or perhaps you just file away mygraph.gph for use later. Stored in the default live format, you can come back to it and change the way it looks by specifying a different scheme, and you can edit it.

If, on the other hand, you specify asis, the graph will forever look just as it looked the instant it was saved. You cannot edit it and you cannot change the scheme. If you send the as-is graph to colleagues, they will see it exactly in the form that you see it.

Whether a graph is saved as-is or live makes no difference in terms of printing. As-is graphs usually require fewer bytes to store and they generally display more quickly, but that is all.

replace specifies that the file may be replaced if it already exists.

Remarks

You may alternatively specify the graph be saved at the instant you draw it by specifying the saving(*filename*[, asis replace]) option; see [G] *saving_option*.

Also See

Complementary: [G] *saving_option*; [G] **gph files**, [G] **graph manipulation**

Title

graph twoway — Twoway graphs

Syntax

<u>graph</u> <u>two</u>way *plot* [if *exp*] [in *range*] [, *twoway_options*]

where *plot* is defined

[( ] *plottype* *varlist* ... , *options* [ )] [||]

and where *plottype* may be

plottype	description
scatter	scatterplot
line	line plot
connected	connected-line plot
scatteri	scatter with immediate arguments
area	line plot with shading
bar	bar plot
spike	spike plot
dropline	dropline plot
dot	dot plot
rarea	range plot with area shading
rbar	range plot with bars
rspike	range plot with spikes
rcap	range plot with capped spikes
rcapsym	range plot with spikes capped with symbols
rscatter	range plot with markers
rline	range plot with lines
rconnected	range plot with lines and markers
mband	median-band line plot
mspline	spline line plot
lowess	LOWESS line plot
lfit	linear prediction plot
qfit	quadratic prediction plot
fpfit	fractional polynomial plot
lfitci	linear prediction plot with CIs
qfitci	quadratic prediction plot with CIs
fpfitci	fractional polynomial plot with CIs

(Continued on next page)

230

plottype	description
function	line plot of function
histogram	histogram plot
kdensity	kernel density plot

For each of the above, see [G] **graph twoway** *plottype*,
where you substitute for *plottype* a word from the left column.

and where *twoway_options* are as defined in [G] ***twoway_options***.

Note that the leading graph is optional. If the first (or only) *plot* is scatter, you may omit twoway
as well, and then the syntax is

<u>sc</u>atter ... [, *scatter_options*] [|| *plot* [*plot* [...]]]

and the same applies to line. The other *plottypes* must be preceded by twoway.

Regardless of how the command is specified, *twoway_options* may be specified among the *scatter_options*, *line_options*, etc., and they will be treated just as if they were specified among the *twoway_options* of the graph twoway command.

Description

twoway is a family of plots all of which fit on numeric y and x scales.

Remarks

Remarks are presented under the headings

Definition
Syntax
Multiple if and in restrictions
twoway and plot options

Definition

Twoway graphs are for showing the relationship between numeric data. Say you have data on life expectancy in the U.S. between 1900 and 1999:

```
. sysuse uslifeexp, clear
(U.S. life expectancy, 1900-1999)
. keep year le
. list in 1/8
```

	year	le
1.	1900	47.3
2.	1901	49.1
3.	1902	51.5
4.	1903	50.5
5.	1904	47.6
6.	1905	48.7
7.	1906	48.7
8.	1907	47.6

We could graph these data as a twoway scatterplot,

```
. twoway scatter le year
```

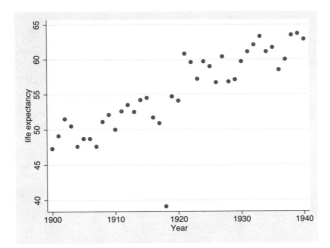

or we could graph these data as a twoway line plot,

```
. twoway line le year
```

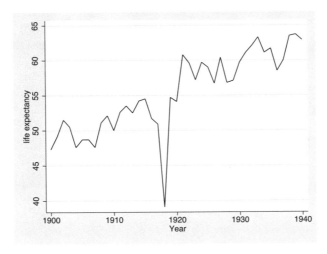

or we could graph these data as a twoway connected plot, marking both the points and connecting them with straight lines,

(Continued on next page)

. twoway connected le year

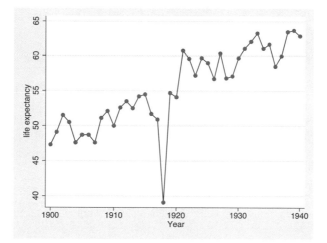

or we could graph these data as a scatterplot and put on top of that the prediction from a linear regression of le on year,

. twoway (scatter le year) (lfit le year)

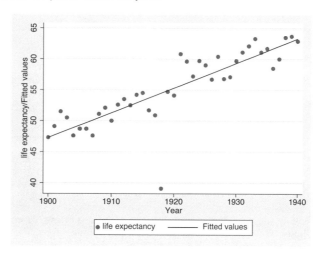

or we could graph these data in lots of other ways.

What is important is that all of the above are examples of twoway graphs. What distinguishes a twoway graph is that it fits onto numeric y and x axes.

Each of what we produced above is called a *graph*. What appeared in the graphs are called *plots*. In the first graph the plottype was a scatter, in the second the plottype was a line, in the third the plottype was connected, and in the fourth there were two plots: a scatter combined with a line plot of a linear fit.

twoway provides lots of different plottypes. Some, such as scatter and line, simply render the data in different ways. Others, such as lfit, transform the data and render that. And still others, such as function, actually make up data to be rendered. This last class makes it easy to overlay $y = x$ lines or $y = f(x)$ functions on your graphs.

By the way, in case you are wondering, there are no errors in the above data. In 1918 there was an outbreak of influenza known as the 1918 Influenza Pandemic that, in the U.S., was the worst epidemic ever known and which killed more citizens than all combat deaths of the 20th century.

Syntax

If you want to graph y1 versus x and y2 versus x, the formal way that you are supposed to type this is

```
. graph twoway (scatter y1 x) (scatter y2 x)
```

If you wanted y1 versus x plotted with solid circles and y2 versus x plotted with hollow circles, formally you would type

```
. graph twoway (scatter y1 x, ms(O)) (scatter y2 x, ms(Oh))
```

If you wanted y1 versus x plotted with solid circles and wanted a line graph for y2 versus x, formally you would type

```
. graph twoway (scatter y1 x, ms(O)) (line y2 x, sort)
```

We included the sort option under the assumption that the data are not already sorted by x.

The above is the formal way to type each of our requests but nobody types that. First, they omit the graph:

```
. twoway (scatter y1 x) (scatter y2 x)
. twoway (scatter y1 x, ms(O)) (scatter y2 x, ms(Oh))
. twoway (scatter y1 x, ms(O)) (line y2 x, sort)
```

Second, most people use the | |-separator notation rather than the ()-binding notation:

```
. twoway scatter y1 x || scatter y2 x
. twoway scatter y1 x, ms(O) || scatter y2 x, ms(Oh)
. twoway scatter y1 x, ms(O) || line y2 x, sort
```

Third, most people now omit the twoway:

```
. scatter y1 x || scatter y2 x
. scatter y1 x, ms(O) || scatter y2 x, ms(Oh)
. scatter y1 x, ms(O) || line y2 x, sort
```

And finally, most people quickly realize that scatter allows plotting more than one y variable against the same x variable:

```
. scatter y1 y2 x
. scatter y1 y2 x, ms(O Oh)
. scatter y1 x, ms(O) || line y2 x, sort
```

Note that the third example did not change: in that example, we are combining a scatterplot and a line plot. Actually, in this particular case, there is a way we can combine that, too:

```
. scatter y1 y2 x, ms(O i) connect(. l)
```

That we can combine scatter and line just happens to be an oddity of the examples we picked. What is important to understand is that there is nothing wrong with any of the above ways of typing our request, and sometimes the wordier syntaxes are the only way to obtain what we want. If we wanted to graph y1 versus x1 and y2 versus x2, the only way to type that is

```
. scatter y1 x1 || scatter y2 x2
```

or to type the equivalent in one of the wordier syntaxes above it. We have to do this because `scatter` (see [G] **graph twoway scatter**) draws a scatterplot against a single x variable. Ergo, if we want two different x variables, we need two different scatters.

In any case, we will often refer to the `graph twoway` command even though, when we give the command, we will seldom type the `graph` and mostly, we will not type the `twoway` either.

Multiple if and in restrictions

Each *plot* may have its own `if` *exp* and `in` *range* restrictions:

```
. twoway (scatter mpg weight if foreign, msymbol(O))
         (scatter mpg weight if !foreign, msymbol(Oh))
```

Multiple *plots* in a single `graph twoway` command draw one graph with multiple things plotted in it. The above will produce a scatter of `mpg` versus `weight` for foreign cars (making the points with solid circles) and a scatter of `mpg` versus `weight` for domestic cars (using hollow circles).

In addition, the `graph twoway` command itself can have `if` *exp* and `in` *range* restrictions:

```
. twoway (scatter mpg weight if foreign, msymbol(O))
         (scatter mpg weight if !foreign, msymbol(Oh)) if mpg>20
```

The `if mpg>20` restriction will apply to both scatters.

We have chosen to show these two examples using the ()-binding notation because it makes the scope of each `if` *exp* so clear. In ||-separator notation, the commands would read

```
. twoway scatter mpg weight if foreign, msymbol(O) ||
         scatter mpg weight if !foreign, msymbol(Oh)
```

and

```
. twoway scatter mpg weight if foreign, msymbol(O) ||
         scatter mpg weight if !foreign, msymbol(Oh) || if mpg>20
```

or even

```
. scatter mpg weight if foreign, msymbol(O) ||
         scatter mpg weight if !foreign, msymbol(Oh)
```

and

```
. scatter mpg weight if foreign, msymbol(O) ||
         scatter mpg weight if !foreign, msymbol(Oh) || if mpg>20
```

You may specify `graph twoway` restrictions only, of course:

```
. twoway (scatter mpg weight) (lfit mpg weight) if !foreign
. scatter mpg weight || lfit mpg weight || if !foreign
```

twoway and plot options

`graph twoway` allows options and the individual *plots* allow options. For instance, `graph twoway` allows the `saving()` option, and `scatter` (see [G] **graph twoway scatter**) allows the `msymbol()` option (which specifies the marker symbol to be used). Nevertheless, you do not have to keep track of which option belongs to which. If you type

```
. scatter mpg weight, saving(mygraph) msymbol(Oh)
```

results will be just the same as if you more formally typed

>   . twoway (scatter mpg weight, msymbol(Oh)), saving(mygraph)

Similarly, you could type

>   . scatter mpg weight, msymbol(Oh) || lfit mpg weight, saving(mygraph)

or

>   . scatter mpg weight, msymbol(Oh) saving(mygraph) || lfit mpg weight

and, either way, results would be as if you typed

>   . twoway (scatter mpg weight, msymbol(Oh))
>   (lfit mpg weight), saving(mygraph)

You may specify a `graph twoway` option "too deeply", but you cannot go the other way. The following is an error:

>   . scatter mpg weight || lfit mpg weight ||, msymbol(Oh) saving(mygraph)

It is an error because you specified a `scatter` option where only a `graph twoway` option may be specified, and given what you typed, there is insufficient information for `graph twoway` to determine for which *plot* you meant the `msymbol()` option. Even when there is sufficient information (say option `msymbol()` were not allowed by `lfit`), it would still be an error. `graph twoway` can reach in and pull out its options, but it cannot take from its options and distribute them back to the individual *plots*.

Also See

Complementary: [G] **graph twoway area**, [G] **graph twoway bar**,
[G] **graph twoway connected**, [G] **graph twoway dot**,
[G] **graph twoway dropline**, [G] **graph twoway fpfit**,
[G] **graph twoway fpfitci**, [G] **graph twoway function**,
[G] **graph twoway histogram**, [G] **graph twoway kdensity**,
[G] **graph twoway lfit**, [G] **graph twoway lfitci**,
[G] **graph twoway line**, [G] **graph twoway lowess**,
[G] **graph twoway mband**, [G] **graph twoway mspline**,
[G] **graph twoway qfit**, [G] **graph twoway qfitci**,
[G] **graph twoway rarea**, [G] **graph twoway rbar**,
[G] **graph twoway rcap**, [G] **graph twoway rcapsym**,
[G] **graph twoway rconnected**, [G] **graph twoway rline**,
[G] **graph twoway rscatter**, [G] **graph twoway scatter**,
[G] **graph twoway scatteri**, [G] **graph twoway spike**

Title

> **graph twoway area** — Twoway line plot with area shading

Syntax

twoway area *yvar xvar* [if *exp*] [in *range*] [, *area_options scatter_options*]

where *area_options* are

area_options	description
vertical	vertical area plot (default)
horizontal	horizontal area plot
base(#)	value to drop to (default = 0)
nodropbase	programmer's option
sort	recommendation: specify this option
bstyle(*areastyle*)	overall look of shaded area
bcolor(*colorstyle*)	outline and fill color
bfcolor(*colorstyle*)	fill color
blstyle(*linestyle*)	overall look of outline
blcolor(*colorstyle*)	outline color
blwidth(*linewidthstyle*)	thickness of outline
blpattern(*linepatternstyle*)	whether outline solid, dashed, etc.

See [G] *areastyle*, [G] *colorstyle*, [G] *linestyle*, [G] *linewidthstyle*, and
[G] *linepatternstyle*.

All options are *rightmost*; see [G] **repeated options**.

and where *scatter_options* refer to any of the options allowed by scatter except that
marker_options, *marker_placement_option*, *marker_label_options*, and *connect_options* are
ignored even if specified; see [G] **graph twoway scatter**.

Description

twoway area displays (*y*,*x*) connected by straight lines and shaded underneath.

Options

vertical and horizontal specify whether a vertical or a horizontal area plot is desired. vertical
is the default. If horizontal is specified, the values recorded in *yvar* are treated as *x* values,
and the values recorded in *xvar* are treated as *y* values. That is, to make horizontal plots, do not
switch the order of the two variables specified.

In the vertical case, shading at each *xvar* value extends up or down from 0 according to the
corresponding *yvar* values. If 0 is not in the range of the *y* axis, shading extends up or down to
the *x* axis.

In the `horizontal` case, shading at each *xvar* value extends left or right from 0 according to the corresponding *yvar* values. If 0 is not in the range of the x axis, shading extends left or right to the y axis.

`base(#)` specifies the value from which the shading should extend. The default is `base(0)` and, in the above description of options `vertical` and `horizontal`, this default was assumed.

`nodropbase` is a programmer's option and is an alternative to `base()`. It specifies that rather than the enclosed area dropping to `base(#)`—or `base(0)`—it drops to the line formed by (y_1, x_1) and (y_N, x_N), where (y_1, x_1) are the y and x values in the first observation being plotted and (y_N, x_N) are the values in the last observation being plotted.

`sort` specifies that the data be sorted by *xvar* before plotting.

`bstyle(`*areastyle*`)` specifies the look of the shaded area. The options listed below allow you to change each attribute, but `bstyle()` provides the starting point.

You need not specify `bstyle()` just because there is something you want to change. You specify `bstyle()` when another style exists that is exactly what you desire or when another style would allow you to specify fewer changes to obtain what you want.

See [G] *areastyle* for a list of available area styles.

`bcolor(`*colorstyle*`)` specifies a single color to be used both to outline the shape and to fill its interior. See [G] *colorstyle* for a list of color choices.

`bfcolor(`*colorstyle*`)` specifies the color to be used to fill the interior. See [G] *colorstyle* for a list of color choices.

`blstyle(`*linestyle*`)` specifies the overall style of the line used to outline the area, which includes its pattern (solid, dashed, etc.), its thickness, and its color. The other options listed below allow you to change the line's attributes, but `blstyle()` is the starting point. See [G] *linestyle* for a list of choices.

`blcolor(`*colorstyle*`)` specifies the color to be used to outline the area. See [G] *colorstyle* for a list of color choices.

`blwidth(`*linewidthstyle*`)` specifies the thickness of the line to be used to outline the area. See [G] *linewidthstyle* for a list of choices.

`blpattern(`*linepatternstyle*`)` specifies whether the line used to outline the area is solid, dashed, etc. See [G] *linepatternstyle* for a list of pattern choices.

scatter_options refer to any of the options allowed by `scatter` except that *marker_options*, *marker_placement_option*, *marker_label_options*, and *connect_options* are ignored even if specified; see [G] **graph twoway scatter**.

Remarks

Remarks are presented under the headings

> *Typical use*
> *Advanced use*
> *Cautions*

Typical use

We have quarterly data recording the U.S. GNP in constant 1996 dollars:

```
. sysuse gnp96, clear
. list in 1/5
```

	date	gnp96
1.	1967q1	3631.6
2.	1967q2	3644.5
3.	1967q3	3672
4.	1967q4	3703.1
5.	1968q1	3757.5

It is our opinion that the area under a curve should only be shaded if the area is meaningful:

```
. sysuse gnp96, clear
. twoway area d.gnp96 date
```

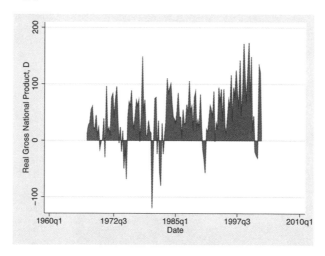

Advanced use

Here is the same graph, dressed to kill:

```
. twoway area d.gnp96 date, xlabel(36(8)164, angle(90))
        ylabel(-100(50)200, angle(0))
        ytitle("Billions of 1996 Dollars")
        xtitle("")
        subtitle("Change in U.S. GNP", position(11))
        note("Source: U.S. Department of Commerce,
                   Bureau of Economic Analysis")
```

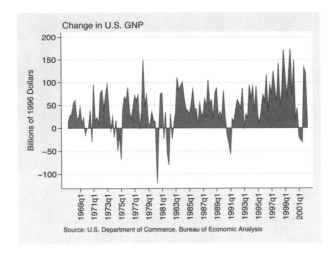

Cautions

Be sure that the data are in the order of *xvar* or specify area's `sort` option. If you do neither, you will get something that looks like modern art:

```
. sysuse gnp96, clear
. generate d = d.gnp96
. generate u = uniform()
. sort u                        (put in random order)
. twoway area d date
```

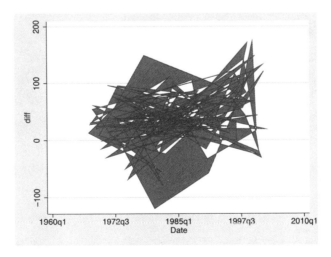

Also See

Complementary: [G] **graph twoway scatter**; [G] **graph twoway dot**,
                       [G] **graph twoway dropline**, [G] **graph twoway histogram**,
                       [G] **graph twoway spike**; [G] **graph bar**

Title

graph twoway bar — Twoway bar plots

Syntax

<u>two</u>way bar *yvar xvar* [if *exp*] [in *range*] [, *bar_options scatter_options*]

where *bar_options* are

bar_options	description
<u>vertical</u>	vertical bar plot (default)
<u>horizontal</u>	horizontal bar plot
base(*#*)	value to drop to (default = 0)
<u>barwidth</u>(*#*)	width of bar in *xvar* units
[G] *barlook_options*	look of bars

See [G] *barlook_options*.
All options are *merged-implicit*; see [G] **repeated options**.

and where *scatter_options* refer to any of the options allowed by scatter, except *marker_options*, *marker_placement_option*, *marker_label_options*, and *connect_options* are ignored even if specified; see [G] **graph twoway scatter**.

Description

twoway bar displays numeric (*y*,*x*) data as bars. twoway bar is useful for drawing bar plots of time-series data or other equally spaced data and is useful as a programming tool. For finely spaced data, also see [G] **graph twoway spike**.

Also see [G] **graph bar** for traditional bar charts and [G] **graph twoway histogram** for histograms.

Options

vertical and horizontal specify whether a vertical or a horizontal bar plot is desired. vertical is the default. If horizontal is specified, the values recorded in *yvar* are treated as *x* values, and the values recorded in *xvar* are treated as y values. That is, to make horizontal plots, do not switch the order of the two variables specified.

In the vertical case, bars are drawn at the specified *xvar* values and extend up or down from 0 according to the corresponding *yvar* values. If 0 is not in the range of the *y* axis, bars extend up or down to the *x* axis.

In the horizontal case, bars are drawn at the specified *xvar* values and extend left or right from 0 according to the corresponding *yvar* values. If 0 is not in the range of the *x* axis, bars extend left or right to the *y* axis.

base(*#*) specifies the value from which the bar should extend. The default is base(0) and, in the above description of options vertical and horizontal, this default was assumed.

barwidth(*#*) specifies the width of the bar in *xvar* units. The default is width(1). When a bar is plotted, it is centered at *x*, so half the width extends below *x* and half above.

barlook_options set the look of the bars. The most important of these options is bcolor(*colorstyle*), which specifies the color of the bars; see [G] *colorstyle* for a list of color choices. See [G] *barlook_options* for information on the other *barlook_options*.

scatter_options refer to any of the options allowed by scatter except that *marker_options*, *marker_placement_option*, *marker_label_options*, and *connect_options* are ignored even if specified; see [G] **graph twoway scatter**.

Remarks

Remarks are presented under the headings

> *Typical use*
> *Advanced use: Overlaying*
> *Advanced use: Population pyramid*
> *Cautions*

Typical use

We have daily data recording the values for the S&P 500 in 2001:

```
. sysuse sp500, clear
(S&P 500)

. list date close change in 1/5
```

	date	close	change
1.	02jan2001	1283.27	.
2.	03jan2001	1347.56	64.29004
3.	04jan2001	1333.34	-14.22009
4.	05jan2001	1298.35	-34.98999
5.	08jan2001	1295.86	-2.48999

We will use the first 57 observations from these data:

(*Continued on next page*)

. twoway bar change date in 1/57

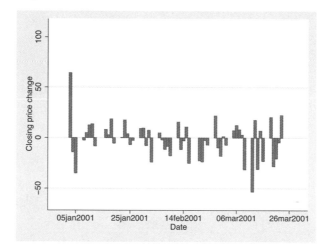

We get a different visual effect if we reduce the width of the bars from 1 day to .6 days:

. twoway bar change date in 1/57, barw(.6)

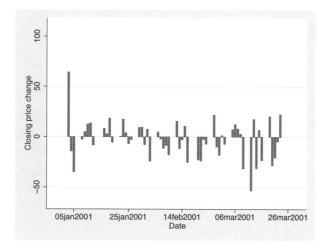

Advanced use: Overlaying

The useful thing about twoway bar is that it can be combined with other twoway plottypes:

```
. twoway line close date || bar change date || in 1/52
```

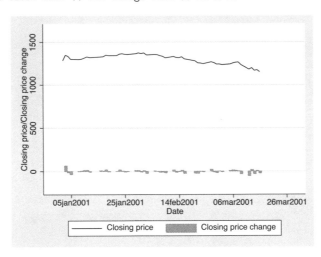

We can improve this graph by typing

```
. twoway
        line close date, yaxis(1)
    ||
        bar change date, yaxis(2)
    ||
    in 1/52,
        ysca(axis(1) r(1000 1400)) ylab(1200(50)1400, axis(1))
        ysca(axis(2) r(-50 300)) ylab(-50 0 50, axis(2))
                ytick(-25 0 25, axis(2) grid)
        legend(off)
        title("S&P 500")
        subtitle("January - March 2001")
        note("Source:  Yahoo!Finance and Commodity Systems, Inc.")
        yline(1150, axis(1) lstyle(foreground))
```

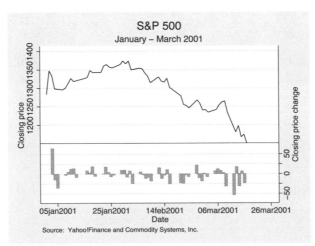

Notice our use of

```
yline(1150, axis(1) lstyle(foreground))
```

The 1150 put the horizontal line at $y = 1150$; axis(1) stated that y should be interpreted according to the left y axis; and lstyle(foreground) specified that the line be drawn in the foreground style.

Instead of lstyle(foreground) we could have coded lcolor(white) or lcolor(black), which one being determined by whether we were producing a graph for display on the screen or for printing in the manuals. Specifying lstyle(foreground) got us around the problem of having to use a different option for the help files and the manual.

Advanced use: Population pyramid

We have the following aggregate data from the U.S. 2000 Census recording total population by age and sex:

```
. sysuse pop2000, clear

. list agegrp maletotal femtotal
```

	agegrp	maletotal	femtotal
1.	Under 5	9,810,733	9,365,065
2.	5 to 9	10,523,277	10,026,228
3.	10 to 14	10,520,197	10,007,875
4.	15 to 19	10,391,004	9,828,886
5.	20 to 24	9,687,814	9,276,187
6.	25 to 29	9,798,760	9,582,576
7.	30 to 34	10,321,769	10,188,619
8.	35 to 39	11,318,696	11,387,968
9.	40 to 44	11,129,102	11,312,761
10.	45 to 49	9,889,506	10,202,898
11.	50 to 54	8,607,724	8,977,824
12.	55 to 59	6,508,729	6,960,508
13.	60 to 64	5,136,627	5,668,820
14.	65 to 69	4,400,362	5,133,183
15.	70 to 74	3,902,912	4,954,529
16.	75 to 79	3,044,456	4,371,357
17.	80 to 84	1,834,897	3,110,470

From this, we produce the following population pyramid:

```
. replace maletotal = -maletotal/1e+6

. replace femtotal = femtotal/1e+6

. twoway
        bar maletotal agegrp, horizontal xvarlab(Males)
   ||
        bar  femtotal agegrp, horizontal xvarlab(Females)
   ||
   , ylabel(1(1)17, angle(horizontal) valuelabel labsize(*.8))
     xtitle("Population in millions") ytitle("")
     xlabel(-10 "10" -7.5 "7.5" -5 "5" -2.5 "2.5" 2.5 5 7.5 10)
     legend(label(1 Males) label(2 Females))
     title("US Male and Female Population by Age")
     subtitle("Year 2000")
     note("Source:  U.S. Census Bureau, Census 2000, Tables 1, 2 and 3", span)
```

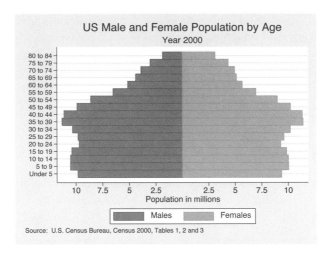

At its heart, the above graph is very simple: We turned the bars sideways and changed the male total to be negative. Our first attempt at the above was simply

```
. sysuse pop2000, clear

. replace maletotal = -maletotal

. twoway bar maletotal agegrp, horizontal ||
        bar  femtotal agegrp, horizontal
```

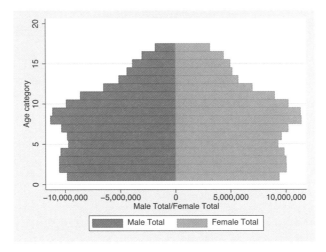

From there, we divided the population totals by one million and added options. Two of the options that we added are worth mentioning.

xlabel(-10 "10" -7.5 "7.5" -5 "5" -2.5 "2.5" 2.5 5 7.5 10) was a clever way to disguise that the bars for males extended in the negative direction. We said to label the values -10, -7.5, -5, -2.5, 2.5, 5, 7.5, and 10, but then we substituted text for the negative numbers to make it appear that they were positive. See [G] *axis_label_options*.

The use of the span suboption to note() aligned the text on the left side of the graph rather than on the plot region. See [G] *textbox_options*.

For another rendition of the pyramid, we tried

```
. sysuse pop2000, clear

. replace maletotal = -maletotal/1e+6

. replace femtotal = femtotal/1e+6

. generate zero = 0

. twoway
        bar maletotal agegrp, horizontal xvarlab(Males)
    ||
        bar  femtotal agegrp, horizontal xvarlab(Females)
    ||
        sc  agegrp zero      , mlabel(agegrp) mlabcolor(black) msymbol(i)
    ||
        , xtitle("Population in millions") ytitle("")
    plotregion(style(none))                                          (note 1)
    ysca(noline) ylabel(none)                                        (note 2)
    xsca(noline titlegap(-3.5))                                      (note 3)
    xlabel(-12 "12" -10 "10" -8 "8" -6 "6" -4 "4" 4(2)12 , tlength(0)
                                                    grid gmin gmax)
    legend(label(1 Males) label(2 Females)) legend(order(1 2))
    title("US Male and Female Population by Age, 2000")
    note("Source:  U.S. Census Bureau, Census 2000, Tables 1, 2 and 3")
```

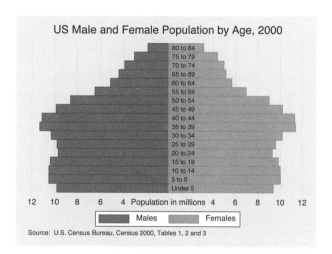

In the above rendition, we moved the labels from the x axis to be inside the bars by overlaying a `scatter` on top of the bars. The points of the scatter we plotted at $y = $ `agegrp` and $x = 0$, and rather than showing the markers, we displayed marker labels containing the desired labelings. See [G] *marker_label_options*.

We also played the following tricks:

1. `plotregion(style(none))` suppressed the outlining of the plot region; see [G] *region_options*.

2. `ysca(noline)` suppressed the drawing of the y axis—see [G] *axis_scale_options*—and `ylabel(none)` suppressed the labeling of it—see [G] *axis_label_options*.

3. `xsca(noline titlegap(-3.5))` suppressed the drawing of the x axis and moved the x-axis title up to be in between its labels; see [G] *axis_scale_options*.

Cautions

It is your responsibility to extend the scale of the axis if that is necessary. Consider using twoway bar to produce a histogram (ignoring the better alternative of using twoway histogram). Assume that you have already aggregated data of the form

x	frequency
1	400
2	800
3	3,000
4	1,800
5	1,100

which you enter into Stata to make variables x and frequency. You type

    . twoway bar frequency x

to make a histogram-style bar chart. The y axis will be scaled to go between 400 and 3,000 (labeled at 500, 1,000, ..., 3,000) and the shortest bar will have zero height. You need to type

    . twoway bar frequency x, ysca(r(0))

Also See

Complementary: [G] **graph twoway scatter**; [G] **graph twoway dot**,

[G] **graph twoway dropline**, [G] **graph twoway histogram**,

[G] **graph twoway spike**; [G] **graph bar**

Title

graph twoway connected — Twoway connected plots

Syntax

<u>tw</u>oway <u>con</u>nected *varlist* $\left[\textit{weight}\right]$ $\left[\text{if } \textit{exp}\right]$ $\left[\text{in } \textit{range}\right]$ $\left[, \textit{scatter\_options}\right]$

where *varlist* has the interpretation

$$y_1 \left[y_2\left[\ldots\right]\right] x$$

and where *scatter_options* are any of the options allowed by graph twoway scatter; see [G] **graph twoway scatter**.

aweights, fweights, and pweights are allowed; see [U] **14.1.6 weight**.

Description

connected draws connected-line plots. In a connected-line plot, the markers are displayed and the points are connected.

connected is a *plottype* as defined in [G] **graph twoway**. Thus, the syntax for connected is

    . graph twoway connected ...
    . twoway connected ...

Being a plottype, line may be combined with other plottypes in the twoway family, as in,

    . twoway (connected ...) (scatter ...) (lfit ...) ...

Options

scatter_options are any of the options allowed by the graph twoway scatter command; see [G] **graph twoway scatter**.

Remarks

connected is, in fact, scatter, the difference being that by default the points are connected:

Default connect() option: connect(l ...)

Thus, you get the same results by typing

    . twoway connected yvar xvar

as typing

    . scatter yvar xvar, connect(l)

You can just as easily turn connected into scatter: Typing

    . scatter yvar xvar

249

is the same as typing

```
. line yvar xvar, connect(none)
```

Also See

Complementary: [G] **graph twoway scatter**

Title

graph twoway dot — Twoway dot plots

Syntax

<u>two</u>way dot *yvar xvar* [if *exp*] [in *range*] [, *dot_options scatter_options*]

where *dot_options* are

dot_options	description
<u>vert</u>ical	vertical bar plot (default)
<u>horizontal</u>	horizontal bar plot
<u>dotextend</u>(yes | no)	whether dots extend beyond point
base(*#*)	value to drop to if dotextend(no)
<u>ndots</u>(*#*)	# of dots in full span of y or x
dstyle(*markerstyle*)	overall marker style of dots
dsymbol(*symbolstyle*)	marker symbol for dots
dcolor(*colorstyle*)	fill and outline color for dots
dfcolor(*colorstyle*)	fill color for dots
dsize(*markersizestyle*)	size of dots
dlstyle(*linestyle*)	overall outline style of dots
dlcolor(*colorstyle*)	outline color for dots
dlwidth(*linewidthstyle*)	thickness of outline for dots
dlpattern(*linepatternstyle*)	outline pattern for dots

See [G] *markerstyle*, [G] *symbolstyle*, [G] *colorstyle*, [G] *markersizestyle*, [G] *linestyle*, [G] *linewidthstyle*, and [G] *linepatternstyle*.

All options are *merged-implicit*; see [G] **repeated options**.

and where *scatter_options* refer to any of the options allowed by scatter except that *connect_options* are ignored even if specified; see [G] **graph twoway scatter**.

Description

twoway dots displays numeric (y,x) data as dot plots. Also see [G] **graph dot** to create dot plots of categorical variables. twoway dots is useful in programming contexts.

Options

vertical and horizontal specify whether a vertical or a horizontal dot plot is desired. vertical is the default. If horizontal is specified, the values recorded in *yvar* are treated as x values, and the values recorded in *xvar* are treated as y values. That is, to make horizontal plots, do not switch the order of the two variables specified.

251

In the vertical case, dots are drawn at the specified *xvar* values and extend up and down.

In the horizontal case, lines are drawn at the specified *xvar* values and extend left and right.

dotextend(yes | no) determines whether the dots extend beyond the y value (or x value if horizontal is specified). dotextend(yes) is the default.

base(#) is relevant only if dotextend(no) is also specified. base() specifies the value from which the dots are to extend. The default is base(0).

ndots(#) specifies the number of dots across a line; ndots(75) is the default. Depending on printer/screen resolution, fewer or more dots can make the graph look better.

dstyle(*markerstyle*) specifies the overall look of the markers used to create the dots, including their shape and color. The other options listed below allow you to change their attributes, but dstyle() provides the starting point.

You need not specify dstyle() just because there is something you want to change. You specify dstyle() when another style exists that is exactly what you desire or when another style would allow you to specify fewer changes to obtain what you want.

See [G] *markerstyle* for a list of available marker styles.

dsymbol(*symbolstyle*) specifies the shape of the marker used for the dot. See [G] *symbolstyle* for a list of symbol choices, although it really makes little sense to change the shape of dots; else why would it be called a dot plot?

dcolor(*colorstyle*) specifies the color of the symbol used for the dot. See [G] *colorstyle* for a list of color choices.

dfcolor(*colorstyle*), dsize(*markersizestyle*), dlstyle(*linestyle*), dlcolor(*colorstyle*), dlwidth(*linewidthstyle*), and dlpattern(*linepatternstyle*) are rarely (never) specified options. They control, respectively, the fill color, size, outline style, outline color, outline width, outline pattern of the marker used for the dots and, if you are really using dots, dots are affected by none of these things. For these options to be useful, you must also specify dsymbol(), and, as we said earlier, why then would it be called a dot plot? In any case, see [G] *colorstyle*, [G] *markersizestyle*, [G] *linestyle*, [G] *linewidthstyle*, and [G] *linepatternstyle* for a list of choices.

scatter_options refer to any of the options allowed by scatter, and most especially the *marker_options* which control how the marker (not the dot) appears. *connect_options*, even if specified, are ignored. See [G] **graph twoway scatter**.

Remarks

twoway dot is of little, if any use. We cannot think of a use for it, but perhaps someday, somewhere, someone will. Understand that we have nothing against the dot plot used with categorical data—see [G] **graph dot** for a very useful command—but the use of the dot plot in a twoway context is bizarre. It is nonetheless included for logical completeness.

In [G] **graph twoway bar**, we graphed the change in the value for the S&P 500. Here is a bit of that data graphed as a dot plot:

```
. sysuse sp500, clear
(S&P 500)
```

. twoway dot change date in 1/45

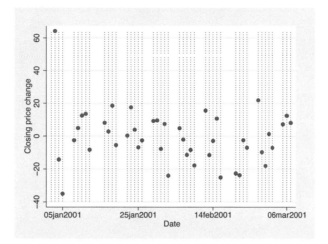

Dot plots are usually presented horizontally,

. twoway dot change date in 1/45, horizontal

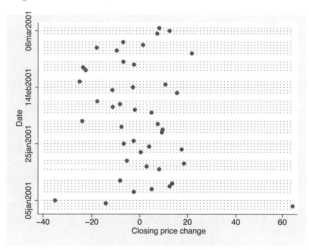

and below we specify the dotextend(n) option to prevent the dots from extending clear across the range of x:

(Continued on next page)

. twoway dot change date in 1/45, horizontal dotext(n)

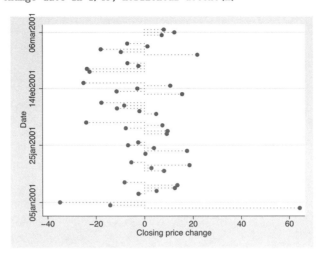

Also See

Complementary: [G] **graph twoway scatter**; [G] **graph dot**

Title

> **graph twoway dropline** — Twoway dropped-line plots

Syntax

<u>two</u>way dropline *yvar xvar* [if *exp*] [in *range*] [, *dropline_options scatter_options*]

where *dropline_options* are

dropline_options	description
<u>vert</u>ical	vertical dropped-line plot (default)
<u>hor</u>izontal	horizontal dropped-line plot
base(#)	value to drop to (default = 0)
<u>bl</u>style(*linestyle*)	overall look of line
<u>blc</u>olor(*colorstyle*)	line color
<u>blw</u>idth(*linewidthstyle*)	thickness of line
<u>blp</u>attern(*linepatternstyle*)	whether line solid, dashed, etc.

See [G] *linestyle*, [G] *colorstyle*, [G] *linewidthstyle*, and [G] *linepatternstyle*.

All options are *merged-implicit*; see [G] **repeated options**.

and where *scatter_options* refer to any of the options allowed by scatter except that *connect_options* are ignored even if specified; see [G] **graph twoway scatter**.

Description

twoway dropline displays numeric (*y*,*x*) data as dropped lines capped with a marker. twoway dropline is useful for drawing plots in which the numbers vary around zero.

Options

vertical and horizontal specify whether a vertical or a horizontal dropped-line plot is desired. vertical is the default. If horizontal is specified, the values recorded in *yvar* are treated as *x* values, and the values recorded in *xvar* are treated as *y* values. That is, to make horizontal plots, do not switch the order of the two variables specified.

In the vertical case, dropped lines are drawn at the specified *xvar* values and extend up or down from 0 according to the corresponding *yvar* values. If 0 is not in the range of the *y* axis, lines extend up or down to the *x* axis.

In the horizontal case, dropped lines are drawn at the specified *xvar* values and extend left or right from 0 according to the corresponding *yvar* values. If 0 is not in the range of the *x* axis, lines extend left or right to the *y* axis.

255

base(*#*) specifies the value from which the lines should extend. The default is base(0) and, in the above description of options vertical and horizontal, this default was assumed.

blstyle(*linestyle*) specifies the overall style of the dropped line, which includes its pattern (solid, dashed, etc.), its thickness, and its color. The other options listed below allow you to change the line's attributes, but blstyle() is the starting point. See [G] *linestyle* for a list of choices.

blcolor(*colorstyle*) specifies the color of the dropped line. See [G] *colorstyle* for a list of color choices.

blwidth(*linewidthstyle*) specifies the thickness of the dropped line. See [G] *linewidthstyle* for a list of choices.

blpattern(*linepatternstyle*) specifies whether the dropped line used is solid, dashed, etc. See [G] *linepatternstyle* for a list of pattern choices.

scatter_options refer to any of the options allowed by scatter, and most especially the *marker_options*, which control how the marker atop the dropped line looks, and the *marker_label_options*, which allow labeling of the individual points. *connect_options*, even if specified, are ignored. See [G] **graph twoway scatter**.

Remarks

Remarks are presented under the headings

> *Typical use*
> *Advanced use*
> *Cautions*

Typical use

We have daily data recording the values for the S&P 500 in 2001:

```
. sysuse sp500, clear
(S&P 500)

. list date close change in 1/5
```

	date	close	change
1.	02jan2001	1283.27	.
2.	03jan2001	1347.56	64.29004
3.	04jan2001	1333.34	-14.22009
4.	05jan2001	1298.35	-34.98999
5.	08jan2001	1295.86	-2.48999

In [G] **graph twoway bar**, we graphed the first 57 observations of these data using bars. Here is the same graph presented as dropped lines:

(*Continued on next page*)

. twoway dropline change date in 1/57, yline(0, lstyle(foreground))

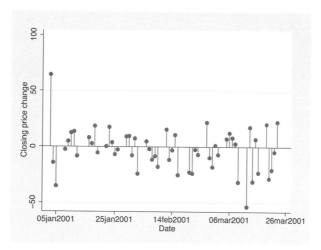

In the above, we specified yline(0) to add a line across the graph at 0, and we specified yline(, lstyle(foreground)) so that the line would have the same color as the foreground. We could have instead specified yline(, lcolor()). For an explanation of why we chose lstyle() over foreground(), see *Advanced use* in [G] **graph twoway bar**.

Advanced use

Dropped-line plots work especially well when the points are labeled. For instance,

```
. sysuse lifeexp, clear
(Life expectancy, 1998)
. keep if region==3
. generate lngnp = ln(gnppc)
. quietly regress le lngnp
. predict r, resid
. twoway dropline r gnp,
        yline(0, lstyle(foreground)) mlabel(country) mlabpos(9)
        ylab(-6(1)6)
        subtitle("Regression of life expectancy on ln(gnp)"
                "Residuals:" " ", pos(11))
        note("Residuals in years; positive values indicate"
            "longer than predicted life expectancy")
```

(*Continued on next page*)

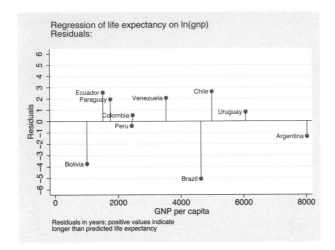

Cautions

See *Cautions* in [G] **graph twoway bar**. What is said there applies equally to `twoway dropline`.

Also See

Complementary: [G] **graph twoway scatter**; [G] **graph twoway spike**

Title

> **graph twoway fpfit** — Twoway fractional-polynomial prediction plots

Syntax

<u>two</u>way fpfit *yvar xvar* [*weight*] [if *exp*] [in *range*] [, *fpfit_options line_options*]

where *fpfit_options* are

fpfit_options	description
<u>estcmd</u>(*estcmd*)	estimation command; default `regress`
<u>esto</u>pts(*fracpoly_options*)	options for `fracpoly` *estcmd*
<u>predo</u>pts(*predict_options*)	options for `predict`

All options are *rightmost*; see [G] **repeated options**.

and where *line_options* are any of the options allowed by `graph twoway line`; see [G] **graph twoway line**.

`aweights`, `fweights`, and `pweights` are allowed. Weights, if specified, affect estimation but not how the weighted results are plotted. See [U] **14.1.6 weight**.

Description

`twoway fpfit` calculates the prediction for *yvar* based on estimation of a fractional polynomial of *xvar*, and plots the resulting curve.

Options

`estcmd(`*estcmd*`)` specifies the estimation command to be used; `estcmd(regress)` is the default.

`estopts(`*fracpoly_options*`)` specifies options to be passed along to `fracpoly` to estimate the fractional polynomial regression from which the curve will be predicted; see [R] **fracpoly**.

`predopts(`*predict_options*`)` specifies options to be passed along to `predict` to obtain the predictions after estimation by `fracpoly regress`. `predopts()` may only be used with `estcmd(regress)`. Predictions in all cases are calculated at all the *xvar* values in the data.

line_options refer to any of the options allowed by `graph twoway line`; see [G] **graph twoway line**.

Remarks

Remarks are presented under the headings

> *Typical use*
> *Cautions*
> *Use with by()*

Typical use

`twoway fpfit` is nearly always used in conjunction with other `twoway` plottypes, such as

```
. sysuse auto, clear
(1978 Automobile Data)
. scatter mpg weight || fpfit mpg weight
```

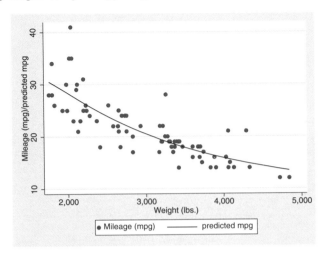

Results are visually the same as typing

```
. fracpoly regress mpg weight
. predict fitted
. scatter mpg weight || line fitted weight
```

Cautions

Be careful (i.e., do not use) `twoway fpfit` when specifying the *axis_scale_options* `yscale(log)` or `xscale(log)` to create log scales. For instance, typing

```
. scatter mpg weight, xscale(log) || fpfit mpg weight
```

will produce a curve that will be fit from a fractional polynomial regression of `mpg` on `weight` rather than `log(weight)`.

Use with by()

`fpfit` may be used with `by()` (as can all the `twoway` plot commands):

. scatter mpg weight || fpfit mpg weight ||, by(foreign, total row(1))

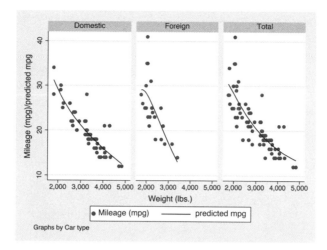

Also See

Complementary: [G] **graph twoway line**, [G] **graph twoway fpfitci**;

[G] **graph twoway lfit**, [G] **graph twoway qfit**;

[G] **graph twoway mband**, [G] **graph twoway mspline**;

[R] **fracpoly**

Title

> **graph twoway fpfitci** — Twoway fractional-polynomial prediction plots with CIs

Syntax

> <u>tw</u>oway fpfitci *yvar* *xvar* [*weight*] [if *exp*] [in *range*] [, *fpfitci_options*]

where *fpfitci_options* are

fpfitci_options	description
fpfit_options	rarely specified
level(#)	confidence level for CI of mean
fitplot(*plottype*)	how to plot fit; line default
ciplot(*plottype*)	how to plot CIs; rarea default
line_options	how to render prediction
rarea_options	how to render CIs

All options are *rightmost*; see [G] **repeated options**.

aweights, fweights, and pweights are allowed. Weights, if specified, affect estimation but not how the weighted results are plotted. See [U] **14.1.6 weight**.

Description

twoway fpfitci calculates the prediction for *yvar* based on estimation of a fractional polynomial of *xvar*, and plots the resulting curve along with the confidence interval of the mean.

Options

fpfit_options refers to any of the options of graph twoway fpfit; see [G] **graph twoway fpfit**. These options are rarely specified.

level(#) specifies the confidence level, in percent, for the confidence interval of the mean; see [R] **level**.

fitplot(*plottype*) is rarely specified. It specifies how the prediction is to be plotted. The default is fitplot(line), meaning that the prediction will be plotted by graph twoway line. See [G] **graph twoway** for a list of *plottype* choices. You may choose any that expect a single y and a single x variable.

ciplot(*plottype*) specifies how the confidence interval is to be plotted. The default is ciplot(rarea), meaning that the prediction will be plotted by graph twoway rarea.

A very reasonable alternative is ciplot(rline), which will substitute lines around the prediction for shading. See [G] **graph twoway** for a list of *plottype* choices. You may choose any that expect two y variables and one x variable.

262

line_options specify how the prediction line is rendered; see [G] **graph twoway line**. If you specify `fitplot()`, then rather than *line_options*, you should specify whatever is appropriate.

rarea_options specify how the confidence interval is rendered; see [G] **graph twoway rarea**. If you specify `ciplot()`, then rather than *rarea_options*, you should specify whatever is appropriate.

Remarks

Remarks are presented under the headings

> *Typical use*
> *Advanced use*
> *Cautions*
> *Use with by()*

Typical use

`twoway fpfitci` by default draws the confidence interval of the predicted mean:

```
. sysuse auto, clear
(1978 Automobile Data)
. twoway fpfitci mpg weight
```

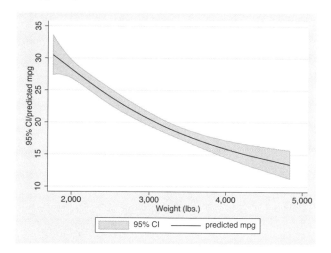

If you specify the `ciplot(rline)` option, rather than shading the confidence interval, it will be designated by lines:

(Continued on next page)

```
. twoway fpfitci mpg weight, ciplot(rline)
```

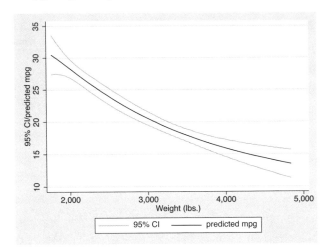

Advanced use

fpfitci can be usefully overlaid with other plots:

```
. sysuse auto, clear
(1978 Automobile Data)
. twoway fpfitci mpg weight || scatter mpg weight
```

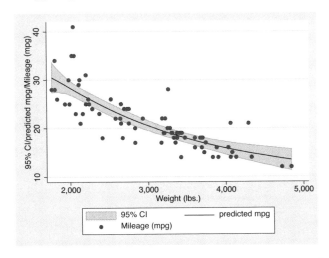

Note that in the above, the shaded area corresponds to the 95% confidence interval for the mean. Of great importance, we typed

```
. twoway fpfitci ... || scatter ...
```

and not

```
. twoway scatter ... || fpfitci ...
```

Had we drawn the scatter diagram first, the confidence interval would have covered up most of the points.

Cautions

Be careful (i.e., do not use) `twoway fpfitci` when specifying the *axis_scale_options* `yscale(log)` or `xscale(log)` to create log scales. For instance, typing

. `twoway fpfitci mpg weight || scatter mpg weight ||, xscale(log)`

will produce a curve that will be fit from a fractional polynomial regression of `mpg` on `weight` rathere than `log(weight)`.

See *Cautions* in [G] **graph twoway lfitci**.

Use with by()

`fpfitci` may be used with `by()` (as can all the `twoway` plot commands):

. `twoway fpfitci  mpg weight  ||`
        `scatter  mpg weight  ||`
   `, by(foreign, total row(1))`

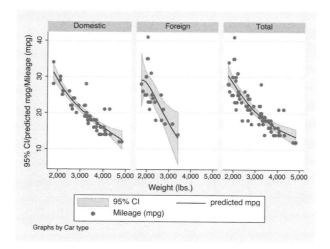

Also See

Complementary: [G] **graph twoway lfitci**, [G] **graph twoway qfitci**;

[G] **graph twoway fpfit**;

[R] **fracpoly**

Title

graph twoway function — Twoway line plot of function

Syntax

twoway function $\big[\,[y] =\big]\ f(x)\ \big[\text{if } exp\big]\ \big[\text{in } range\big]$

$\big[\,,\ function\_options\ line\_options\ \big]$

where *function_options* are

function_options	description
range(# #)	plot over $x = $ # to #
range(*varname*)	plot over $x = $ min to max of *varname*
n(#)	evaluate at # points; default $= 300$
droplines(*numlist*)	draw lines to axis at specified x values
base(#)	base value for dropline(); default $= 0$
horizontal	draw plot horizontally
yvarformat(%*fmt*)	display format for y
xvarformat(%*fmt*)	display format for x

All options are *rightmost*; see [G] **repeated options**.

Note that if *exp* and in *range* play no role unless option range(*varname*) is specified.
In the above syntax diagram, $f(x)$ stands for an *expression* in terms of x.

Description

twoway function plots $y = f(x)$, where $f(x)$ is some function of x. That is, you literally type

. twoway function y=sqrt(x)

It makes no difference whether y and x are variables in your data.

Options

range(# #) and range(*varname*) specify the range of values for x. In the first syntax, range() is a pair of numbers identifying the minimum and maximum. In the second syntax, range() is a variable name and the range used will be obtained from the minimum and maximum values of the variable. If range() is not specified, range(0 1) is assumed.

n(#) specifies the number of points at which $f(x)$ is to be evaluated. The default is n(300).

droplines(*numlist*) adds dropped lines from the function down to, or up to, the axis (or $y = $ base() if base() is specified) at each x value specified in *numlist*.

266

base(*#*) specifies the base for the droplines(). The default is base(0). This option does not affect the range of the axes, so you may also want to specify the *line_option* yscale(range(*#*)) as well; see [G] *axis_scale_options*.

horizontal specifies that the roles of y and x be interchanged and that the graph be plotted horizontally rather than vertically (that the plotted function be reflected along the identity line).

yvarformat(*%fmt*) and xvarformat(*%fmt*) specify the display format to be used for y and x. These formats are used when labeling the axes; see [G] *axis_label_options*.

line_options are any of the options allowed by graph twoway line; see [G] **graph twoway line**.

Remarks

Remarks are presented under the headings

> *Typical use*
> *Advanced use 1*
> *Advanced use 2*

Typical use

You wish to plot the function $y = \exp(-x/6)\sin(x)$ over the range 0 to 4π:

. twoway function y=exp(-x/6)*sin(x), range(0 12.57)

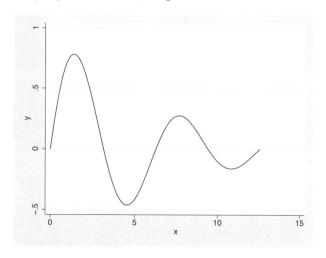

A better rendition of the graph above is

```
. twoway function y=exp(-x/6)*sin(x), range(0 12.57)
        yline(0, lstyle(foreground))
        xlabel(0 3.14 "pi" 6.28 "2 pi" 9.42 "3 pi" 12.57 "4 pi")
        plotregion(style(none))
        xsca(noline)
```

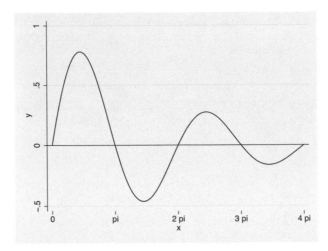

yline(0, lstyle(foreground)) added a line at $y = 0$; lstyle(foreground) gave the line the same style as used for the axes. See [G] ***added_line_options***.

xlabel(0 3.14 "pi" 6.28 "2 pi" 9.42 "3 pi" 12.57 "4 pi") labeled the x axis with the numeric values given and substituted text for the numeric values; see [G] ***axis_label_options***.

plotregion(style(none)) suppressed the border around the plot region; see [G] ***region_options***.

xsca(noline) suppressed the drawing of the x-axis line; see [G] ***axis_scale_options***.

Advanced use 1

The following graph appears in many introductory textbooks:

```
. twoway
        function y=normden(x), range(-4 -1.96) bcolor(gs12) recast(area)
    ||  function y=normden(x), range(1.96 4)    bcolor(gs12) recast(area)
    ||  function y=normden(x), range(-4 4) clstyle(foreground)
    ||,
        plotregion(style(none))
        ysca(off) xsca(noline)
        legend(off)
        xlabel(-4 "-4 sd" -3 "-3 sd" -2 "-2 sd" -1 "-1 sd" 0 "mean"
                1  "1 sd"  2  "2 sd"  3  "3 sd"  4  "4 sd"
          , grid gmin gmax)
        xtitle("")
```

(Continued on next page)

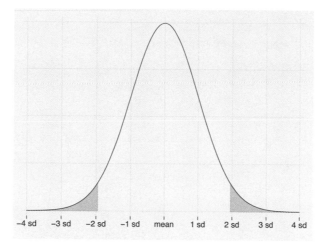

We drew the graph in three parts: the shaded area on the left, the shaded area on the right, and then the overall function. To obtain the shaded areas, we used the *advanced_option* recast(area) so that, rather than the function being plotted by graph twoway line, it was plotted by graph twoway rarea; see [G] *advanced_options* and [G] **graph twoway rarea**. Concerning the overall function, we drew it last so that its darker foreground-colored line would not get covered up by the shaded areas.

Advanced use 2

function plots may be overlaid with other twoway plots. For instance, function is one way to add $y = x$ lines to a plot:

```
. sysuse sp500, clear
(S&P 500)

. scatter open close, msize(*.25) mcolor(*.6) ||
  function y=x, range(close) yvarlab("y=x") clwidth(*1.5)
```

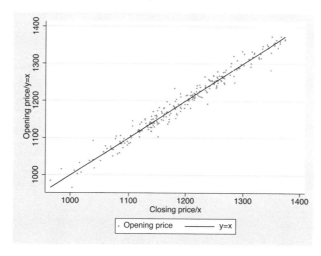

In the above, we specified the *advanced_option* yvarlab("y=x") so that the variable label of y would be treated as "y=x" in the construction of the legend; see [G] *advanced_options*. We specified msize(*.25) to make the marker symbols smaller, and we specified mcolor(*.6) to make them dimmer; see [G] *relativesize* and [G] *colorstyle*.

Also See

Complementary: [G] **graph twoway line**

Title

graph twoway histogram — Histogram plots

Syntax

<u>two</u>way <u>histo</u>gram *varname* [*weight*] [if *exp*] [in *range*]

[, [*discrete_options* | *continuous_options*] *common_options*]

where *discrete_options* are

discrete_options	description
<u>discrete</u>	specify data are discrete
width(#)	width of bins in *varname* units
start(#)	theoretical minimum value

and where *continuous_options* are

continuous_options	description
bin(#)	# of bins
width(#)	width of bins in *varname* units
start(#)	lower limit of first bin

and where *common_options* are

common_options	description
<u>dens</u>ity	draw as density (default)
<u>frac</u>tion	draw as fractions
<u>freq</u>uency	draw as frequencies
gap(#)	reduce width of bars, $0 \leq \# < 100$
rbar_options	rendition of bars
scatter_options	look of graph

See [G] **graph twoway rbar** and [G] **graph twoway scatter**.
The *scatter_options*, *marker_options*, *marker_placement_option*,
marker_label_options, and *connect_options* are ignored if specified.

fweights are allowed; see [U] **14.1.6 weight**.

Description

twoway histogram draws histograms of *varname*. Also see [R] **histogram** for an easier-to-use alternative.

Options

Options for use in the discrete case

discrete specifies that *varname* is discrete and that you want each unique value of *varname* to be given its own bin (bar of histogram).

width(*#*) is rarely specified in the discrete case; it specifies the width of the bins. The default is width(*d*), where *d* is the observed minimum difference between the unique values of *varname*.

Specify width() if you are concerned that your data are sparse. For example, *varname* could in theory take on the values 1, 2, 3, ..., 9, but due to sparseness, perhaps only the values 2, 4, 7, and 8 are observed. In this case, the default width calculation would produce width(2) and you would want to specify width(1).

start(*#*) is also rarely specified in the discrete case; it specifies the theoretical minimum value of *varname*. The default is start(*m*), where *m* is the observed minimum value.

As with width(), specify start() when you are concerned about sparseness. In the previous example, you would also want to specify start(1). Note that start() does nothing more than add white space to the left side of the graph.

start(), if specified, must be less than or equal to *m*, or else an error will be issued.

Options for use in the continuous case

bin(*#*) and width(*#*) are alternatives that specify how the data are to be aggregated into bins. bin() specifies the number of bins (from which the width can be derived), and width() specifies the bin width (from which the number of bins can be derived).

If neither option is specified, results are as if bin(*k*) were specified, where

$$k = \min\left(\sqrt{N}, 10 \times \frac{\ln(N)}{\ln(10)}\right)$$

and where *N* is the number of nonmissing observations of *varname*.

start(*#*) specifies the theoretical minimum of *varname*. The default is start(*m*), where *m* is the observed minimum value of *varname*.

Specify start() when you are concerned about sparse data. For instance, you know that *varname* can go down to 0, but you are concerned that 0 may not be observed.

start(), if specified, must be less than or equal to *m*, or else an error will be issued.

Options for use in both the discrete and continuous cases

density, fraction, and frequency are alternatives that specify whether you want the histogram scaled to density, fractional, or frequency units. density is the default.

density scales the height of the bars so that the sum of their areas equals 1.

fraction scales the height of the bars so that the sum of their heights equals 1.

frequency scales the height of the bars so that each bar's height is equal to the number of observations in the category, and thus the sum of the heights is equal to the total number of nonmissing observations of *varname*.

gap(#) specifies that the bar width be reduced by # percent. gap(0) is the default; histogram sets the width so that adjacent bars just touch. If you want gaps between the bars, specify, for instance, gap(5).

Also see *rbar_options* below for another way to set the display width of the bars.

rbar_options are any of the labels allowed by graph twoway rbar; see [G] **graph twoway rbar**. (It is *rbar_options* and not *bar_options* that are allowed because graph twoway histogram secretly adds a line at 0 to ensure that the bars extend all the way down to 0.)

One of the more useful *rbar_options* is barwidth(#), which specifies the width of the bars in *varname* units. By default, histogram draws the bars so that adjacent bars just touch. If you want gaps between the bars, do not specify histogram's width() option—which would change how the histogram is calculated—specify the *rbar_option* barwidth() or the histogram option gap(), both of which affect only how the bar is rendered.

Another useful *rbar_option* is bcolor(*colorstyle*), which sets the color of the bars. This is especially useful when combined with the *colorstyle* *# syntax to adjust the intensity. bcolor(*.5) leaves the color unchanged, but dims it, which can be especially useful when overlaying other plots on top of the histogram.

scatter_options refer to any of the options allowed by scatter except that *marker_options*, *marker_placement_option*, *marker_label_options*, and *connect_options* are ignored even if specified; see [G] **graph twoway scatter**.

Remarks

Remarks are presented under the headings

> *Relationship between graph twoway histogram and histogram*
> *Typical use*
> *Use with by()*
> *History*

Relationship between graph twoway histogram and histogram

graph twoway histogram—documented here—and histogram—documented in
[R] **histogram**—are almost the same command. Comparing the two, histogram has the advantages that

1. it allows overlaying of a normal density or a kernel estimate of the density, and

2. if a density estimate is overlaid, it scales the density to reflect the scaling of the bars.

For your information, histogram is implemented in terms of graph twoway histogram.

Typical use

When you do not specify otherwise, graph twoway histogram assumes the variable is continuous:

```
. sysuse lifeexp, clear
(Life expectancy, 1998)
. twoway histogram le
```

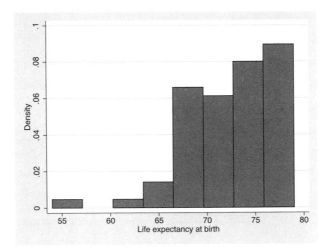

Even with a continuous variable, you may specify the discrete option to see the individual values:

```
. twoway histogram le, discrete
```

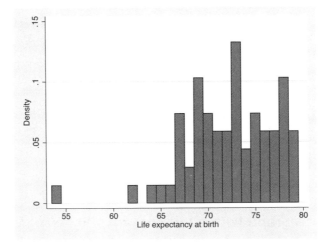

Use with by()

graph twoway histogram may be used with by():

```
. sysuse lifeexp, clear
(Life expectancy, 1998)
. twoway histogram le, discrete by(region, total)
```

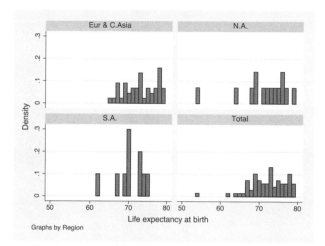

Graphs by Region

In this case, specifying frequency is a good way to show both the distribution and the overall contribution to the total:

```
. twoway histogram le, discrete freq by(region, total)
```

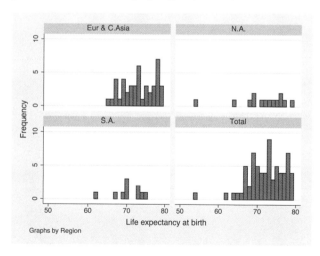

Graphs by Region

The height of the bars reflects the number of countries. In this case—and in all the above examples—we would do better by obtaining population data on the countries and then typing

```
. twoway histogram le [fw=pop], discrete freq by(region, total)
```

so that bar height reflected total population.

History

According to Beniger and Robyn, (1978, 4) although A. M. Guerry published a histogram in 1833, the word "histogram" was first used by Karl Pearson in 1895.

References

Beniger, J. R. and D. L. Robyn. 1978. Quantitative graphics in statistics: a brief history. *The American Statistician* 32: 1–11.

Guerry, A. M. 1833. *Essai sur la Statistique Morale de la France.* Paris.

Pearson, K. 1895. Contributions to the mathematical theory of evolution.—II. Skew variation in homogeneous material. *Philosophical Transactions of the Royal Society of London*, A, 186: 343–414.

Also See

Complementary: [R] **histogram**;

[G] **graph twoway kdensity**

Title

graph twoway kdensity — Kernel density plots

Syntax

<u>tw</u>oway kdensity *varname* [if *exp*] [in *range*] [, *kdensity_options line_options*]

where *kdensity_options* are

kdensity_options	description
<u>w</u>idth(*#*)	smoothing parameter
<u>epan</u>	use Epanechnikov kernel (default)
<u>bi</u>weight	use biweight kernel
<u>cos</u>ine	use cosine trace
<u>gauss</u>	use Gaussian kernel
<u>parzen</u>	use Parzen kernel
<u>rect</u>angle	use rectangular kernel
<u>tri</u>angle	use triangular kernel
<u>r</u>ange(*# #*)	range for plot, minimum and maximum
<u>r</u>ange(*varname*)	range for plot obtained from *varname*
n(*#*)	number of points to evaluate
area(*#*)	rescaling parameter
<u>ho</u>rizontal	graph horizontally

Description

graph twoway kdensity plots a kernel density estimate for *varname* using graph twoway line.

Options

width(*#*) and epan, biweight, ..., triangle specify how the kernel density estimate is to be obtained and are in fact the same options as specified with the command kdensity; see [R] **kdensity**.

width(*#*) specifies the smoothing parameter.

epan, biweight, ..., triangle are alternatives and specify the kernel-weight function to be used. epan (Epanechnikov) is the default.

See [R] **kdensity** for more information about these options.

All the other graph twoway kdensity options modify how the result is displayed, not how it is obtained.

277

range(*# #*) and range(*varname*) specify the range of values at which the kernel density estimates are to be plotted. The default is range(*m M*), where *m* and *M* are the minimum and maximum of the *varname* specified on the graph twoway kdensity command.

range(*# #*) specifies a pair of numbers to be used as the minimum and maximum.

range(*varname*) specifies another variable for which its minimum and maximum are to be used.

n(*#*) specifies the number of points at which the estimate is evaluated. The default is n(300).

area(*#*) specifies a multiplier by which the density estimates are adjusted before being plotted. The default is area(1). area() is useful when overlaying a density estimate on top of a histogram that is itself not scaled as a density. For instance, if you wished to scale the density estimate as a frequency, area() would be specified as the total number of nonmissing observations.

horizontal specifies that the result be plotted horizontally (i.e, reflected along the identity line).

line_options affect the rendition of the plotted kernel density estimate and include all the other common graph twoway options; see [G] **graph twoway line**.

Remarks

graph twoway kdensity *varname* uses the kdensity command to obtain an estimate of the density of *varname* and uses graph twoway line to plot the result.

Remarks are presented under the headings

> *Typical use*
> *Use with by*

Typical use

The density estimate is often graphed on top of the histogram:

```
. sysuse lifeexp, clear
(Life expectancy, 1998)

. tw histogram lexp, bcolor(*.5) || kdensity lexp
```

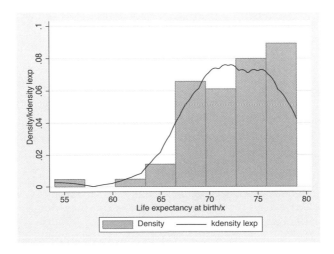

Notice our use of graph twoway histogram's bcolor(*.5) option to dim the bars and make the line stand out; see [G] *colorstyle*.

Notice also the y and x axes titles: "Density/kdensity lexp" and "Life expectancy at birth/x". The "kdensity lexp" and "x" were contributed by the twoway kdensity. When you overlay graphs, you nearly always need to respecify the axes titles using the *axis_title_options* ytitle() and xtitle(); see [G] *axis_title_options*.

Use with by

graph twoway kdensity may be used with by():

```
. sysuse lifeexp, clear
(Life expectancy, 1998)

. tw histogram lexp, bcolor(*.5) || kdensity lexp ||, by(region)
```

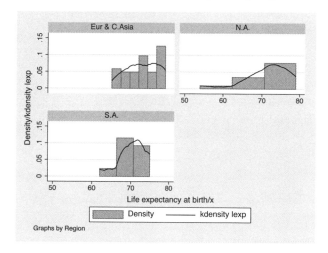

Also See

Complementary: [R] **kdensity**;

[G] **graph twoway histogram**

Title

graph twoway lfit — Twoway linear prediction plots

Syntax

<u>two</u>way lfit *yvar xvar* [*weight*] [if *exp*] [in *range*] [, *lfit_options line_options*]

where *lfit_options* are

lfit_options	description
<u>range</u>(*# #*)	range over which predictions calculated
n(*#*)	number of prediction points
atobs	calculate predictions at *xvar*
<u>est</u>opts(*regress_options*)	options for regress
<u>pred</u>opts(*predict_options*)	options for predict

All options are *rightmost*; see [G] **repeated options**.

and where *line_options* are any of the options allowed by graph twoway line; see [G] **graph twoway line**.

aweights, fweights, and pweights are allowed. Weights, if specified, affect estimation but not how the weighted results are plotted. See [U] **14.1.6 weight**.

Description

twoway lfit calculates the prediction for *yvar* based on a linear regression of *yvar* on *xvar*, and plots the resulting line.

Options

range(*# #*) specifies the *x* range over which predictions are calculated. The default is range(. .), meaning the minimum and maximum values of *xvar*. range(0 10) would make the range 0 to 10, range(. 10) would make the range the minimum to 10, and range(0 .) would make the range 0 to the maximum.

n(*#*) specifies the number of points at which predictions over range() are to be calculated. The default is n(3).

atobs is an alternative to n(). It specifies that the predictions be calculated at the *xvar* values. atobs is the default if predopts() is specified and any statistic other than the xb is requested.

estopts(*regress_options*) specifies options to be passed along to regress to estimate the linear regression from which the line will be predicted; see [R] **regress**. If this option is specified, commonly specified is estopts(nocons).

predopts(*predict_options*) specifies options to be passed along to predict to obtain the predictions after estimation by regress.

line_options refer to any of the options allowed by graph twoway line; see [G] **graph twoway line**.

Remarks

Remarks are presented under the headings

Typical use
Cautions
Use with by()

Typical use

twoway lfit is nearly always used in conjunction with other twoway plottypes, such as

```
. sysuse auto, clear
(1978 Automobile Data)
. scatter mpg weight || lfit mpg weight
```

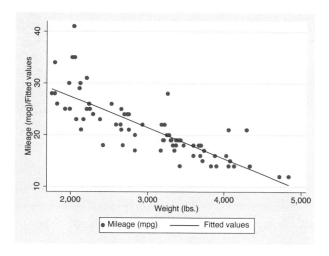

Results are visually the same as typing

```
. regress mpg weight
. predict fitted
. scatter mpg weight || line fitted weight
```

Cautions

Be careful (i.e., do not use) twoway lfit when specifying the *axis_scale_options* yscale(log) or xscale(log) to create log scales. For instance,

. scatter mpg weight, xscale(log) || lfit mpg weight

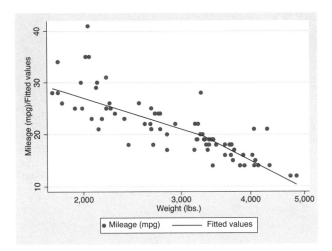

The line is not straight because the regression estimated for the prediction was for mpg on weight, not mpg on log(weight). (The default for n() is 3 so that, if you make this mistake, you will spot it.)

Use with by()

lfit may be used with by() (as can all the twoway plot commands):

. scatter mpg weight || lfit mpg weight ||, by(foreign, total row(1))

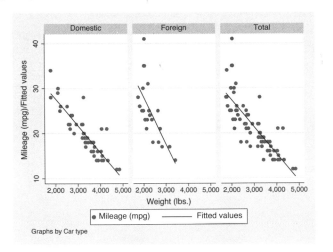

Also See

Complementary: [G] **graph twoway line**, [G] **graph twoway qfit**,
                  [G] **graph twoway fpfit**, [G] **graph twoway mband**,
                  [G] **graph twoway mspline**; [G] **graph twoway lfitci**;
                  [R] **regress**

Title

> **graph twoway lfitci** — Twoway linear prediction plots with CIs

Syntax

<u>tw</u>oway lfitci *yvar xvar* [*weight*] [if *exp*] [in *range*] [, *lfitci_options*]

where *lfitci_options* are

lfitci_options	description
lfit_options	rarely specified
stdp	CIs from SE of prediction (default)
stdf	CIs from SE of forecast
stdr	CIs from SE of residual (rarely specified)
level(*#*)	confidence level for CI
<u>fitp</u>lot(*plottype*)	how to plot fit; line default
<u>cip</u>lot(*plottype*)	how to plot CIs; rarea default
line_options	how to render prediction
rarea_options	how to render CIs

All options are *rightmost*; see [G] **repeated options**.

aweights, fweights, and pweights are allowed. Weights, if specified, affect estimation but not how the weighted results are plotted. See [U] **14.1.6 weight**.

Description

twoway lfitci calculates the prediction for *yvar* based on a linear regression of *yvar* on *xvar*, and plots the resulting line along with a confidence interval.

Options

lfit_options refers to any of the options of graph twoway lfit; see [G] **graph twoway lfit**. These options are rarely specified.

stdp, stdf, and stdr determine the basis for the confidence interval. stdp is the default.

stdp states that the confidence interval is to be the confidence interval of the mean.

stdf states that the confidence interval is to be the confidence interval for an individual forecast, which includes both the uncertainty of the mean prediction and the residual.

stdr states that the confidence interval is to be based only on the standard error of the residual.

level(*#*) specifies the confidence level, in percent, for the confidence interval; see [R] **level**.

fitplot(*plottype*) is rarely specified. It specifies how the prediction is to be plotted. The default is fitplot(line), meaning that the prediction will be plotted by graph twoway line. See [G] **graph twoway** for a list of *plottype* choices. You may choose any that expect a single y and a single x variable.

ciplot(*plottype*) specifies how the confidence interval is to be plotted. The default is ciplot(rarea), meaning that the prediction will be plotted by graph twoway rarea.

A very reasonable alternative is ciplot(rline), which will substitute lines around the prediction for shading. See [G] **graph twoway** for a list of *plottype* choices. You may choose any that expect two y variables and one x variable.

line_options specify how the prediction line is rendered; see [G] **graph twoway line**. If you specify fitplot(), then rather than *line_options*, you should specify whatever is appropriate.

rarea_options specify how the confidence interval is rendered; see [G] **graph twoway rarea**. If you specify ciplot(), then rather than *rarea_options*, you should specify whatever is appropriate.

Remarks

Remarks are presented under the headings

> *Typical use*
> *Advanced use*
> *Cautions*
> *Use with by()*

Typical use

twoway lfitci by default draws the confidence interval of the predicted mean:

```
. sysuse auto, clear
(1978 Automobile Data)

. twoway lfitci mpg weight
```

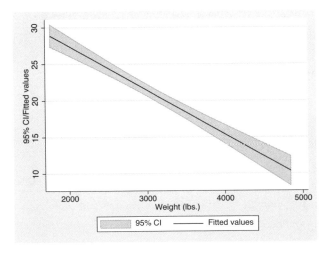

If you specify the ciplot(rline) option, rather than shading the confidence interval, it will be designated by lines:

. twoway lfitci mpg weight, ciplot(rline)

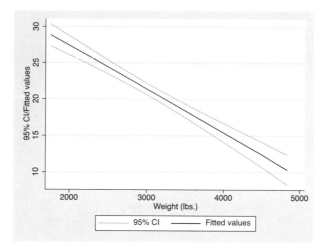

Advanced use

lfitci can be usefully overlaid with other plots:

```
. sysuse auto, clear
(1978 Automobile Data)
. twoway lfitci mpg weight, stdf || scatter mpg weight
```

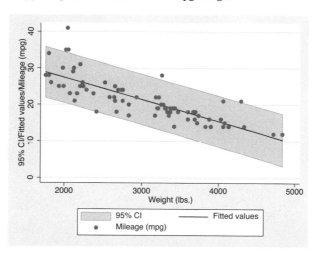

Note that in the above, we specified stdf so as to obtain a confidence interval based on the standard error of the forecast rather than the standard error of the mean. This is more useful for identifying outliers.

Note that we typed

```
. twoway lfitci ... || scatter ...
```

and not

```
. twoway scatter ... || lfitci ...
```

Had we drawn the scatter diagram first, the confidence interval would have covered up most of the points.

Cautions

Be careful (i.e., do not use) twoway lfitci when specifying the *axis_scale_options* yscale(log) or xscale(log) to create log scales. For instance,

. twoway lfitci mpg weight, stdf || scatter mpg weight ||, xscale(log)

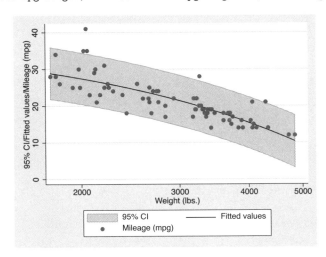

The result may look pretty but, if you think about it, it is not what you want. The prediction line is not straight because the regression estimated for the prediction was for mpg on weight, not mpg on log(weight).

Use with by()

lfitci may be used with by() (as can all the twoway plot commands):

(Continued on next page)

```
. twoway lfitci  mpg weight, stdf ||
        scatter mpg weight      ||
   , by(foreign, total row(1))
```

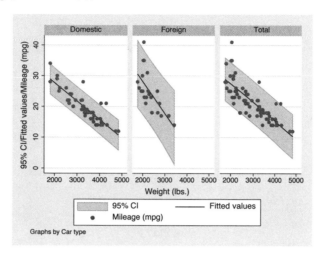

Also See

Complementary: [G] **graph twoway qfitci**, [G] **graph twoway fpfitci**;

[G] **graph twoway lfit**;

[R] **regress**

Title

graph twoway line — Twoway line plots

Syntax

$$\left[\underline{\text{tw}}\text{oway}\right] \text{ line } \textit{varlist} \left[\text{if } \textit{exp}\right] \left[\text{in } \textit{range}\right] \left[,\ \textit{line\_options}\right]$$

where *varlist* has the interpretation

$$y_1 \left[y_2\left[\dots\right]\right]\ x$$

and where *line_options* are

line_options	description
scatter_options	any of the options allowed by graph twoway scatter with the exception of *marker_options* *marker_placement_option* *marker_label_options* which will be ignored if specified

See [G] **graph twoway scatter**.

Description

line draws line plots.

line is both a command and it is a *plottype* as defined in [G] **graph twoway**. Thus, the syntax for line is

```
. graph twoway line ...
. twoway line ...
. line ...
```

Being a plottype, line may be combined with other plottypes in the twoway family, as in,

```
. twoway (line ...) (scatter ...) (lfit ...) ...
```

which can equivalently be written

```
. line ... || scatter ... || lfit ... || ...
```

Options

line_options are any of the options allowed by the graph twoway scatter command except that *marker_options*, *marker_placement_option*, and *marker_label_options* will be ignored if specified; see [G] **graph twoway scatter**.

Remarks

Remarks are presented under the headings

Oneway equivalency of line and scatter
Typical use
Advanced use
Cautions

Oneway equivalency of line and scatter

line, in fact, is scatter, the differences being that by default the marker symbols are not displayed and the points are connected:

Default msymbol() option: msymbol(none ...)

Default connect() option: connect(l ...)

Thus, you get the same results typing

    . line yvar xvar

as typing

    . scatter yvar xvar, msymbol(none) connect(l)

You can use scatter in place of line, but you may not use line in place of scatter. Typing

    . line yvar xvar, msymbol(O) connect(none)

will not achieve the same results as

    . scatter yvar xvar

because line, while it allows you to specify the *marker_option* msymbol(), ignores its setting.

Typical use

line draws line charts:

    . sysuse uslifeexp, clear
    (U.S. life expectancy, 1900-1999)
    . line le year

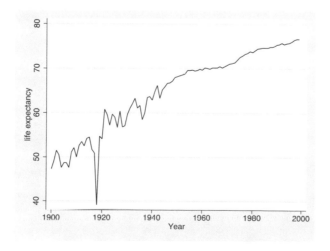

Line charts work well with time-series data. With other datasets, lines are often used to show predicted values and confidence intervals:

```
. sysuse auto, clear
(1978 Automobile Data)
. quietly regress mpg weight
. predict hat
. predict stdf, stdf
. generate lo = hat - 1.96*stdf
. generate hi = hat + 1.96*stdf
. scatter mpg weight || line hat lo hi weight, pstyle(p2 p3 p3) sort
```

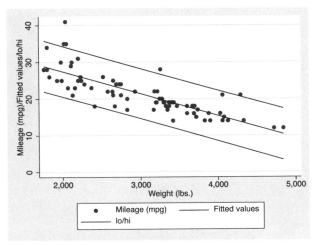

Do not forget to include the `sort` option when the data are not in the order of the x variable, as they are not above. We also included `pstyle(p2 p3 p3)` to give the lower and upper confidence limit lines the same look; see *Styles and composite style* under *Remarks* in [G] **graph twoway scatter**.

Because `line` is `scatter`, we can use any of the options allowed by `scatter`. Below we return to the U.S. life expectancy data and graph black and white male life expectancies along with the difference, specifying lots of options to create an informative and visually pleasing graph:

```
. sysuse uslifeexp, clear
(U.S. life expectancy, 1900-1999)
. generate diff = le_wm - le_bm
. label var diff "Difference"
```

(Continued on next page)

```
     .   line le_wm year, yaxis(1 2) xaxis(1 2)
      || line le_bm year
      || line diff  year
      || lfit diff  year
      ||,
         ylabel(0(5)20, axis(2) gmin angle(horizontal))
         ylabel(0 20(10)80,    gmax angle(horizontal))
         ytitle("", axis(2))
         xlabel(1918, axis(2)) xtitle("", axis(2))
         ytitle("Life expectancy at birth (years)")
         title("White and black life expectancy")
         subtitle("USA, 1900-1999")
         note("Source: National Vital Statistics, Vol 50, No. 6"
    "(1918 dip caused by 1918 Influenza Pandemic)")
```

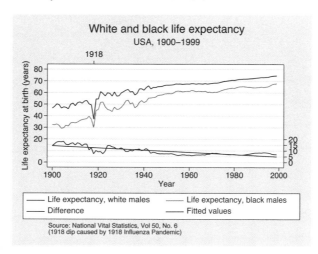

See [G] **graph twoway scatter**.

Advanced use

The above graph would look better if we shortened the descriptive text used in the keys. Below we add

```
legend(label(1 "White males") label(2 "Black males"))
```

to our previous command:

```
     .   line le_wm year, yaxis(1 2) xaxis(1 2)
      || line le_bm year
      || line diff  year
      || lfit diff  year
      ||,
         ylabel(0(5)20, axis(2) gmin angle(horizontal))
         ylabel(0 20(10)80,    gmax angle(horizontal))
         ytitle("", axis(2))
         xlabel(1918, axis(2)) xtitle("", axis(2))
         ytitle("Life expectancy at birth (years)")
         title("White and black life expectancy")
         subtitle("USA, 1900-1999")
         note("Source: National Vital Statistics, Vol 50, No. 6"
              "(1918 dip caused by 1918 Influenza Pandemic)")
         legend(label(1 "White males") label(2 "Black males"))
```

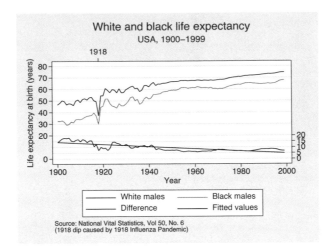

We might also consider moving the legend to the right of the graph, which we can do by adding

```
legend(col(1) pos(3))
```

resulting in

```
.    line le_wm year, yaxis(1 2) xaxis(1 2)
|| line le_bm year
|| line diff   year
|| lfit diff   year
||,
    ylabel(0(5)20, axis(2) gmin angle(horizontal))
    ylabel(0 20(10)80,    gmax angle(horizontal))
    ytitle("", axis(2))
    xlabel(1918, axis(2)) xtitle("", axis(2))
    ytitle("Life expectancy at birth (years)")
    title("White and black life expectancy")
    subtitle("USA, 1900-1999")
    note("Source: National Vital Statistics, Vol 50, No. 6"
"(1918 dip caused by 1918 Influenza Pandemic)")
    legend(label(1 "White males") label(2 "Black males"))
    legend(col(1) pos(3))
```

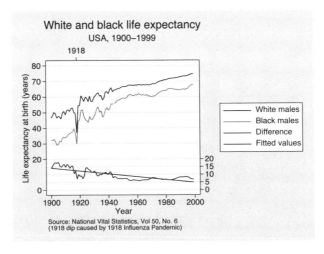

See [G] *legend_option* for more information on dealing with legends.

Cautions

Be sure that the data are in the order of the x variable or specify line's sort option. If you do neither, you will get something that looks like the scribblings of a child:

```
. sysuse auto, clear
(1978 Automobile Data)

. line mpg weight
```

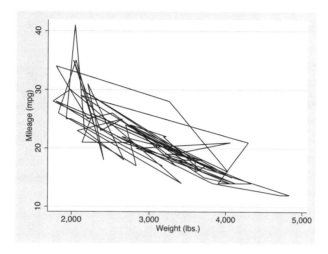

Also See

Complementary: [G] **graph twoway scatter**; [G] **graph twoway fpfit**,

[G] **graph twoway lfit**, [G] **graph twoway mband**,

[G] **graph twoway mspline**, [G] **graph twoway qfit**

Title

> **graph twoway lowess** — Local linear smooth plots

Syntax

<u>two</u>way lowess *yvar* *xvar* [if *exp*] [in *range*] [, *lowess_options* *line_options*]

where *lowess_options* are

lowess_options	description
<u>bw</u>idth(#)	smoothing parameter
<u>me</u>an	use running-mean smoothing
<u>nowe</u>ight	use unweighted smoothing
<u>lo</u>git	transform the smooth to logits
<u>ad</u>just	adjust smooth's mean to equal *yvar*'s mean

Description

graph twoway lowess plots a lowess smooth of *yvar* on *xvar* using graph twoway line.

Options

bwidth(#) specifies the bandwidth. bwidth(.8) is the default. Centered subsets of $N*$bwidth() observations, N = number of observations, are used for calculating smoothed values for each point in the data except for end points, where smaller, uncentered subsets are used. The greater the bwidth(), the greater the smoothing.

mean specifies running-mean smoothing; the default is running-line least-squares smoothing.

noweight prevents the use of Cleveland's (1979) tricube weighting function; the default is to use the weighting function.

logit transforms the smoothed *yvar* into logits.

adjust adjusts by multiplication the mean of the smoothed *yvar* to equal the mean of *yvar*. This is useful when smoothing binary (0/1) data.

line_options affect the rendition of the plotted kernel density estimate and include all the other common graph twoway options; see [G] **graph twoway line**.

Remarks

graph twoway lowess *yvar* *xvar* uses the lowess command—see [R] **lowess**—to obtain a local linear smooth of *yvar* on *xvar* and uses graph twoway line to plot the result.

Remarks are presented under the headings

> *Typical use*
> *Use with by*

Typical use

The local linear smooth is often graphed on top of the data, possibly with other regression lines:

```
. sysuse auto, clear
(1978 Automobile Data)

. twoway scatter mpg weight, mcolor(*.6) ||
         lfit    mpg weight    ||
         lowess  mpg weight
```

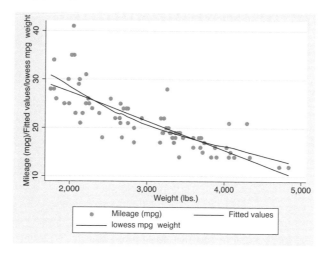

Notice our use of mcolor(*.6) to dim the points and thus make the lines stand out; see [G] *colorstyle*.

Notice also the y axis title: "Mileage (mpg)/Fitted values/lowess mpg weight". The "Fitted values" was contributed by twoway lfit and "lowess mpg weight" by twoway lowess. When you overlay graphs, you nearly always need to respecify the axes titles using the *axis_title_options* ytitle() and xtitle(); see [G] *axis_title_options*.

Use with by

graph twoway lowess may be used with by():

```
. sysuse auto, clear
(1978 Automobile Data)
```

(Continued on next page)

```
. twoway scatter mpg weight, mcolor(*.6) ||
         lfit    mpg weight ||
         lowess  mpg weight ||, by(foreign)
```

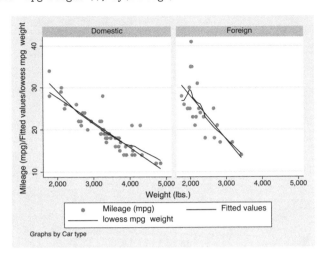

Reference

Cleveland, W. S. 1979. Robust locally weighted regression and smoothing scatterplots. *Journal of the American Statistical Association* 74: 829–836.

Also See

Complementary:	[R] **lowess**;
	[G] **graph twoway mspline**

Title

> **graph twoway mband** — Twoway median-band plots

Syntax

twoway mband *yvar* *xvar* [if *exp*] [in *range*] [, *mband_options* *line_options*]

where *mband_options* are

mband_options	description
bands(#)	number of bands

All options are *rightmost*; see [G] **repeated options**.

and where *line_options* are any of the options allowed by graph twoway line; see [G] **graph twoway line**.

Description

twoway mband calculates cross medians and then graphs the cross medians as a line plot.

Options

bands(#) specifies the number of bands on which the calculation is to be based. The default is $\max(10, \text{round}(10 \times \log 10(N)))$, where N is the number of observations.

In a median-band plot, the x axis is divided into # equal-width intervals and then the median of y and the median of x are calculated in each interval. It is these cross medians that mband graphs as a line plot.

Remarks

Remarks are presented under the headings

> Typical use
> Use with by()

Typical use

Median bands provide a convenient but crude way to show the tendency in the relationship between y and x:

```
. sysuse auto, clear
(1978 Automobile Data)
```

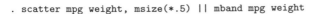

. scatter mpg weight, msize(*.5) || mband mpg weight

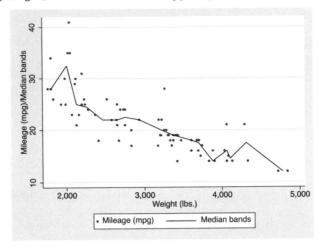

The important part of the above is "mband mpg weight". On the scatter, we specified msize(*.5) to make the marker symbols half their normal size; see [G] *relativesize*.

Use with by()

mband may be used with by() (as can all the twoway plot commands):

. scatter mpg weight, ms(oh) ||
  mband mpg weight ||, by(foreign, total row(1))

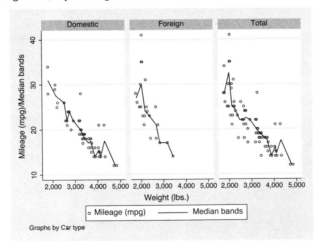

In the above, we specified ms(oh) so as to use hollow symbols; see [G] *symbolstyle*.

Also See

Complementary: [G] **graph twoway line**, [G] **graph twoway mspline**,

[G] **graph twoway lfit**, [G] **graph twoway qfit**,

[G] **graph twoway fpfit**

Title

graph twoway mspline — Twoway median-spline plots

Syntax

<u>tw</u>oway mspline *yvar* *xvar* [if *exp*] [in *range*] [, *mspline_options* *line_options*]

where *mspline_options* are

mspline_options	description
<u>b</u>ands(#)	number of cross-median knots
n(#)	number of points between knots

All options are *rightmost*; see [G] **repeated options**.

and where *line_options* are any of the options allowed by graph twoway line; see [G] **graph twoway line**.

Description

twoway mspline calculates cross medians and then uses the cross medians as knots to fit a cubic spline. The resulting spline is graphed as a line plot.

Options

bands(#) specifies the number of bands for which cross medians should be calculated. The default is $\max(10, \text{round}(10 \times \log10(N)))$, where N is the number of observations.

The x axis is divided into # equal-width intervals and then the median of y and the median of x are calculated in each interval. It is these cross medians to which a cubic spline is then fit.

n(#) specifies the number of points between the knots for which the cubic spline should be evaluated. n(10) is the default. n() does not affect the result that is calculated but it does affect how smooth the result appears.

Remarks

Remarks are presented under the headings

> *Typical use*
> *Cautions*
> *Use with by()*

Typical use

Median splines provide a convenient way to show the relationship between y and x:

```
. sysuse auto, clear
(1978 Automobile Data)

. scatter mpg weight, msize(*.5) || mspline mpg weight
```

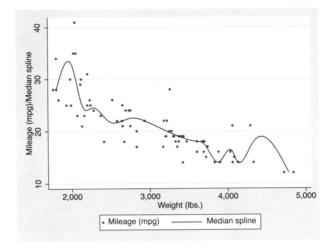

The important part of the above is "mspline mpg weight". On the scatter, we specified msize(*.5) to make the marker symbols half their normal size; see [G] *relativesize*.

Cautions

The graph shown above illustrates a common problem with this technique: it tracks wiggles that may not be real and can introduce wiggles if too many bands are chosen. An improved version of the graph above would be

```
. scatter mpg weight, msize(*.5) || mspline mpg weight, bands(8)
```

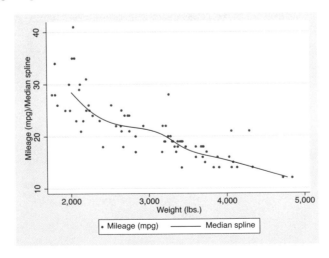

Use with by()

mband may be used with by() (as can all the twoway plot commands):

```
. scatter mpg weight, msize(*.5) ||
  mspline mpg weight, bands(8)   ||, by(foreign, total row(1))
```

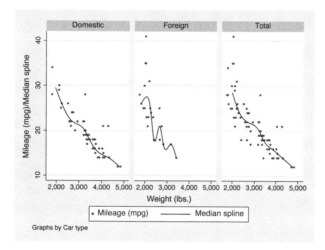

Also See

Complementary: [G] **graph twoway line**, [G] **graph twoway mband**,
[G] **graph twoway lfit**, [G] **graph twoway qfit**,
[G] **graph twoway fpfit**

Title

> **graph twoway qfit** — Twoway quadratic prediction plots

Syntax

> t̲woway qfit *yvar xvar* [*weight*] [if *exp*] [in *range*] [, *qfit_options line_options*]

where *qfit_options* are

qfit_options	description
r̲ange(# #)	range over which predictions calculated
n(#)	number of prediction points
atobs	calculate predictions at *xvar*
est̲opts(*regress_options*)	options for regress
predopts(*predict_options*)	options for predict

All options are *rightmost*; see [G] **repeated options**.

and where *line_options* are any of the options allowed by graph twoway line; see [G] **graph twoway line**.

aweights, fweights, and pweights are allowed. Weights, if specified, affect estimation but not how the weighted results are plotted. See [U] **14.1.6 weight**.

Description

twoway qfit calculates the prediction for *yvar* based on a linear regression of *yvar* on *xvar* and *xvar*^{2}, and plots the resulting curve.

Options

range(# #) specifies the x range over which predictions are calculated. The default is range(. .), meaning the minimum and maximum values of *xvar*. range(0 10) would make the range 0 to 10, range(. 10) would make the range the minimum to 10, and range(0 .) would make the range 0 to the maximum.

n(#) specifies the number of points at which predictions over range() are to be calculated. The default is n(100).

atobs is an alternative to n(). It specifies that the predictions be calculated at the *xvar* values. atobs is the default if predopts() is specified and any statistic other than xb is requested.

estopts(*regress_options*) specifies options to be passed along to regress to estimate the linear regression from which the curve will be predicted; see [R] **regress**. If this option is specified, commonly specified is estopts(nocons).

predopts(*predict_options*) specifies options to be passed along to predict to obtain the predictions after estimation by regress.

line_options refer to any of the options allowed by graph twoway line; see [G] **graph twoway line**.

302

Remarks

Remarks are presented under the headings

> *Typical use*
> *Cautions*
> *Use with by()*

Typical use

`twoway qfit` is nearly always used in conjunction with other `twoway` plottypes, such as

```
. sysuse auto, clear
(1978 Automobile Data)
. scatter mpg weight || qfit mpg weight
```

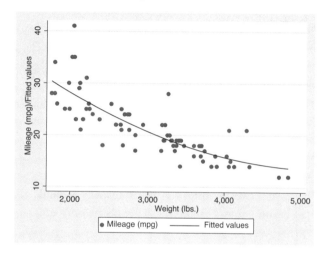

Results are visually the same as typing

```
. generate tempvar = weight^2
. regress mpg weight tempvar
. predict fitted
. scatter mpg weight || line fitted weight
```

Cautions

Be careful (i.e., do not use) `twoway qfit` when specifying the *axis_scale_options* `yscale(log)` or `xscale(log)` to create log scales. For instance,

```
. scatter mpg weight, xscale(log) || qfit mpg weight
```

produces something that is not a parabola because the regression estimated for the prediction was for `mpg` on `weight` and `weight^2`, not `mpg` on `log(weight)` and `log(weight)^2`.

Use with by()

qfit may be used with by() (as can all the twoway plot commands):

. scatter mpg weight || qfit mpg weight ||, by(foreign, total row(1))

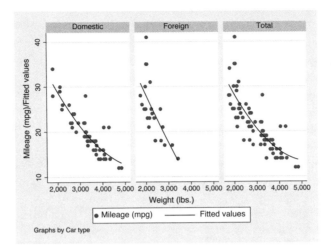

Also See

Complementary: [G] **graph twoway line**, [G] **graph twoway lfit**,
[G] **graph twoway fpfit**, [G] **graph twoway mband**,
[G] **graph twoway mspline**; [G] **graph twoway qfitci**;
[R] **regress**

Title

graph twoway qfitci — Twoway quadratic prediction plots with CIs

Syntax

twoway qfitci *yvar* *xvar* [*weight*] [if *exp*] [in *range*] [, *qfitci_options*]

where *qfitci_options* are

qfitci_options	description
qfit_options	rarely specified
stdp	CIs from SE of prediction (default)
stdf	CIs from SE of forecast
stdr	CIs from SE of resid. (rarely specified)
level(*#*)	confidence level for CI
fitplot(*plottype*)	how to plot fit; line default
ciplot(*plottype*)	how to plot CIs; rarea default
line_options	how to render prediction
rarea_options	how to render CIs

All options are *rightmost*; see [G] **repeated options**.

aweights, fweights, and pweights are allowed. Weights, if specified, affect estimation but not how the weighted results are plotted. See [U] **14.1.6 weight**.

Description

twoway qfitci calculates the prediction for *yvar* based on a regression of *yvar* on *xvar* and $xvar^2$, and plots the resulting line along with a confidence interval.

Options

qfit_options refers to any of the options of graph twoway qfit; see [G] **graph twoway qfit**. These options are rarely specified.

stdp, stdf, and stdr determine the basis for the confidence interval. stdp is the default.

stdp states that the confidence interval is to be the confidence interval of the mean.

stdf states that the confidence interval is to be the confidence interval for an individual forecast, which includes both the uncertainty of the mean prediction and the residual.

stdr states the confidence interval is to be based only on the standard error of the residual.

level(*#*) specifies the confidence level, in percent, for the confidence interval; see [R] **level**.

`fitplot(`*plottype*`)` is rarely specified. It specifies how the prediction is to be plotted. The default is `fitplot(line)`, meaning that the prediction will be plotted by `graph twoway line`. See [G] **graph twoway** for a list of *plottype* choices. You may choose any that expect a single y and a single x variable.

`ciplot(`*plottype*`)` specifies how the confidence interval is to be plotted. The default is `ciplot(rarea)`, meaning that the prediction will be plotted by `graph twoway rarea`.

A very reasonable alternative is `ciplot(rline)`, which will substitute lines around the prediction for shading. See [G] **graph twoway** for a list of *plottype* choices. You may choose any that expect two y variables and one x variable.

line_options specify how the prediction line is rendered; see [G] **graph twoway line**. If you specify `fitplot()`, then rather than *line_options*, you should specify whatever is appropriate.

rarea_options specify how the confidence interval is rendered; see [G] **graph twoway rarea**. If you specify `ciplot()`, then rather than *rarea_options*, you should specify whatever is appropriate.

Remarks

Remarks are presented under the headings

> *Typical use*
> *Advanced use*
> *Cautions*
> *Use with by()*

Typical use

`twoway qfitci` by default draws the confidence interval of the predicted mean:

```
. sysuse auto, clear
(1978 Automobile Data)
. twoway qfitci mpg weight
```

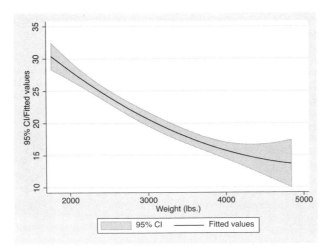

If you specify the `ciplot(rline)` option, rather than shading the confidence interval, it will be designated by lines:

```
. twoway qfitci mpg weight, ciplot(rline)
```

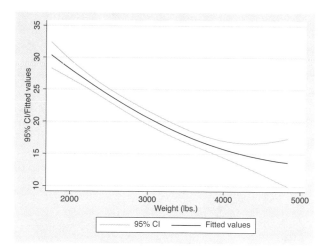

Advanced use

`qfitci` can be usefully overlaid with other plots:

```
. sysuse auto, clear
(1978 Automobile Data)
. twoway qfitci mpg weight, stdf || scatter mpg weight
```

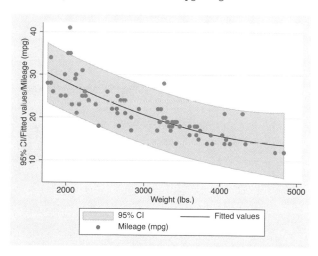

Note that in the above, we specified `stdf` so as to obtain a confidence interval based on the standard error of the forecast rather than the standard error of the mean. This is more useful for identifying outliers.

Also note that we typed

```
. twoway qfitci ... || scatter ...
```

and not

```
. twoway scatter ... || qfitci ...
```

Had we drawn the scatter diagram first, the confidence interval would have covered up most of the points.

Cautions

Be careful (i.e., do not use) `twoway qfitci` when specifying the *axis_scale_options* `yscale(log)` or `xscale(log)` to create log scales. For instance,

```
. twoway qfitci mpg weight, stdf || scatter mpg weight ||, xscale(log)
```

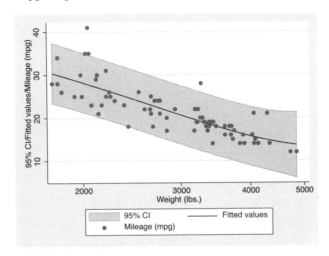

The result may look pretty but, if you think about it, it is not what you want. The prediction line is not a parabola because the regression estimated for the prediction was for mpg on `weight` and `weight^2`, not mpg on `log(weight)` and `log(weight)^2`.

Use with by()

`qfitci` may be used with `by()` (as can all the `twoway` plot commands):

(*Continued on next page*)

```
. twoway qfitci  mpg weight, stdf ||
     scatter mpg weight      ||
     , by(foreign, total row(1))
```

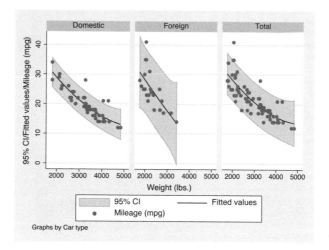

Also See

Complementary:	[G] **graph twoway lfitci**, [G] **graph twoway fpfitci**;
	[G] **graph twoway qfit**;
	[R] **regress**

Title

graph twoway rarea — Range plot with area shading

Syntax

twoway rarea *y1var y2var xvar* [if *exp*] [in *range*] [, *rarea_options scatter_options*]

where *rarea_options* are

rarea_options	description
vertical	vertical area plot (default)
horizontal	horizontal area plot
sort	recommendation: specify this option
bstyle(*areastyle*)	overall look of shaded area
bcolor(*colorstyle*)	outline and fill color
bfcolor(*colorstyle*)	fill color
blstyle(*linestyle*)	overall look of outline
blcolor(*colorstyle*)	outline color
blwidth(*linewidthstyle*)	thickness of outline
blpattern(*linepatternstyle*)	whether outline solid, dashed, etc.

See [G] *areastyle*, [G] *colorstyle*, [G] *linestyle*, [G] *linewidthstyle*, and [G] *linepatternstyle*.

All options are *rightmost*; see [G] **repeated options**.

and where *scatter_options* refer to any of the options allowed by scatter except that *marker_options*, *marker_placement_option*, *marker_label_options*, and *connect_options* are ignored even if specified; see [G] **graph twoway scatter**.

Description

A range plot has two *y* variables, such as high and low daily stock prices or upper and lower 95% confidence limits.

twoway rarea plots range as a shaded area.

Also see [G] **graph twoway area** for area plots filled to the axis.

Options

vertical and horizontal specify whether the high and low *y* values are to be presented vertically (the default) or horizontally.

In the default vertical case, *y1var* and *y2var* record the minimum and maximum (or maximum and minimum) *y* values to be graphed against each *xvar* value.

If `horizontal` is specified, the values recorded in *y1var* and *y2var* are plotted in the x direction and *xvar* is treated as the y value.

`sort` specifies that the data be sorted by *xvar* before plotting.

`bstyle(`*areastyle*`)` specifies the look of the shaded area. The options listed below allow you to change each attribute, but `bstyle()` provides the starting point.

You need not specify `bstyle()` just because there is something you want to change. You specify `bstyle()` when another style exists that is exactly what you desire or when another style would allow you to specify fewer changes to obtain what you want.

See [G] *areastyle* for a list of available area styles.

`bcolor(`*colorstyle*`)` specifies a single color to be used both to outline the shape and to fill its interior. See [G] *colorstyle* for a list of color choices.

`bfcolor(`*colorstyle*`)` specifies the color to be used to fill the interior. See [G] *colorstyle* for a list of color choices.

`blstyle(`*linestyle*`)` specifies the overall style of the line used to outline the area, which includes its pattern (solid, dashed, etc.), its thickness, and its color. The other options listed below allow you to change the line's attributes, but `blstyle()` is the starting point. See [G] *linestyle* for a list of choices.

`blcolor(`*colorstyle*`)` specifies the color to be used to outline the area. See [G] *colorstyle* for a list of color choices.

`blwidth(`*linewidthstyle*`)` specifies the thickness of the line to be used to outline the area. See [G] *linewidthstyle* for a list of choices.

`blpattern(`*linepatternstyle*`)` specifies whether the line used to outline the area is solid, dashed, etc. See [G] *linepatternstyle* for a list of pattern choices.

scatter_options refer to any of the options allowed by `scatter` except that *marker_options*, *marker_placement_option*, *marker_label_options*, and *connect_options* are ignored even if specified; see [G] **graph twoway scatter**.

Remarks

Remarks are presented under the headings

> *Typical use*
> *Advanced use*
> *Cautions*

Typical use

We have daily data recording the values for the S&P 500 in 2001:

```
. sysuse sp500, clear
(S&P 500)
```

```
. list date high low close in 1/5
```

	date	high	low	close
1.	02jan2001	1320.28	1276.05	1283.27
2.	03jan2001	1347.76	1274.62	1347.56
3.	04jan2001	1350.24	1329.14	1333.34
4.	05jan2001	1334.77	1294.95	1298.35
5.	08jan2001	1298.35	1276.29	1295.86

We will use the first 57 observations from these data:

```
. twoway rarea high low date in 1/57
```

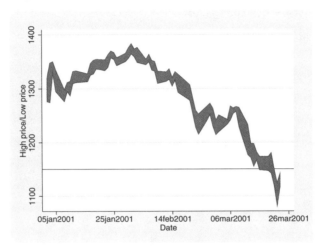

Advanced use

rarea works particularly well when the upper and lower limits are smooth functions and when the area is merely shaded rather than given an eye-catching color:

```
. sysuse auto, clear
(1978 Automobile Data)
. quietly regress mpg weight
. predict hat
. predict s, stdf
. generate low = hat - 1.96*s
. generate hi  = hat + 1.96*s
```

(Continued on next page)

```
. twoway
      rarea low hi weight, sort bcolor(gs14) ||
      scatter  mpg weight
```

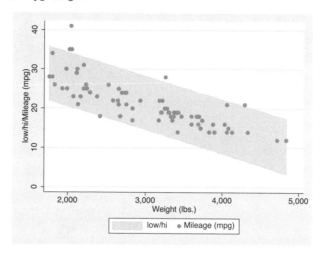

Notice our use of option `bcolor()` to change the color of the shaded area. Also note that we graphed the shaded area first and then the scatter. Typing

```
. twoway scatter ... || rarea ...
```

would not have produced the desired result because the shaded area would have covered up the scatterplot.

Also see [G] **graph twoway lfitci**.

Cautions

Be sure that the data are in the order of *xvar* or specify `rarea`'s `sort` option. If you do neither, you will get something that looks like modern art; see *Cautions* in [G] **graph twoway area** for an example.

Also See

Complementary: [G] **graph twoway area**; [G] **graph twoway rbar**,

[G] **graph twoway rspike**, [G] **graph twoway rcap**,

[G] **graph twoway rcapsym**, [G] **graph twoway rline**,

[G] **graph twoway rconnected**, [G] **graph twoway rscatter**

Title

graph twoway rbar — Range plot with bars

Syntax

twoway rbar *y1var y2var xvar* [if *exp*] [in *range*] [, *rbar_options scatter_options*]

where *rbar_options* are

rbar_options	description
vertical	vertical bars (default)
horizontal	horizontal bars
barwidth(*#*)	width of bar in *xvar* units
mwidth	use msize() rather than barwidth()
msize(*markersizestyle*)	width of bar in relative size units
barlook_options	look of bars

See [G] *markersizestyle* and [G] *barlook_options*.
All options are *rightmost*; see [G] **repeated options**.

and where *scatter_options* refer to any of the options allowed by scatter except that *marker_options*, *marker_placement_option*, *marker_label_options*, and *connect_options* are ignored even if specified; see [G] **graph twoway scatter**.

Description

A range plot has two *y* variables, such as high and low daily stock prices or upper and lower 95% confidence limits.

twoway rbar plots a range using bars to connect the high and low values.

Also see [G] **graph bar** for more traditional bar charts.

Options

vertical and horizontal specify whether the high and low *y* values are to be presented vertically (the default) or horizontally.

In the default vertical case, *y1var* and *y2var* record the minimum and maximum (or maximum and minimum) *y* values to be graphed against each *xvar* value.

If horizontal is specified, the values recorded in *y1var* and *y2var* are plotted in the *x* direction and *xvar* is treated as the *y* value.

barwidth(*#*) specifies the width of the bar in *xvar* units. The default is width(1). When a bar is plotted, it is centered at *x*, so half the width extends below *x* and half above.

314

mwidth and msize(*markersizestyle*) change how the width of the bars is specified. Usually, the width of the bars is determined by the barwidth() option documented below. If mwidth is specified, barwidth() becomes irrelevant and the bar width switches to being determined by msize(). This all has to do with the units in which the width of the bar is specified.

By default, bar widths are specified in the units of *xvar* and, if option barwidth() is not specified, the default width is 1 *xvar* unit.

mwidth() specifies that you wish bar widths to be measured in relative size units; see [G] *relativesize*. When you specify mwidth, the default changes from being 1 *xvar* unit to the default width of a marker symbol.

If you also specify msize(), the width of the bar is modified to be the relative size specified.

barlook_options set the look of the bars. The most important of these options is bcolor(*colorstyle*), which specifies the color of the bars; see [G] *colorstyle* for a list of color choices. See [G] *barlook_options* for information on the other *barlook_options*.

scatter_options refer to any of the options allowed by scatter except that *marker_options*, *marker_placement_option*, *marker_label_options*, and *connect_options* are ignored even if specified; see [G] **graph twoway scatter**.

Remarks

Remarks are presented under the headings

> *Typical use*
> *Advanced use*

Typical use

We have daily data recording the values for the S&P 500 in 2001:

```
. sysuse sp500, clear
(S&P 500)

. list date high low close in 1/5
```

	date	high	low	close
1.	02jan2001	1320.28	1276.05	1283.27
2.	03jan2001	1347.76	1274.62	1347.56
3.	04jan2001	1350.24	1329.14	1333.34
4.	05jan2001	1334.77	1294.95	1298.35
5.	08jan2001	1298.35	1276.29	1295.86

We will use the first 57 observations from these data:

```
. twoway rbar high low date in 1/57, barwidth(.6)
```

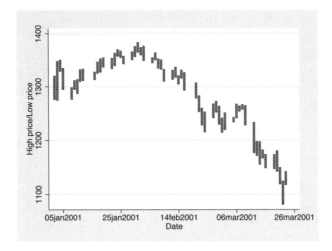

We specified `barwidth(.6)` to reduce the width of the bars. By default, bars are 1 x unit wide (meaning 1 day in our data. That default resulted in the bars touching. `barwidth(.6)` reduced the width of the bars to .6 days.

Advanced use

The useful thing about `twoway rbar` is that it can be combined with other `twoway` plottypes:

```
. twoway rbar high low date, barwidth(.6) bcolor(gs7) ||
      line close date || in 1/57
```

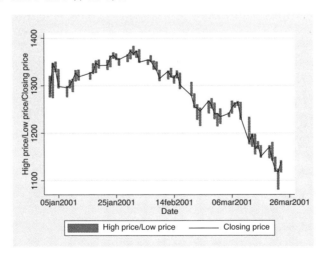

There are two things to note in the example above: our specification of `bcolor(gs7)` and that we specified that the range bars be drawn first, followed by the line. We specified `bcolor(gs7)` to tone down the bars: By default, the bars were too bright, making the line plot of close versus date all but invisible. Concerning the ordering, we typed

```
. twoway rbar high low date, barwidth(.6) bcolor(gs7) ||
       line close date || in 1/57
```

so that the bars would be drawn first and then the line drawn over them. Had we specified

```
. twoway line close date ||
       rbar high low date, barwidth(.6) bcolor(gs7) || in 1/57
```

the bars would have been placed on top of the line and thus would have occulted the line.

Also See

Complementary: [G] **graph twoway bar**; [G] **graph twoway rarea**,
                       [G] **graph twoway rspike**, [G] **graph twoway rcap**,
                       [G] **graph twoway rcapsym**, [G] **graph twoway rline**,
                       [G] **graph twoway rconnected**, [G] **graph twoway rscatter**

Title

> **graph twoway rcap** — Range plot with capped spikes

Syntax

twoway rcap *y1var y2var xvar* [if *exp*] [in *range*] [, *rcap_options scatter_options*]

where *rcap_options* are

rcap_options	description
vertical	vertical spikes (default)
horizontal	horizontal spikes
msize(*markersizestyle*)	width of cap *(sic)*
blstyle(*linestyle*)	overall look of spike
blcolor(*colorstyle*)	spike color
blwidth(*linewidthstyle*)	thickness of spike
blpattern(*linepatternstyle*)	whether spike solid, dashed, etc.

See [G] *markersizestyle*, [G] *linestyle*, [G] *colorstyle*, [G] *linewidthstyle*, and [G] *linepatternstyle*.

All options are *merged-implicit*; see [G] **repeated options**.

and where *scatter_options* refer to any of the options allowed by scatter except that *marker_options*, *marker_placement_option*, *marker_label_options*, and *connect_options* are ignored even if specified; see [G] **graph twoway scatter**.

Description

A range plot has two *y* variables, such as high and low daily stock prices or upper and lower 95% confidence limits.

twoway rcap plots a range using capped spikes (I-beams) to connect the high and low values.

Options

vertical and horizontal specify whether the high and low *y* values are to be presented vertically (the default) or horizontally.

In the default vertical case, *y1var* and *y2var* record the minimum and maximum (or maximum and minimum) *y* values to be graphed against each *xvar* value.

If horizontal is specified, the values recorded in *y1var* and *y2var* are plotted in the *x* direction and *xvar* is treated as the *y* value.

msize(*markersizestyle*) specifies the width of the cap. Option msize() is in fact twoway scatter's *marker_option* that sets the size of the marker symbol, but in this case msymbol() is borrowed to set the cap width. See [G] *markersizestyle* for a list of size choices.

blstyle(*linestyle*) specifies the overall style of the line used to draw the spike, which includes its pattern (solid, dashed, etc.), its thickness, and its color. The other options listed below allow you to change the line's attributes, but blstyle() is the starting point. See [G] *linestyle* for a list of choices.

blcolor(*colorstyle*) specifies the color of the line used to draw the spike. See [G] *colorstyle* for a list of color choices.

blwidth(*linewidthstyle*) specifies the thickness of the line used to draw the spike. See [G] *linewidthstyle* for a list of choices.

blpattern(*linepatternstyle*) specifies whether the line used to draw the spike is solid, dashed, etc. See [G] *linepatternstyle* for a list of pattern choices.

scatter_options refer to any of the options allowed by scatter except that *marker_options*, *marker_placement_option*, *marker_label_options*, and *connect_options* are ignored even if specified; see [G] **graph twoway scatter**.

Remarks

Remarks are presented under the headings

> *Typical use*
> *Advanced use*

Typical use

We have daily data recording the values for the S&P 500 in 2001:

```
. sysuse sp500, clear
(S&P 500)

. list date high low close in 1/5
```

	date	high	low	close
1.	02jan2001	1320.28	1276.05	1283.27
2.	03jan2001	1347.76	1274.62	1347.56
3.	04jan2001	1350.24	1329.14	1333.34
4.	05jan2001	1334.77	1294.95	1298.35
5.	08jan2001	1298.35	1276.29	1295.86

We will use the first 37 observations from these data:

```
. twoway rcap high low date in 1/37
```

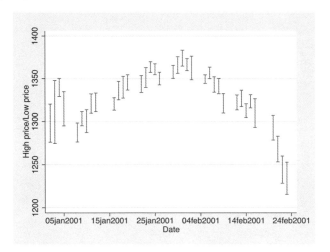

Advanced use

twoway rcap looks best when combined with horizontal lines:

```
. sysuse sp500, clear
(S&P 500)

. generate month = month(date)

. sort month

. by month: egen lo = min(volume)

. by month: egen hi = max(volume)

. format lo hi %10.0gc

. summarize volume
```

Variable	Obs	Mean	Std. Dev.	Min	Max
volume	248	12320.68	2585.929	4103	23308.3

```
. by month: keep if _n==_N
(236 observations deleted)

. twoway rcap lo hi month,
    xlabel(1 "J"  2 "F"  3 "M"  4 "A"  5 "M"  6 "J"
                 7 "J"  8 "A"  9 "S" 10 "O" 11 "N" 12 "D")
    xtitle("Month of 2001")
    ytitle("High and Low Volume")
    yaxis(1 2) ylabel(12321 "12,321 (mean)", axis(2) angle(0))
    ytitle("", axis(2))
    yline(12321, lstyle(foreground))
    msize(*2)
    title("Volume of the S&P 500", margin(b+2.5))
    note("Source:  Yahoo!Finance and Commodity Systems Inc.")
```

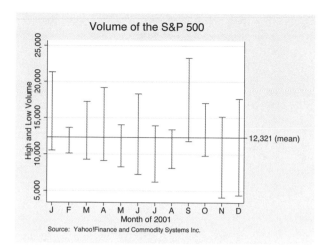

Also See

Complementary: [G] **graph twoway rarea**, [G] **graph twoway rbar**,
[G] **graph twoway rspike**, [G] **graph twoway rcapsym**,
[G] **graph twoway rline**, [G] **graph twoway rconnected**,
[G] **graph twoway rscatter**

Title

> **graph twoway rcapsym** — Range plot with spikes capped with marker symbols

Syntax

twoway rcapsym *y1var y2var xvar* [if *exp*] [in *range*] [, *rcapsym_options*

scatter_options]

where *rcapsym_options* are

rcapsym_options	description
vertical	vertical spikes (default)
horizontal	horizontal spikes
marker_options	look of marker symbols
marker_label_options	labels for markers (of limited use)
blstyle(*linestyle*)	overall look of spike
blcolor(*colorstyle*)	spike color
blwidth(*linewidthstyle*)	thickness of spike
blpattern(*linepatternstyle*)	whether spike solid, dashed, etc.

See [G] *marker_options*, [G] *marker_label_options*, [G] *linestyle*, [G] *colorstyle*, [G] *linewidthstyle*, and [G] *linepatternstyle*.

All options are *merged-implicit*; see [G] **repeated options**.

and where *scatter_options* refer to any of the options allowed by scatter except that *marker_placement_option* and *connect_options* are ignored even if specified; see [G] **graph twoway scatter**.

Description

A range plot has two *y* variables, such as high and low daily stock prices or upper and lower 95% confidence limits.

twoway rcapsym plots a range using spikes capped with marker symbols.

Options

vertical and horizontal specify whether the high and low *y* values are to be presented vertically (the default) or horizontally.

In the default vertical case, *y1var* and *y2var* record the minimum and maximum (or maximum and minimum) *y* values to be graphed against each *xvar* value.

If horizontal is specified, the values recorded in *y1var* and *y2var* are plotted in the *x* direction and *xvar* is treated as the *y* value.

marker_options specify the symbol to be used at the ends of the spikes. The same symbol is used on both ends. Of these options, option msize(*markersize*) specifies the size of the symbol. See [G] ***marker_options*** for a list of symbol choices and [G] ***markersizestyle*** for a list of size choices.

marker_label_options specify if and how the markers are to be labeled. Because the same marker label would be used to label both ends of the spike, these options are of limited use in this case. See [G] ***marker_label_options***.

blstyle(*linestyle*) specifies the overall style of the line used to draw the spike, which includes its pattern (solid, dashed, etc.), its thickness, and its color. The other options listed below allow you to change the line's attributes, but blstyle() is the starting point. See [G] ***linestyle*** for a list of choices.

blcolor(*colorstyle*) specifies the color of the line used to draw the spike. See [G] ***colorstyle*** for a list of color choices.

blwidth(*linewidthstyle*) specifies the thickness of the line used to draw the spike. See [G] ***linewidthstyle*** for a list of choices.

blpattern(*linepatternstyle*) specifies whether the line used to draw the spike is solid, dashed, etc. See [G] ***linepatternstyle*** for a list of pattern choices.

scatter_options refer to any of the options allowed by scatter except that *marker_options*, *marker_placement_option*, *marker_label_options*, and *connect_options* are ignored even if specified; see [G] **graph twoway scatter**.

Remarks

We have daily data recording the values for the S&P 500 in 2001:

```
. sysuse sp500, clear
(S&P 500)

. list date high low close in 1/5
```

	date	high	low	close
1.	02jan2001	1320.28	1276.05	1283.27
2.	03jan2001	1347.76	1274.62	1347.56
3.	04jan2001	1350.24	1329.14	1333.34
4.	05jan2001	1334.77	1294.95	1298.35
5.	08jan2001	1298.35	1276.29	1295.86

We will use the first 37 observations from these data:

(Continued on next page)

. twoway rcapsym high low date in 1/37

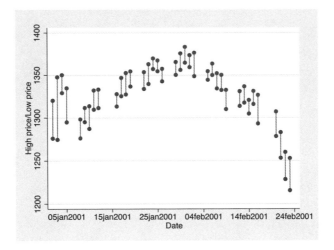

Also See

Complementary: [G] **graph twoway rarea**, [G] **graph twoway rbar**,

[G] **graph twoway rspike**, [G] **graph twoway rcap**,

[G] **graph twoway rline**, [G] **graph twoway rconnected**,

[G] **graph twoway rscatter**

Title

graph twoway rconnected — Range plot with connected lines

Syntax

<u>two</u>way <u>rcon</u>nected *y1var y2var xvar* [if *exp*] [in *range*]

[, *rconnected_options scatter_options*]

where *rconnected_options* are

rconnected_options	description
<u>vert</u>ical	vertical plot (default)
<u>hor</u>izontal	horizontal plot
<u>c</u>onnect(*connectstyle*)	how to connect points
sort[(*varlist*)]	how to order data before connecting
<u>cmis</u>sing(y|n)	whether missing values are ignored
marker_options	look of marker symbols
marker_label_options	labels for markers (of limited use)
<u>bls</u>tyle(*linestyle*)	overall look of line
<u>blc</u>olor(*colorstyle*)	line color
<u>blw</u>idth(*linewidthstyle*)	thickness of line
<u>blp</u>attern(*linepatternstyle*)	whether line solid, dashed, etc.

See [G] *connectstyle*, [G] *marker_options*, [G] *marker_label_options*, [G] *linestyle*, [G] *colorstyle*, [G] *linewidthstyle*, and [G] *linepatternstyle*.

All options are *merged-implicit*; see [G] **repeated options**.

and where *scatter_options* refer to any of the options allowed by scatter except that *marker_placement_option* and *connect_options* (except the *connect_options* listed above) are ignored even if specified; see [G] **graph twoway scatter**.

Description

A range plot has two y variables, such as high and low daily stock prices or upper and lower 95% confidence limits.

twoway rconnected plots the upper and lower ranges using connected lines.

Options

vertical and horizontal specify whether the high and low y values are to be presented vertically (the default) or horizontally.

In the default vertical case, *y1var* and *y2var* record the minimum and maximum (or maximum and minimum) y values to be graphed against each *xvar* value.

If horizontal is specified, the values recorded in *y1var* and *y2var* are plotted in the x direction and *xvar* is treated as the y value.

connect(*connectstyle*), sort, sort(*varlist*), cmissing(y), and cmissing(n) affect how the line between the points is constructed. See [G] **connect_options** for a description of these options, but ignore the descriptions of the options clstyle(), clcolor(), clwidth(), and clpattern(), which also appear there. Those options affect the look of the line and, in the case of graph twoway rconnected, are ignored in favor of the options blstyle(), blcolor(), blwidth(), and blpattern() documented below.

marker_options specify the symbol to be used to mark the points. The same symbol is used for both lines. See [G] **marker_options** for a list of symbol choices.

marker_label_options specify if and how the markers are to be labeled. Because the same marker label would be used to label both lines, these options are of limited use in this case. See [G] **marker_label_options**.

blstyle(*linestyle*) specifies the overall style of the line used to connect the points, which includes its pattern (solid, dashed, etc.), its thickness, and its color. The other options listed below allow you to change the line's attributes, but blstyle() is the starting point. See [G] **linestyle** for a list of choices.

blcolor(*colorstyle*) specifies the color of the line used to connect the points. See [G] **colorstyle** for a list of color choices.

blwidth(*linewidthstyle*) specifies the thickness of the line used to connect the points. See [G] **linewidthstyle** for a list of choices.

blpattern(*linepatternstyle*) specifies whether the line used to connect the points is solid, dashed, etc. See [G] **linepatternstyle** for a list of pattern choices.

scatter_options refer to any of the options allowed by scatter except that *marker_options*, *marker_placement_option*, *marker_label_options*, and the *connect_options* clstyle(), clcolor(), clwidth(), and clpattern() are ignored even if specified; see [G] **graph twoway scatter**.

Remarks

Visually, there is no difference between

> . twoway rconnected *y1var y2var xvar*

and

> . twoway connected *y1var xvar* || connected *y2var xvar*, pstyle(p1)

Note that the two connected lines are presented in the same overall style, meaning symbol selection and color and line color, thickness, and pattern.

Also See

Complementary: [G] **graph twoway rarea**, [G] **graph twoway rbar**,

[G] **graph twoway rspike**, [G] **graph twoway rcap**,

[G] **graph twoway rcapsym**, [G] **graph twoway rline**,

[G] **graph twoway rscatter**

Title

> **graph twoway rline** — Range plot with lines

Syntax

twoway <u>rl</u>ine *y1var y2var xvar* [*if exp*] [*in range*] [, *rline_options scatter_options*]

where *rline_options* are

rline_options	description
<u>v</u>ertical	vertical plot (default)
<u>h</u>orizontal	horizontal plot
<u>c</u>onnect(*connectstyle*)	how to connect points
<u>sort</u> [(*varlist*)]	how to order data before connecting
<u>cmis</u>sing(y|n)	whether missing values are ignored
<u>bls</u>tyle(*linestyle*)	overall look of lines
<u>blc</u>olor(*colorstyle*)	color of lines
<u>blw</u>idth(*linewidthstyle*)	thickness of lines
blpattern(*linepatternstyle*)	whether lines solid, dashed, etc.

See [G] **connectstyle**, [G] **linestyle**, [G] **colorstyle**, [G] **linewidthstyle**, and [G] **linepatternstyle**.

All options are *merged-implicit*; see [G] **repeated options**.

and where *scatter_options* refer to any of the options allowed by scatter except that *marker_placement_option* and *connect_options* (except the *connect_options* listed above) are ignored even if specified; see [G] **graph twoway scatter**.

Description

A range plot has two y variables, such as high and low daily stock prices or upper and lower 95% confidence limits.

twoway rline plots the upper and lower ranges using lines.

Options

vertical and horizontal specify whether the high and low y values are to be presented vertically (the default) or horizontally.

In the default vertical case, *y1var* and *y2var* record the minimum and maximum (or maximum and minimum) y values to be graphed against each *xvar* value.

If `horizontal` is specified, the values recorded in *y1var* and *y2var* are plotted in the x direction and *xvar* is treated as the y value.

`connect(`*connectstyle*`)`, `sort`, `sort(`*varlist*`)`, `cmissing(y)`, and `cmissing(n)` affect how the line between the points is constructed. See [G] ***connect_options*** for a description of these options, but ignore the descriptions of the options `clstyle()`, `clcolor()`, `clwidth()`, and `clpattern()`, which also appear there. Those options affect the look of the line and, in the case of `graph twoway rline`, are ignored in favor of the options `blstyle()`, `blcolor()`, `blwidth()`, and `blpattern()` documented below.

`blstyle(`*linestyle*`)` specifies the overall style of the line used to connect the points, which includes its pattern (solid, dashed, etc.), its thickness, and its color. The other options listed below allow you to change the line's attributes, but `blstyle()` is the starting point. See [G] ***linestyle*** for a list of choices.

`blcolor(`*colorstyle*`)` specifies the color of the line used to connect the points. See [G] ***colorstyle*** for a list of color choices.

`blwidth(`*linewidthstyle*`)` specifies the thickness of the line used to connect the points. See [G] ***linewidthstyle*** for a list of choices.

`blpattern(`*linepatternstyle*`)` specifies whether the line used to connect the points is solid, dashed, etc. See [G] ***linepatternstyle*** for a list of pattern choices.

scatter_options refer to any of the options allowed by `scatter` except that *marker_options*, *marker_placement_option*, *marker_label_options*, and the *connect_options* `clstyle()`, `clcolor()`, `clwidth()`, and `clpattern()` are ignored even if specified; see [G] **graph twoway scatter**.

Remarks

Visually, there is no difference between

```
. twoway rline y1var y2var xvar
```

and

```
. twoway line y1var xvar || line y2var xvar, pstyle(p1)
```

Note that the two lines are presented in the same overall style, meaning color, thickness, and pattern.

Also See

Complementary: [G] **graph twoway rarea**, [G] **graph twoway rbar**,

[G] **graph twoway rspike**, [G] **graph twoway rcap**,

[G] **graph twoway rcapsym**, [G] **graph twoway rconnected**,

[G] **graph twoway rscatter**

Title

---
graph twoway rscatter — Range plot with markers

Syntax

>twoway rscatter *y1var y2var xvar* [if *exp*] [in *range*]
>
>>[, *rscatter_options scatter_options*]

where *rscatter_options* are

rscatter_options	description
<u>vert</u>ical	vertical plot (default)
<u>hor</u>izontal	horizontal plot
marker_options	look of marker symbols
marker_label_options	labels for markers (of limited use)

See [G] ***marker_options*** and [G] ***marker_label_options***.

All options are *merged-implicit*; see [G] **repeated options**.

and where *scatter_options* refer to any of the options allowed by scatter.

Description

A range plot has two *y* variables, such as high and low daily stock prices or upper and lower 95% confidence limits.

twoway rscatter plots the upper and lower ranges as scatters.

Options

vertical and horizontal specify whether the high and low *y* values are to be presented vertically (the default) or horizontally.

In the default vertical case, *y1var* and *y2var* record the minimum and maximum (or maximum and minimum) *y* values to be graphed against each *xvar* value.

If horizontal is specified, the values recorded in *y1var* and *y2var* are plotted in the *x* direction and *xvar* is treated as the *y* value.

marker_options specify the symbol to be used to mark the points. The same symbol is used for both lines. See [G] ***marker_options*** for a list of symbol choices.

marker_label_options specify if and how the markers are to be labeled. Because the same marker label would be used to label both lines, these options are of limited use in this case. See [G] ***marker_label_options***.

scatter_options refer to any of the options allowed by scatter; see [G] **graph twoway scatter**.

Remarks

Visually, there is no difference between

. twoway rscatter *y1var y2var xvar*

and

. twoway scatter *y1var xvar* || scatter *y2var xvar*, pstyle(p1)

Note that the two scatters are presented in the same overall style, meaning that the markers (symbol shape and color) are the same.

Also See

Complementary: [G] **graph twoway rarea**, [G] **graph twoway rbar**,
[G] **graph twoway rspike**, [G] **graph twoway rcap**,
[G] **graph twoway rcapsym**, [G] **graph twoway rline**,
[G] **graph twoway rconnected**

Title

graph twoway rspike — Range plot with spikes

Syntax

twoway rspike *y1var y2var xvar* [if *exp*] [in *range*] [, *rspike_options*

    scatter_options]

where *rspike_options* are

rspike_options	description
<u>vertical</u>	vertical spikes (default)
<u>horizontal</u>	horizontal spikes
<u>bl</u>style(*linestyle*)	overall look of spike
<u>blc</u>olor(*colorstyle*)	spike color
<u>blw</u>idth(*linewidthstyle*)	thickness of spike
<u>blp</u>attern(*linepatternstyle*)	whether spike solid, dashed, etc.

See [G] *linestyle*, [G] *colorstyle*, [G] *linewidthstyle*, and [G] *linepatternstyle*.

All options are *merged-implicit*; see [G] **repeated options**.

and where *scatter_options* refer to any of the options allowed by scatter except that
marker_options, *marker_placement_option*, *marker_label_options*, and *connect_options* are
ignored even if specified; see [G] **graph twoway scatter**.

Description

A range plot has two y variables, such as high and low daily stock price or upper and lower 95%
confidence limits.

twoway rspike plots a range using spikes to connect the high and low values.

Also see [G] **graph twoway spike** for another style of spike chart.

Options

vertical and horizontal specify whether the high and low y values are to be presented vertically
(the default) or horizontally.

In the default vertical case, *y1var* and *y2var* record the minimum and maximum (or maximum
and minimum) y values to be graphed against each *xvar* value.

If horizontal is specified, the values recorded in *y1var* and *y2var* are plotted in the x direction
and *xvar* is treated as the y value.

blstyle(*linestyle*) specifies the overall style of the line used to draw the spike, which includes its pattern (solid, dashed, etc.), its thickness, and its color. The other options listed below allow you to change the line's attributes, but blstyle() is the starting point. See [G] *linestyle* for a list of choices.

blcolor(*colorstyle*) specifies the color of the line used to draw the spike. See [G] *colorstyle* for a list of color choices.

blwidth(*linewidthstyle*) specifies the thickness of the line used to draw the spike.
See [G] *linewidthstyle* for a list of choices.

blpattern(*linepatternstyle*) specifies whether the line used to draw the spike is solid, dashed, etc.
See [G] *linepatternstyle* for a list of pattern choices.

scatter_options refer to any of the options allowed by scatter except that *marker_options*, *marker_placement_option*, *marker_label_options*, and *connect_options* are ignored even if specified; see [G] **graph twoway scatter**.

Remarks

Remarks are presented under the headings

> Typical use
> Advanced use
> Advanced use 2

Typical use

We have daily data recording the values for the S&P 500 in 2001:

```
. sysuse sp500, clear
(S&P 500)

. list date high low close in 1/5
```

	date	high	low	close
1.	02jan2001	1320.28	1276.05	1283.27
2.	03jan2001	1347.76	1274.62	1347.56
3.	04jan2001	1350.24	1329.14	1333.34
4.	05jan2001	1334.77	1294.95	1298.35
5.	08jan2001	1298.35	1276.29	1295.86

We will use the first 57 observations from these data:

(Continued on next page)

```
. twoway rspike high low date in 1/57
```

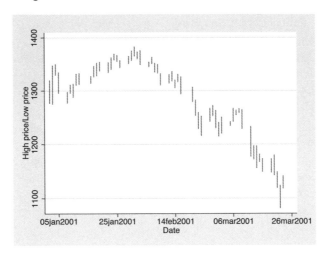

Advanced use

twoway rspike can be usefully combined with other twoway plottypes:

```
. twoway rspike high low date, bcolor(gs11) ||
        line close date || in 1/57
```

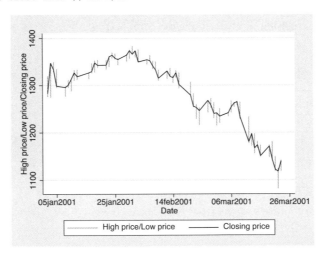

We specified bcolor(gs11) to tone down the spikes and so give the line plot more prominence.

Advanced use 2

A popular financial graph is

```
. sysuse sp500, clear
(S&P 500)
. replace volume = volume/1000
```

```
. twoway
        rspike hi low date ||
        line   close date ||
        bar    volume date, barw(.25) yaxis(2) ||
   in 1/57
  , ysca(axis(1) r(900 1400))
    ysca(axis(2) r( 9   45))
    ytitle("                            Price -- High, Low, Close")
    ytitle(" Volume (millions)", axis(2) astext just(left))
    legend(off)
    subtitle("S&P 500", margin(b+2.5))
    note("Source:  Yahoo!Finance and Commodity Systems, Inc.")
```

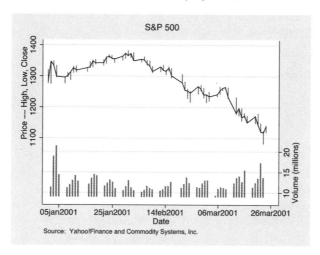

Also See

Complementary: [G] **graph twoway spike**; [G] **graph twoway rarea**,

[G] **graph twoway rbar**, [G] **graph twoway rcap**,

[G] **graph twoway rcapsym**, [G] **graph twoway rline**,

[G] **graph twoway rconnected** [G] **graph twoway rscatter**

Title

> **graph twoway scatter** — Twoway scatterplots

Syntax

$$\big[\underline{\text{two}}\text{way}\big]\ \underline{\text{sc}}\text{atter } \textit{varlist } \big[\textit{weight}\big]\ \big[\underline{\text{if}}\ \textit{exp}\big]\ \big[\underline{\text{in}}\ \textit{range}\big]\ \big[\,,$$

> *marker_options*
> *marker_placement_option*
> *marker_label_options*
> *connect_options*
> *axis_selection_options*
> *composite_style_option*
> *twoway_options* $\big]$

where *varlist* has the interpretation

$$y_1\ \big[\,y_2\big[\,\ldots\,\big]\,\big]\ x$$

and where *marker_options* are

msymbol(*symbolstylelist*)	shape of marker
mcolor(*colorstylelist*)	color of marker, inside and out
msize(*markersizestylelist*)	size of marker
mfcolor(*colorstylelist*)	inside or "fill" color
mlcolor(*colorstylelist*)	color of outline
mlwidth(*linewidthstylelist*)	thickness of outline
mlpattern(*linepatternstylelist*)	outline solid, dashed, etc.
mlstyle(*linestylelist*)	overall style of outline
mstyle(*markerstylelist*)	overall style of marker

See [G] **marker_options**.

and *marker_placement_option* is

jitter(*relativesizelist*)	perturb location of point

See *Jittered markers* under *Remarks* below.

(Continued on next page)

and *marker_label_options* are

mlabel(*varlist*)	specify marker variables
mlabposition(*clockposlist*)	where to locate label
mlabvposition(*varname*)	where to locate label 2
mlabgap(*relativesizelist*)	gap between marker and label
mlabangle(*anglestylelist*)	angle of label
mlabsize(*textsizestylelist*)	size of label
mlabcolor(*colorstylelist*)	color of label
mlabtextstyle(*textstylelist*)	overall style of text
mlabstyle(*markerlabelstylelist*)	overall style of label

See [G] ***marker_label_options***.

and *connect_options* are

connect(*connectstylelist*)	whether and how to connect points
sort $\lceil$ (*varlist*) $\rceil$	how to order data before connecting
cmissing({ y | n } ...)	whether missing values are ignored
clpattern(*linepatternstylelist*)	whether line solid, dashed, etc.
clwidth(*linewidthstylelist*)	thickness of line
clcolor(*colorstylelist*)	color of line
clstyle(*linestylelist*)	overall style of line

See [G] ***connect_options***.

and where *axis_selection_options* are

yaxis(# $\lceil$ # ... $\rceil$)	which y axis to use
xaxis(# $\lceil$ # ... $\rceil$)	which x axis to use

See [G] ***axis_selection_options***.

and where *composite_style_option* is

pstyle(*pstylelist*)	all the ...style() options above

See *Styles and composite styles* under *Remarks* below.

and where *twoway_options* include

added_line_options	draw lines at specified y or x values
added_text_option	display text at specified (y,x value)
axis_options	labels, ticks, grids, log scales
title_options	titles, subtitles, notes, captions
legend_option	legend explaining what means what
scale(#)	resize text and markers
region_options	outlining, shading, aspect ratio
scheme(*schemename*)	overall look
by(*varlist*, ...)	repeat for subgroups
nodraw	suppress display of graph
name(*name*, ...)	specify name for graph
saving(*filename*, ...)	save graph in file
advanced_options	difficult to explain

See [G] *twoway_options*.

aweights, fweights, and pweights are allowed; see [U] **14.1.6 weight**.

Description

scatter draws scatterplots and is the mother of all the twoway plottypes, such as line, lfit, etc.

scatter is both a command and it is a *plottype* as defined in [G] **graph twoway**. Thus, the syntax for scatter is

```
. graph twoway scatter ...
. twoway scatter ...
. scatter ...
```

Being a plottype, scatter may be combined with other plottypes in the twoway family, as in,

```
. twoway (scatter ...) (line ...) (lfit ...) ...
```

which can equivalently be written

```
. scatter ... || line ... || lfit ... || ...
```

Options

marker_options specify how the points on the graph are to be designated. Markers are the ink used to mark where points are on a plot. Markers have shape, color, and size, and other characteristics. See [G] *marker_options* for a description of markers and the options that specify them.

msymbol(O D S T + X o d s t smplus x) is the default. msymbol(i) will suppress the appearance of the marker altogether.

jitter(*relativesize*) adds spherical random noise to the data before plotting. This is useful when plotting data which otherwise would result in points plotted on top of each other. See *Jittered markers* under *Remarks* below.

Commonly specified are jitter(5) or jitter(6); jitter(0) is the default. See [G] *relativesize* for a description of relative sizes.

marker_label_options specify labels to appear next to or in place of the markers. For instance, if you were plotting country data, marker labels would allow you to have "Argentina", "Bolivia", ..., appear next to each point and, with a small amount of data, that might be desirable. See [G] *marker_label_options* for a description of marker labels and the options that control them.

By default, no marker labels are displayed. If you wish to display marker labels in place of the markers, specify mlabposition(0) and msymbol(i).

connect_options specify how the points are to be connected. The default is not to connect the points.

connect() specifies whether points are to be connected and, if so, how the line connecting them is to be shaped. The line between each pair of points can connect them directly or in stairstep fashion.

sort specifies that the data should be sorted by the x variable before the points are connected. Unless you are after a special effect or your data are already sorted, do not forget to specify this option. If you are after a special effect, and if the data are not already sorted, you can specify sort(*varlist*) to specify exactly how the data should be sorted. Understand that specifying sort or sort(*varlist*) when it is not necessary will slow Stata down a little. You must specify sort if you wish to connect points, and you must specify the *twoway_option* by() with total.

cmissing(y) and cmissing(n) specify whether missing values are ignored when points are connected; whether the line should have a break in it. The default is cmissing(y), meaning that there will be no breaks.

clpattern() specifies how the style of the line is to be drawn: solid, dashed, etc.

clwidth() specifies the width of the line.

clcolor() specifies the color of the line.

clstyle() specifies the overall style of the line.

See [G] *connect_options* for more information on these and related options. See [G] **lines** for an overview of lines.

axis_selection_options are for use when you have multiple x or y axes.
See [G] *axis_selection_options* for more information.

pstyle(*pstyle*) specifies the overall style of the plot and is a composite of mstyle(), mlabstyle(), clstyle(), connect(), and cmissing(). The default is pstyle(p1) for the first plot, pstyle(p2) for the second, and so on. See *Styles and composite styles* under *Remarks* below.

twoway_options include

added_line_options, which specify that horizontal or vertical lines be drawn on the graph; see [G] *added_line_options*. If your interest is in drawing grid lines through the plot region, see *axis_options* below.

added_text_option, which specifies text to be displayed on the graph (inside the plot region); see [G] *added_text_option*.

axis_options, which allows you to specify labels, ticks, and grids. These options also allow you to obtain logarithmic scales; see [G] *axis_options*.

title_options, which allows you to specify titles, subtitles, notes, and captions; see [G] *title_options*.

legend_option, which allows specifying the legend explaining the symbols and line styles used; see [G] *legend_option*.

scale(*#*), which makes all the text and markers on a graph larger or smaller (scale(1) means no change); see [G] *scale_option*.

region_options, which allows you to control the aspect ratio and to specify that the graph be outlined, or given a background shading; see [G] *region_options*.

scheme(*schemename*), which specifies the overall look of the graph; see [G] *scheme_option*.

by(*varlist*, ...), which allows drawing multiple graphs for each subgroup of the data; see [G] *by_option*.

nodraw, which prevents the graph from being displayed; see [G] *nodraw_option*.

name(*name*), which allows you to save the graph in memory under a name different from Graph; see [G] *name_option*.

saving(*filename*[, asis replace]), which allows you to save the graph to disk; see [G] *saving_option*.

other options, which allows you to suppress the display of the graph, to name the graph, etc.

See [G] *twoway_options*.

Remarks

Remarks are presented under the headings

> *Typical use*
> *Scatter syntax*
> *The overall look for the graph*
> *The size and aspect ratio of the graph*
> *Titles*
> *Axis titles*
> *Axis labels and ticking*
> *Grid lines*
> *Added lines*
> *Axis range*
> *Log scales*
> *Multiple axes*
> *Markers*
> *Weighted markers*
> *Jittered markers*
> *Connected lines*
> *Graphs by groups*
> *Saving graphs*
> *Appendix: Styles and composite styles*

Typical use

The scatter plottype by default individually marks the location of each point:

```
. sysuse uslifeexp2, clear
(U.S. life expectancy, 1900-1940)
. scatter le year
```

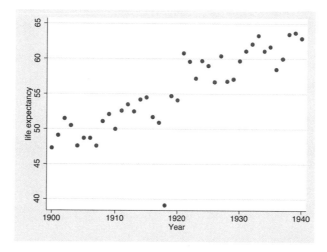

With the specification of options, you can produce the same effect as `twoway connected`,

```
. scatter le year, connect(l)
```

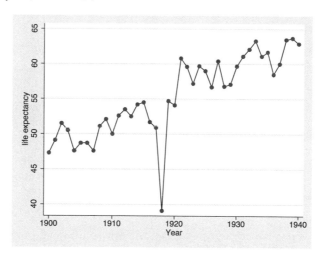

or `twoway line`:

. scatter le year, connect(1) msymbol(i)

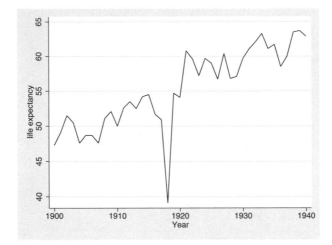

In fact, all the other twoway plottypes eventually work their way back to executing scatter. scatter literally is the mother of all twoway graphs in Stata.

Scatter syntax

See [G] **graph twoway** for an overview of graph twoway syntax. In the particular case of graph twoway scatter, the only thing to know is that if more than two variables are specified, all but the last are given the interpretation of being *y* variables. For example,

. scatter *y1var y2var xvar*

would plot *y1var* versus *xvar* and overlay that with a plot of *y2var* versus *xvar*, and so is the same as typing

. scatter *y1var xvar* || scatter *y2var xvar*

If, using the multiple-variable syntax, you specify scatter-level options (i.e., all options except *twoway_options* as defined in the syntax diagram), you specify arguments for *y1var, y2var, ...*, separated by spaces. That is, you might type

. scatter *y1var y2var xvar*, ms(O i) c(. l)

ms() and c() are abbreviations for the msymbol() and connect() options. In any case, the results from the above are the same as if you typed

. scatter *y1var xvar*, ms(O) c(.) || scatter *y2var xvar*, ms(i) c(l)

There need not be a one-to-one correspondence between options and *y* variables when you use the multiple-variable syntax. If you typed

. scatter *y1var y2var xvar*, ms(O) c(l)

then options ms() and c() will have default values for the second scatter, and if you typed

. scatter *y1var y2var xvar*, ms(O S i) c(l l l)

the extra options for the nonexistent third variable would be ignored.

If you wish to specify the default for one of the y variables, you may specify period (.):

. scatter *y1var y2var xvar*, ms(. O) c(. l)

There are other shorthands available to make specifying multiple arguments easier; see [G] *stylelists*.

Since multiple variables are interpreted as multiple *y* variables, to produce graphs containing multiple *x* variables, you must chain together separate `scatter` commands:

```
. scatter yvar x1var, ...  ||  . scatter yvar x2var, ...
```

The overall look for the graph

The overall look of the graph is mightily affected by the scheme, and note that there is a `scheme()` option that will allow you to specify which scheme to use. We showed earlier the results of `scatter le year`. Here is the same graph repeated using the `economist` scheme:

```
. sysuse uslifeexp2, clear
(U.S. life expectancy, 1900-1940)

. scatter le year,
      title("Scatter plot")
      subtitle("Life expectancy at birth, U.S.")
      note("1")
      caption("Source: National Vital Statistics Report,
      Vol. 50 No. 6")
      scheme(economist)
```

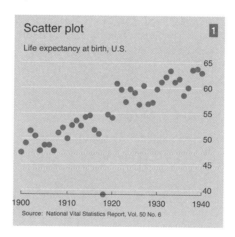

See [G] **schemes**.

The size and aspect ratio of the graph

The size and aspect ratio of the graph are controlled by the *region_options* `ysize(#)` and `xsize(#)`, which specify the height and width in inches of the graph. For instance,

```
. scatter yvar xvar, xsize(4) ysize(4)
```

would produce a 4 × 4 inch square graph. See [G] *region_options*.

Titles

By default, no titles appear on the graph, but the *title_options* title(), subtitle(), note(), caption(), and legend() allow you to specify the titles that you wish to appear, as well as to control their position and size. For instance,

. scatter *yvar xvar*, title("My title")

would draw the graph and include the title "My title" (without the quotes) at the top. Multiple-line titles are allowed. Typing

. scatter *yvar xvar*, title("My title" "Second line")

would create a two-line title. The above, however, would probably look better as a title followed by a subtitle:

. scatter *yvar xvar*, title("My title") subtitle("Second line")

In any case, see [G] *title_options*.

Axis titles

Titles do, by default, appear on the y and x axes. The axes are titled with the variable names being plotted or, if the variables have variable labels, with their variable labels. The *axis_title_options* ytitle() and xtitle() allow you to override that. If you specify

. scatter *yvar xvar*, ytitle("")

the title on the y axis would disappear. If you specify

. scatter *yvar xvar*, ytitle("Rate of change")

the y-axis title would become "Rate of change". As with all titles, multiple-line titles are allowed:

. scatter *yvar xvar*, ytitle("Time to event" "Rate of change")

See [G] *axis_title_options*.

Axis labels and ticking

By default, approximately five major ticks and labels are placed on each axis. The *axis_label_options* ylabel() and xlabel() allow you to control that. Typing

. scatter *yvar xvar*, ylabel(#10)

would put approximately ten labels and ticks on the y axis. Typing

. scatter *yvar xvar*, ylabel(0(1)9)

would put exactly ten labels and would put them at the values 0, 1, ..., 9.

ylabel() and xlabel() have other features, and options are also provided for minor labels and minor ticks; see [G] *axis_label_options*.

Grid lines

Assuming you use a member of the s2 family of schemes—see [G] **schemes**—grid lines are included in y but not x, by default. You can specify option xlabel(,grid) to add x grid lines, and you can specify ylabel(,nogrid) to suppress y grid lines.

Grid lines are considered an extension of ticks and are specified as suboptions inside the *axis_label_options* ylabel() and xlabel(). For instance,

```
. sysuse auto, clear
(1978 Automobile Data)

. scatter mpg weight, xlabel(,grid)
```

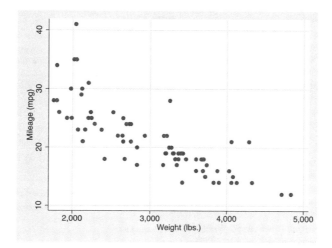

In the above example, the grid lines are placed at the same values as the default ticks and labels, but you can control that, too. See [G] *axis_label_options*.

Added lines

Lines may be added to the graph for emphasis using the *added_line_options* yline() and xline(); see [G] *added_line_options*.

Axis range

The extent or range of an axis is set according to all the things that appear on it—the data being plotted and the values on the axis being labeled or ticked. In the graph that just appeared above,

```
. sysuse auto, clear
(1978 Automobile Data)
```

(Continued on next page)

. scatter mpg weight

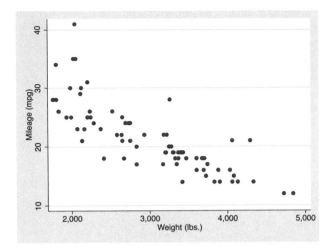

variable mpg varies between 12 and 41 and yet the y axis extends from 10 to 41. The axis was extended to include $10 < 12$ because the value 10 was labeled. Variable weight varies between 1,760 and 4,840; the x axis extends from 1,760 to 5,000. This axis was extended to include $5,000 > 4,840$ because the value 5,000 was labeled.

You can prevent axes from being extended by specifying the ylabel(minmax) and xlabel(minmax) options. minmax specifies that only the minimum and maximum are to be labeled:

. scatter mpg weight, ylabel(minmax) xlabel(minmax)

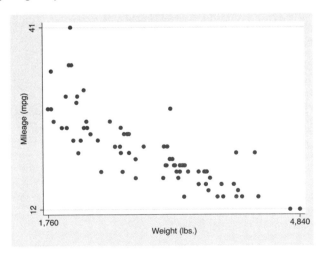

In other cases, you may wish to widen the range of an axis. This you can do by specifying the range() descriptor of the *axis_scale_options* yscale() or xscale(). For instance,

. scatter mpg weight, xscale(range(1000 5000))

would widen the x axis to include 1,000 to 5,000. We typed out the name of the option, but most people would type

. scatter mpg weight, xscale(r(1000 5000))

`range()` can widen, but never narrow, the extent of an axis. Typing

> . scatter mpg weight, xscale(r(1000 4000))

would not omit cars with `weight`> 4000 from the plot. If that is your desire, type

> . scatter mpg weight if weight<=4000

See [G] *axis_scale_options* for more information on `range()`, `yscale()`, and `xscale()`; see [G] *axis_label_options* for more information on `ylabel(minmax)` and `xlabel(minmax)`.

Log scales

By default, arithmetic scales for the axes are used. Log scales can be obtained by specifying the `log` suboption of `yscale()` and `xscale()`. For instance,

> . sysuse lifeexp, clear
> (Life expectancy, 1998)
> . scatter lexp gnppc, xscale(log) xlab(,g)

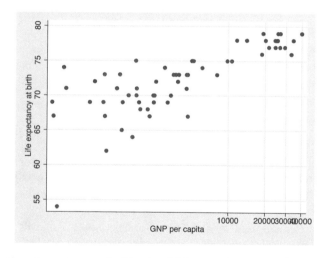

The important option above is `xscale(log)`, which caused `gnppc` to be presented on a log scale.

We included `xlab(,g)` (abbreviated forms of `xlabel(,grid)`) to obtain x grid lines. Note that the values 30,000 and 40,000 are overprinted. We could improve the graph by typing

> . generate gnp000 = gnppc/1000
> . label var gnp000 "GNP per capita, thousands of dollars"

(Continued on next page)

. scatter lexp gnp000, xsca(log) xlab(.5 2.5 10(10)40, grid)

See [G] *axis_options*.

Multiple axes

Graphs may have more than one y and more than one x axis. There are two reasons to do this: you include an extra axis so that you have an extra place to label special values or so that you may plot multiple variables on different scales. In either case, specify the yaxis() or xaxis() options. See [G] *axis_selection_options*.

Markers

Markers are the ink used to mark where points are on the plot. Many people think of markers in terms of their shape (circles, diamonds, etc.), but there are other properties including, most importantly, their color and size. The shape of the marker is specified by the msymbol() option, its color by the mcolor() option, and its size by the msize() option.

By default, solid circles are used for the first y variable, solid diamonds for the second, solid squares for the third, and the list continues; see *marker_options* under *Options* for the remaining details, if you care. In any case, when you type

. scatter *yvar xvar*

results are as if you typed

. scatter *yvar xvar*, msymbol(O)

You can vary the symbol used by specifying other msymbol() arguments. Similarly, you can vary the color and size of the symbol by specifying the mcolor() and msize() options. See [G] *marker_options*.

In addition to the markers themselves, you can request that the individual points be labeled. These marker labels are numbers or text that appear beside the marker symbol—or in place of it—to identify the points. See [G] *marker_label_options*.

Weighted markers

If weights are specified—see [U] **14.1.6 weight**—the size of the marker is scaled according to the size of the weights. aweights, fweights, and pweights are allowed and all are treated the same; iweights are not allowed because scatter would not know what to do with negative values. Weights affect the size of the marker and nothing else about the plot.

Below, we use U.S. state-averaged data to graph the divorce rate in a state versus the state's median age. We scale the symbols to be proportional to the population size:

```
. sysuse census, clear
(1980 Census data by state)

. generate drate = divorce / pop18p
. label var drate "Divorce rate"
. scatter drate medage [w=pop18p] if state!="Nevada", msymbol(Oh)
    note("Stata data excluding Nevada"
        "Area of symbol proportional to state's population aged 18+")
```

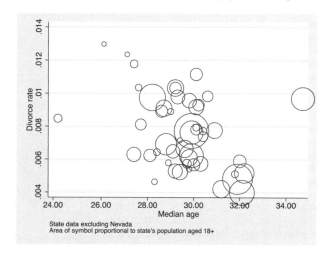

Note our use of the msymbol(Oh) option. Hollow scaled markers look much better than solid ones.

scatter scales the symbols so that the sizes are a fair representation when the weights represent population weights. To wit: If all the weights except one are 1,000 and the exception is 999, the symbols will all be of almost equal size. The weight 999 observation will not be a dot and the weight 1,000 observation giant circles as would be the result if the exception had weight 1.

When weights are specified, option msize() (which also affects the size of the marker), if specified, is ignored.

Jittered markers

scatter will add spherical random noise to your data before plotting if you specify jitter(#), where # represents the size of the noise as a percent of the graphical area. This can be useful for creating graphs of categorical data when, were the data not jittered, many of the points would be on top of each other, making it impossible to tell whether the plotted point represented one or 1,000 observations.

For instance, in a variation on `auto.dta` used below, `mpg` is recorded in units of 5 mpg and `weight` is recorded in units of 500 pounds. A standard scatter has considerable overprinting:

```
. sysuse autornd, clear
(1978 Automobile Data)

. scatter mpg weight
```

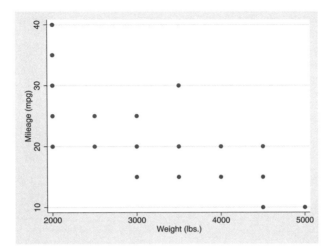

There are 74 points in the graph, even though it appears because of overprinting as if there are only 19. Jittering solves that problem:

```
. scatter mpg weight, jitter(7)
```

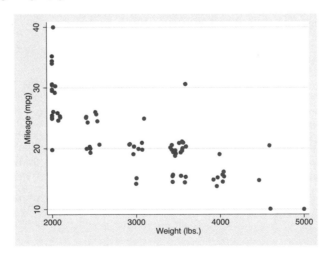

Connected lines

The `connect()` option allows you to connect the points of a graph. The default is not to connect the points.

If you want connected points, you probably want to specify `connect(1)`—which is usually abbreviated `c(1)`. The `1` means that the points are to be connected with straight lines. There are other ways points can be connected (such as a stairstep fashion), but usually `c(l)` is the right choice. The command

> `. scatter` *yvar xvar*`, c(l)`

will plot *yvar* versus *xvar*, marking the points in the usual way, and drawing straight lines between the points. It is common also to specify the `sort` option,

> `. scatter` *yvar xvar*`, c(l) sort`

because otherwise points are connected in the order of the data. If the data are already in the order of *xvar*, then the `sort` is unnecessary. You can also omit the `sort` when creating special effects.

`connect()` is often specified with the `msymbol(i)` option to suppress the display of the individual points:

> `. scatter` *yvar xvar*`, c(l) sort m(i)`

See [G] *connect_options*.

Graphs by groups

Option `by()` specifies that graphs are to be drawn separately for each of the different groups, and then the results arrayed into a single display. Below we use country data and group the results by region of the world:

```
. sysuse lifeexp, clear
(Life expectancy, 1998)
. scatter lexp gnppc, by(region)
```

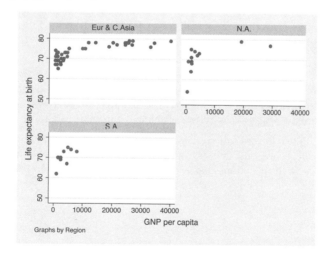

Variable `region` is a numeric variable taking on values 1, 2, and 3. Separate graphs were drawn for each value of region. The graphs were titled "Eur & C. Asia", "N.A.", and "S.A." because numeric variable `region` had been assigned a value label, but results would have been the same had variable `region` been a string directly containing "Eur & C. Asia", "N.A.", and "S.A.".

See [G] *by_option* for more information on this useful option.

Saving graphs

To save a graph to disk for later printing or reviewing, include the `saving()` option,

    . scatter ..., ... saving(*filename*)

or use the `graph save` command afterwards:

    . scatter ...
    . graph save *filename*

See [G] *saving_option* and [G] **graph save**. Also see [G] **gph files** for information on how files such as *filename*.gph can be put to subsequent use.

Appendix: Styles and composite styles

There are a lot of options that end in the word style including, to name a few, `mstyle()`, `mlabstyle()`, and `clstyle()`. Option `mstyle()`, for instance, is described as setting the "overall look" of a marker. What does that mean?

How something looks—a marker, a marker label, a line—is specified by lots of detail options. In the case of markers, option `msymbol()` specifies its shape, `mcolor()` specifies its color, `msize()` specifies its size, and so on.

A *style* specifies a composite of related option settings. Were you to type option `mstyle(p1)`, you would be specifying a whole set of values for `msymbol()`, `mcolor()`, `msize()`, and all the other m*() options. `p1` is called the name of a style, and `p1` contains the settings.

Concerning `mstyle()`, and all the other options ending in the word style, throughout this manual you will read statements such as

Option *whateverstyle()* specifies the overall look of *whatever*, such as its *(insert list here)*. The other options allow you to change the attributes of a *whatever*, but *whateverstyle()* is the starting point.

You need not specify *whateverstyle()* just because there is something you want to change about the look of a *whatever* and, in fact, most people seldom specify the *whateverstyle()* option. You specify *whateverstyle()* when another style exists that is exactly what you desire or when another style would allow you to specify fewer changes to obtain what you want.

Styles actually come in two flavors called *composite styles* and *detail styles*, and the above statement applies only to composite styles and appears only in manual entries concerning composite styles. Composite styles are specified in options that end in the word style. The following are examples of composite styles,

    mstyle(*symbolstyle*)
    mlstyle(*linestyle*)
    mlabstyle(*markerlabelstyle*)
    clstyle(*linestyle*)
    pstyle(*pstyle*)

and the following are examples of detail styles:

    mcolor(*colorstyle*)
    mlwidth(*linewidthstyle*)
    mlabsize(*textsizestyle*)
    clpattern(*linepatternstyle*)

In the above examples, distinguish carefully between option names such as mcolor() and option arguments such as *colorstyle*. *colorstyle* is an example of a detail style because it appears in the option mcolor(), and the option name does not end in the word style.

Detail styles specify precisely how an attribute of something looks and composites specify an overall look in terms of detail-style values. Composite styles are invariably described as specifying the "overall look".

Composite styles sometimes contain other composite styles as members. For instance, when you specify the mstyle() option—which specifies the "overall look" of markers—you are also specifying an mlstyle()—which specifies the "overall look" of the lines that outline the shape of the markers. That does not mean you cannot specify the mlstyle() option, too. It just means that specifying mstyle() implies an mlstyle(). The order in which you specify the options does not matter. You can type

    . scatter ..., ... mstyle(...) ... mlstyle(...) ...

or

    . scatter ..., ... mlstyle(...) ... mstyle(...) ...

and, either way, mstyle() will be set as you specify, and then mlstyle() will be reset as you wish. The same applies for the mixing of composite-style and detail-style options. Option mstyle() implies an mcolor() value. Even so, you may type

    . scatter ..., ... mstyle(...) ... mcolor(...) ...

or

    . scatter ..., ... mcolor(...) ... mstyle(...) ...

and the outcome will be the same.

The grandest composite style of them all is pstyle(*pstyle*). It contains all the other composite styles and scatter (twoway, in fact) makes great use of this grand style. When you type

    . scatter *y1var y2var xvar*, ...

results are as if you typed

    . scatter *y1var y2var xvar*, pstyle(p1 p2) ...

That is, *y1var* versus *xvar* is plotted using pstyle(p1) and *y2var* versus *xvar* is plotted using pstyle(p2). It is the pstyle(p1) that sets all the defaults—which marker symbols are used, what color they are, etc.

The same applies if you type

    . scatter *y1var xvar*, ... || scatter *y2var xvar*, ...

y1var versus *xvar* is plotted using pstyle(p1) and *y2var* versus *xvar* is plotted using pstyle(p2), just as if you had typed

    . scatter *y1var xvar*, pstyle(p1) ... || scatter *y2var xvar*, pstyle(p2) ...

The same applies if you mix scatter with other plottypes:

```
. scatter y1var xvar, ... || line y2var xvar, ...
```

is equivalent to

```
. scatter y1var xvar, pstyle(p1) ... || line y2var xvar, pstyle(p2) ...
```

and

```
. twoway (..., ...) (..., ...), ...
```

is equivalent to

```
. twoway (..., pstyle(p1) ...) (..., pstyle(p2) ...), ...
```

which is why we said that it is `twoway`, and not just `scatter`, that exploits `scheme()`.

You can put this to use. Pretend that you have a dataset on husbands and wives and it contains the variables

`hinc`	husband's income
`winc`	wife's income
`hed`	husband's education
`wed`	wife's education

You wish to draw a graph of income versus education, drawing no distinctions between husbands and wives. You type

```
. scatter hinc hed || scatter winc wed
```

You intend to treat husbands and wives the same in the graph but, in the above, they are treated differently because `msymbol(O)` will be used to mark the points of `hinc` versus `hed` and `msymbol(D)` will be used to designate `winc` versus `wed`. The color of the symbols will be different, too.

You could address that problem in many different ways. You could specify the `msymbol()` and `mcolor()` options along with whatever other detail options are necessary to make the two scatters appear the same. Being knowledgeable, you realize you do not have to do that. There is, you know, a composite style that specifies this. So you get out your manuals, flip through, and discover that the relevant composite style for the marker symbols is `mstyle()`.

Easiest of all, however, would be to remember that `pstyle()` contains all the other styles. Rather than resetting `mstyle()`, just reset `pstyle()`, and whatever needs to be set to make the two plots the same will be set. Type

```
. scatter hinc hed || scatter winc wed, pstyle(p1)
```

or, if you prefer,

```
. scatter hinc hed, pstyle(p1) || scatter winc wed, pstyle(p1)
```

You do not need to specify `pstyle(p1)` for the first plot, however, because that is the default.

As another example, you have a dataset containing

`mpg`	Mileage ratings of cars
`weight`	Each car's weight
`prediction`	A predicted mileage rating based on weight

You wish to draw the graph

```
. scatter mpg weight || line prediction weight
```

but you wish the appearance of the line to "match" that of the markers used to plot mpg versus weight. You could go digging to find out which option controlled the line style and color, and then dig some more to figure out which line style and color goes with the markers used in the first plot, but much easier is simply to type

```
. scatter mpg weight || line prediction weight, pstyle(p1)
```

Also See

Complementary: [G] **graph twoway**; [G] *marker_options*, [G] *marker_label_options*,

[G] *connect_options*, [G] *axis_selection_options*, [G] *twoway_options*

Title

> **graph twoway scatteri** — scatter with immediate arguments

Syntax

<u>tw</u>oway scatteri *immediate_values* $\left[\ ,\ scatter\_options\ \right]$

where *immediate_values* is one or more of

$\#_y\ \#_x\ \left[\ (\#_{clockpos})\ \right]\ \left[\ "text\ for\ label"\ \right]$

See [G] ***clockpos*** for a description of $\#_{clockpos}$.

Description

scatteri is an immediate version of graph twoway scatter; see [U] **22 Immediate commands**. scatteri is intended for programmer use but can be useful interactively.

Options

scatter_options are as defined in [G] **graph twoway scatter**, with the following modifications:

If "*text for label*" is specified among any of the immediate arguments, marker mlabel() is assumed.

If ($\#_{clockpos}$) is specified among any of the immediate arguments, option mlabvposition() is assumed.

Remarks

Immediate commands are commands that obtain data from numbers typed as arguments. Typing

```
. twoway scatteri 1 1   2 2, any_options
```

produces the same graph as typing

```
. clear
. input y x

            y           x
  1. 1 1
  2. 2 2
  3. end
. twoway scatter y x, any_options
```

twoway scatteri does not modify the data in memory.

scatteri is intended for programmer use but can be used interactively. In [G] ***added_text_option*** we demonstrated the use of option text() to add text to a graph:

```
. twoway qfitci  mpg weight, stdf ||
         scatter mpg weight, ms(O)
                  text(41 2040 "VW Diesel", place(e))
                  text(28 3260 "Plymouth Arrow", place(e))
                  text(35 2050 "Datsun 210 and Subaru", place(e))
```

Below we use `scatteri` to obtain similar results:

```
. twoway qfitci  mpg weight, stdf ||
         scatter mpg weight, ms(O) ||
         scatteri 41 2040 (3) "VW Diesel"
                  28 3260 (3) "Plymouth Arrow"
                  35 2050 (3) "Datsun 210 and Subaru"
                  , msymbol(i)
```

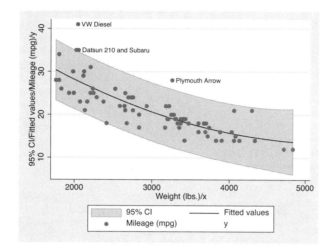

Note that we translated `text(..., place(e))` to the (3), 3 o'clock being the *clockpos* notation for the east *compassdirstyle*. Since labels are by default positioned at 3 o'clock, we could omit the (3) altogether:

```
. twoway qfitci  mpg weight, stdf ||
         scatter mpg weight, ms(O) ||
         scatteri 41 2040 "VW Diesel"
                  28 3260 "Plymouth Arrow"
                  35 2050 "Datsun 210 and Subaru"
                  , msymbol(i)
```

Note that we specified `msymbol(i)` option to suppress the display of the marker symbol.

❏ Technical Note

Programmers: Note carefully `scatter`'s *advanced_option* `recast()`; see [G] *advanced_options*. It can be used to good effect, such as using `scatteri` to add areas, bars, spikes, and dropped lines.

❏

Also See

Complementary: [G] **graph twoway scatter**

Title

graph twoway spike — Twoway spike plots

Syntax

twoway spike *yvar xvar* [if *exp*] [in *range*] [, *spike_options scatter_options*]

where *spike_options* are

spike_options	description
vertical	vertical spike plot (default)
horizontal	horizontal spike plot
base(*#*)	value to drop to (default=0)
blstyle(*linestyle*)	overall look of spike
blcolor(*colorstyle*)	spike color
blwidth(*linewidthstyle*)	thickness of spike
blpattern(*linepatternstyle*)	whether spike solid, dashed, etc.

See [G] *linestyle*, [G] *colorstyle*, [G] *linewidthstyle*, and [G] *linepatternstyle*.
All options are *merged-implicit*; see [G] **repeated options**.

and where *scatter_options* refer to any of the options allowed by scatter except that
marker_options, *marker_placement_option*, *marker_label_options*, and *connect_options* are
ignored even if specified; see [G] **graph twoway scatter**.

Description

twoway spike displays numerical (*y*,*x*) data as spikes. twoway spike is useful for drawing spike
plots of time-series data or other equally spaced data and is useful as a programming tool. For sparse
data, also see [G] **graph bar**.

Options

vertical and horizontal specify whether a vertical or a horizontal spike plot is desired. vertical
is the default. If horizontal is specified, the values recorded in *yvar* are treated as *x* values,
and the values recorded in *xvar* are treated as *y* values. That is, to make horizontal plots, do not
switch the order of the two variables specified.

In the vertical case, spikes are drawn at the specified *xvar* values and extend up or down from
0 according to the corresponding *yvar* values. If 0 is not in the range of the *y* axis, spikes extend
up or down to the *x* axis.

In the horizontal case, spikes are drawn at the specified *xvar* values and extend left or right
from 0 according to the corresponding *yvar* values. If 0 is not in the range of the *x* axis, spikes
extend left or right to the *y* axis.

base(#) specifies the value from which the spike should extend. The default is base(0) and, in the above description of options vertical and horizontal, this default was assumed.

blstyle(*linestyle*) specifies the overall style of the line used to draw the spike, which includes its pattern (solid, dashed, etc.), its thickness, and its color. The other options listed below allow you to change the line's attributes, but blstyle() is the starting point. See [G] *linestyle* for a list of choices.

blcolor(*colorstyle*) specifies the color of the line used to draw the spike. See [G] *colorstyle* for a list of color choices.

blwidth(*linewidthstyle*) specifies the thickness of the line used to draw the spike.
See [G] *linewidthstyle* for a list of choices.

blpattern(*linepatternstyle*) specifies whether the line used to draw the spike is solid, dashed, etc. See [G] *linepatternstyle* for a list of pattern choices.

scatter_options refer to any of the options allowed by scatter except that *marker_options*, *marker_placement_option*, *marker_label_options*, and *connect_options* are ignored even if specified; see [G] **graph twoway scatter**.

Remarks

Remarks are presented under the headings

> *Typical use*
> *Advanced use*
> *Cautions*

Typical use

We have daily data recording the values for the S&P 500 in 2001:

```
. sysuse sp500, clear
(S&P 500)

. list date high low close in 1/5
```

	date	high	low	close
1.	02jan2001	1320.28	1276.05	1283.27
2.	03jan2001	1347.76	1274.62	1347.56
3.	04jan2001	1350.24	1329.14	1333.34
4.	05jan2001	1334.77	1294.95	1298.35
5.	08jan2001	1298.35	1276.29	1295.86

In [G] **graph twoway bar**, we graphed the first 57 observations of these data using bars. Here is the same graph presented as spikes:

. twoway spike change date in 1/57

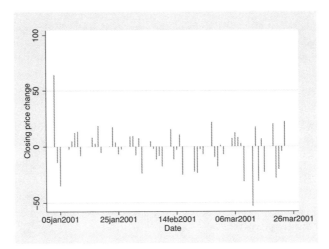

Spikes are especially useful when there is a lot of data. Below, we graph the data for the entire year:

. twoway spike change date

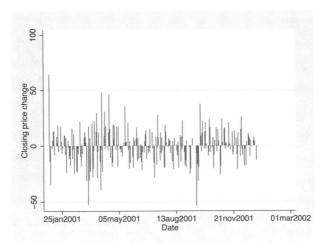

(*Continued on next page*)

Advanced use

The useful thing about `twoway spike` is that it can be combined with other `twoway` plottypes:

```
. twoway line close date || spike change date
```

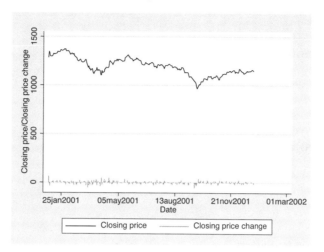

We can improve this graph by typing

```
. twoway
    line close date, yaxis(1)
  ||
    spike change date, yaxis(2)
  ||,
    ysca(axis(1) r(700  1400)) ylab(1000(100)1400, axis(1))
    ysca(axis(2) r(-50 300)) ylab(-50 0 50, axis(2))
          ytick(-25 0 25, axis(2) grid)
    legend(off)
    title("S&P 500")
    subtitle("January - December 2001")
    note("Source:  Yahoo!Finance and Commodity Systems, Inc.")
    yline(950, axis(1) lstyle(foreground))
```

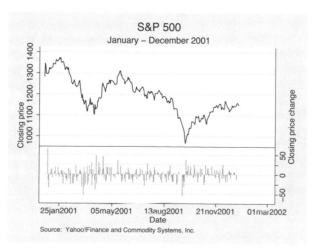

Concerning our use of

```
yline(950, axis(1) lstyle(foreground))
```

see *Advanced use* under [G] **graph twoway bar**.

Cautions

See *Cautions* under [G] **graph twoway bar**. What is said there applies equally to `twoway spike`.

Also See

Complementary: [G] **graph twoway scatter**; [G] **graph twoway bar**,
[G] **graph twoway dot**, [G] **graph twoway dropline**

Title

> **graph use** — Display graph stored on disk

Syntax

graph use *filename* $\big[$, nodraw name(*name*$\big[$, replace$\big]$) scheme(*schemename*) $\big]$

Description

graph use displays (draws) the graph previously saved in a .gph file and, if the graph was stored in live format, loads it.

If *filename* is specified without an extension, .gph is assumed.

Options

nodraw specifies the graph is not to be displayed. If the graph was stored in live format, it is still loaded; otherwise, graph use does nothing. See [G] ***nodraw_option***.

name(*name*$\big[$, replace$\big]$) specifies the name under which the graph is to be stored in memory, assuming the graph was saved in live format. name(Graph, replace) is the default. If the graph is not stored in live format, the graph can only be displayed, not loaded, and the name() is irrelevant. See [G] ***name_option***.

scheme(*schemename*) specifies the scheme to be used; see [G] **schemes**. If scheme() is not specified, the default is the *schemename* recorded in the graph being loaded.

Remarks

Graphs can be saved at the time you draw them either by specifying the saving() option or subsequently using the graph save command; see [G] ***saving_option*** and graph save. Modern graphs are saved in live format or as-is format; see [G] **gph files**. Regardless of how the graph was saved or the format in which it was saved, graph use can redisplay the graph; simply type

        . graph use *filename*

Remarks are presented under the headings

        Typical use
        Advanced use

Typical use

In a prior session, you drew a graph by typing

        . twoway qfitci mpg weight, stdf ||
                 scatter mpg weight ||
          , by(foreign, total row(1)) saving(cigraph)

The result of this was to create file cigraph.gph. At a later date, you can see the contents of the file by typing

        . graph use cigraph

You might now print a copy of the graph.

Advanced use

In the above example, the graph was saved in live format. Thus, when you typed graph use cigraph, in addition to being redisplayed, the graph was loaded under the name Graph. You might now use this graph with the graph combine command. Had you typed

> . graph use cigraph, name(ci)

the graph would have been loaded under the name ci. See [G] *name_option* for more information on the use of graphs stored in memory.

Also See

Complementary: [G] **graph combine**; [G] **graph save**, [G] *saving_option*;
                    [G] **gph files**, [G] *name_option*

Title

gridstyle — Choices for overall look of grid lines

Syntax

gridstyle may be

gridstyle	description
default	determined by scheme
major	determined by scheme; default or bolder
minor	determined by scheme; default or fainter
dot	dotted line

Other *gridstyles* may be available; type

```
. graph query gridstyle
```

to obtain the full list installed on your computer.

Description

Grids are lines that extend from an axis across the plot region. *gridstyle* specifies the overall look of grids. See [G] *axis_label_options*.

Remarks

Remarks are presented under the headings

What is a grid?
What is a gridstyle?
You do not need to specify a gridstyle
Turning off and on the grid

What is a grid?

Grids are lines that extend from an axis across the plot region.

What is a gridstyle?

Grids are defined by three attributes:

1. Whether the grid lines extend into the plot region's margin.

2. Whether the grid lines close to the axes are to be drawn.

3. The line style of the grid, which includes the line's thickness, color, and whether solid, dashed, etc.; see [G] *linestyle*.

The *gridstyle* specifies all three of these attributes.

365

You do not need to specify a gridstyle

The *gridstyle* is specified in the options named

$\big\{\,$ y $|$ x $\big\}\,\big\{\,$ label $|$ tick $|$ mlabel $|$ mtick $\big\}\,(\,\ldots$ gstyle(*gridstyle*) $\ldots\,)$

Correspondingly, there are other $\big\{\,$ y $|$ x $\big\}\big\{\,$ label $|$ tick $|$ mlabel $|$ mtick $\big\}\,()$ suboptions that allow you to specify the individual attributes; see [G] *axis_label_options*.

You specify the *gridstyle* when a style exists that is exactly what you desire or when another style would allow you to specify fewer changes to obtain what you want.

Turning off and on the grid

Whether grid lines are included by default is a function of the scheme; see [G] **schemes**. Regardless of the default, whether grid lines are included is controlled not by the *gridstyle*, but by the $\big\{\,$ y $|$ x $\big\}\big\{\,$ label $|$ tick $|$ mlabel $|$ mtick $\big\}\,()$ suboptions grid and nogrid.

Grid lines are nearly always associated with the ylabel() and/or xlabel() options. Specify $\big\{\,$ y $|$ x $\big\}$ label(,grid) or $\big\{\,$ y $|$ x $\big\}$ label(,nogrid). See [G] *axis_label_options*.

Also See

Complementary: [G] *axis_label_options*

Title

justificationstyle — Choices for how text is justified

Syntax

justificationstyle may be

justificationstyle	description
`left`	left-justified
`center`	centered
`right`	right-justified

Other *justificationstyles* may be available; type

```
. graph query justificationstyle
```

to obtain the full list installed on your computer.

Description

justificationstyle specifies how the text is "horizontally" aligned in the textbox. Choices include `left`, `right`, and `center`. Think of the textbox as being horizontal even if it is vertical when specifying this option.

justificationstyle is specified in the `justification()` option nested within another option, such as `title()`:

```
. graph ..., title("Line 1" "Line 2", justification(justificationstyle)) ...
```

See [G] *textbox_options* for more information on textboxes.

In some cases, you will see that a *justificationstylelist* is allowed. A *justificationstylelist* is a sequence of *justificationstyles* separated by spaces. Shorthands are allowed to make specifying the list easier; see [G] *stylelists*.

Remarks

justificationstyle typically affects the alignment of multiline text within a textbox and not the justification of the placement of the textbox itself; see *Justification* in [G] *textbox_options*.

Also See

Complementary: [G] *textbox_options*, [G] *alignmentstyle*

367

Title

legend_option — Option for specifying legend

Syntax

The *legend_option* is

legend_option	description
legend([*contents*] [*location*])	legend contents and location

legend() is *merged-implicit*; see [G] **repeated options**.

where *contents* and *location* specify the contents and the location of the legend.

contents and *location* are defined

contents	description
order(*orderinfo*)	which keys appear and their order
label(*labelinfo*)	override text for a key
holes(*numlist*)	positions in legend to leave blank
all	generate keys for all symbols
style(*legendstyle*)	overall style of legend
cols(#)	# of keys per line
rows(#)	or # of rows
[no]colfirst	"1, 2, 3" in row 1 or in column 1?
[no]textfirst	symbol-text or text-symbol?
stack	symbol/text vertically stacked
rowgap(*relativesize*)	gap between lines
colgap(*relativesize*)	gap between columns
symplacement(*compassdirstyle*)	alignment/justification of key's symbol
keygap(*relativesize*)	gap between symbol-text
symysize(*relativesize*)	height for key's symbol
symxsize(*relativesize*)	width for key's symbol
textwidth(*relativesize*)	width for key's descriptive text
forcesize	always respect symysize(), symxsize(), and textwidth()
bmargin(*marginstyle*)	outer margin around legend
textbox_options	other text characteristics
title_options	title, subtitle, notes, captions
region(*roptions*)	borders and background shading

See [G] *legendstyle*, [G] *relativesize*, [G] *compassdirstyle*, [G] *marginstyle*, [G] *textbox_options*, and [G] *title_options*.

location	description
off or on	suppress or force display of legend
position(*clockpos*)	where legend appears
ring(*ringpos*)	where legend appears (detail)
span	"centering" of legend
at(#)	allowed with by() only

See *Where legends appear* under *Remarks* below and see *Positioning of titles* in [G] ***title_options*** for definitions of *clockpos* and *ringpos*.

orderinfo, the argument allowed by legend(order()), is defined

$$\{ \, \# \mid - \, \} \; \big[\, \texttt{"text"} \; \big[\, \texttt{"text"} \; \ldots \, \big] \, \big]$$

labelinfo, the argument allowed by legend(label()), is defined

$$\# \; \texttt{"text"} \; \big[\, \texttt{"text"} \; \ldots \, \big]$$

roptions, the arguments allowed by legend(region()), include

roptions	description
style(*areastyle*)	overall style of region
color(*colorstyle*)	line + fill color of region
fcolor(*colorstyle*)	fill color of region
lstyle(*linestyle*)	overall style of border
lcolor(*colorstyle*)	color of border
lwidth(*linewidthstyle*)	thickness of border
lpattern(*linepatternstyle*)	whether border is solid, dashed, etc.
margin(*marginstyle*)	margin between border and contents of legend

See [G] ***areastyle***, [G] ***colorstyle***, [G] ***linestyle***, [G] ***linewidthstyle***, [G] ***linepatternstyle***, and [G] ***marginstyle***.

Description

The legend() option allows you to control the look, contents, and placement of the legend. A sample legend is

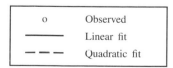

The above legend has three *keys*. Each key is composed of a *symbol* and *descriptive text* describing the symbol, where symbol is understood to mean whatever the symbol might be, be it a marker, a line, or a color swatch.

Options

legend(*contents*, *location*) defines the contents of a legend along with how it is to look and whether and where it is to be displayed.

Content suboptions for use with legend()

order(*orderinfo*) specifies which keys are to appear in the legend and the order in which they are to appear.

order(*# #* ...) is the usual syntax. order(1 2 3) would specify that key 1 is to appear first in the legend, followed by key 2, followed by key 3. order(1 2 3) is the default if there are three keys. If there were four keys, order(1 2 3 4) would be the default, and so on. If there were four keys and you specified order(1 2 3), then the fourth key would not appear in the legend. If you specified order(2 1 3), first key 2 would appear, followed by key 1, followed by key 3.

A dash specifies that text is to be inserted into the legend. For instance, order(1 2 - "*text*" 3) specifies key 1 is to appear first, followed by key 2, followed by the text *text*, followed by key 3. Imagine that the default key were

o	Observed
——	Linear
– – –	Quadratic

Specifying order(1 - "Predicted:" 2 3) would produce

o	Observed
	Predicted:
——	Linear
– – –	Quadratic

and specifying order(1 - " " "Predicted:" 2 3) would produce

o	Observed
	Predicted:
——	Linear
– – –	Quadratic

Note carefully the specification of a blank for the first line of the text insertion: we typed " " and not "". "" would insert nothing.

You may also specify quoted text after # to override the descriptive text associated with a symbol. Specifying order(1 "Observed 1992" - " " "Predicted" 2 3) would change "Observed" in the above to "Observed 1992". It is considered better style, however, to use the label() suboption to relabel symbols.

label(*#* "*text*" ["*text*" ...]) specifies the descriptive text to be displayed next to the #th key. Multiline text is allowed. Specifying label(1 "Observed 1992") would change the descriptive text associated with the first key to be "Observed 1992". Specifying label(1 "Observed" "1992-1993") would change the descriptive text to contain two lines, "Observed" followed by "1992–1993".

Note that the descriptive text of only one key may be changed per `label()` suboption. Specify multiple `label()` suboptions when you wish to change the text of multiple keys.

`holes(`*numlist*`)` specifies where gaps should appear in the presentation of the keys. `holes()` has an effect only if the keys are being presented in more than one row and more than one column.

Consider a case in which the default key is

o	Observed	————	Linear fit
— — —	Quadratic fit		

Specifying `holes(2)` would result in

o	Observed		
————	Linear fit	— — —	Quadratic fit

In this case, `holes(2)` would have the same effect as specifying `order(1 - " " 2 3)` and, as a matter of fact, there is always an `order()` command that will achieve the same result as `holes()`. `order()` has the added advantage of working in all cases.

`all` specifies that keys are to be generated for all the plots of the graph, even when the same symbol is repeated. The default is to generate keys only when the symbols are different, which is determined by the overall style. For example, in

    . scatter ylow yhigh x, pstyle(p1 p1) || ...

there would be only one key generated for the variables ylow and yhigh because they share the style p1. That single key's descriptive text would indicate that the symbol corresponded to both variables. If, on the other hand, you typed

    . scatter ylow yhigh x, pstyle(p1 p1) legend(all) || ...

then separate keys would be generated for ylow and yhigh.

In the above example, do not confuse our use of the *scatter_option* `pstyle()` with `legend()`'s suboption `legend(style())`. The *scatter_option* `pstyle()` sets the overall style for the rendition of the points. `legend()`'s `style()` suboption is documented directly below.

`style(`*legendstyle*`)` specifies the overall look of the legend—whether it is presented horizontally or vertically, how many keys appear across the legend if presented horizontally, etc. The options listed below allow you to change each attribute of the legend, but `style()` is the starting point.

You need not specify `style()` just because there is something you want to change. You specify `style()` when another style exists that is exactly what you desire or when another style would allow you to specify fewer changes to obtain what you want.

See [G] *legendstyle* for a list of available legend styles.

`cols(`*#*`)` and `rows(`*#*`)` are alternatives; they specify in how many columns or rows (lines) the keys are to be presented. The usual default is `cols(2)`, which means legends are to take two columns:

o	Observed	————	Linear fit
— — —	Quadratic fit		

`cols(1)` would force a vertical arrangement,

o	Observed
————	Linear fit
— — —	Quadratic fit

and rows(1) would force a horizontal arrangement:

o Observed	———— Linear fit	— — — Quadratic fit

colfirst and nocolfirst determine whether, when the keys are presented in multiple columns, keys are to read down or to read across, whether you get this

o Observed	— — — Quadratic fit	
———— Linear fit		

or this

o Observed	———— Linear fit
— — — Quadratic fit	

The usual default is nocolfirst, and so colfirst is the option.

textfirst and notextfirst specify whether the keys are presented as descriptive text followed by the symbol or the symbol followed by descriptive text. The usual default is notextfirst, and thus textfirst is the option. textfirst produces keys that look like this,

Observed o	Linear fit ————
Quadratic fit — — —	

and textfirst cols(1) produces

Observed   o
Linear fit   ————
Quadratic fit   — — —

stack specifies that the symbol-text is to be presented vertically with the symbol on top (or with the descriptive text on top if textfirst is also specified). legend(stack) would produce

o	————————
Observed	Linear fit
— — — — — — — —	
Quadratic fit	

legend(stack symplacement(left) symxsize(13) forcesize rowgap(4)) would produce

o	————
Observed	Linear fit
— — —	
Quadratic fit	

stack tends to be used to produce single-column keys. legend(cols(1) stack symplacement(left) symxsize(13) forcesize rowgap(4)) produces

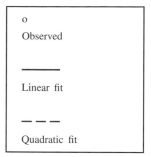

This is the real use of stack: to produce narrow, vertical keys.

rowgap(*relativesize*) and colgap(*relativesize*) specify the distance between lines and the distance between columns. The defaults are rowgap(1.4) and colgap(4.9). See [G] *relativesize*.

symplacement(*compassdirstyle*) specifies how symbols are justified in the key. The default is symplacement(center), meaning that they are vertically and horizontally centered. The two most commonly specified alternatives are symplacement(right) (right alignment) and symplacement(left) (left alignment). See [G] *compassdirstyle* for other alignment choices.

keygap(*relativesize*), symysize(*relativesize*), symxsize(*relativesize*), and
textwidth(*relativesize*) specify the height and width to be allocated for the key and the key's symbols and descriptive text:

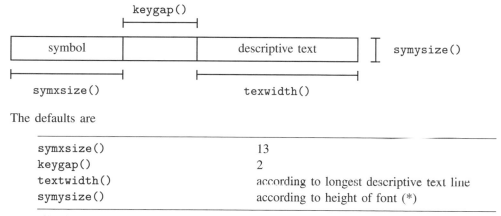

The defaults are

symxsize()	13
keygap()	2
textwidth()	according to longest descriptive text line
symysize()	according to height of font (*)

(*) The size of the font is set by the *textbox_option* size(*relativesize*);
     see *textbox_options* below.

Markers are placed in the symbol area, centered according to symplacement().

Lines are placed in the symbol area vertically according to symplacement() and horizontally are drawn to length symxsize().

Color swatches fill the symysize() × symxsize() area.

See [G] *relativesize* for information on specifying relative sizes.

forcesize causes the sizes specified by symysize() and symxsize() to be respected. If forcesize is not specified, once all the symbols have been placed for all the keys, the symbol area is compressed (or expanded) to be no larger than necessary to contain the symbols.

bmargin(*marginstyle*) specifies the outer margin around the legend. That is, it specifies how close other things appearing near to the legend can get. Also see suboption margin() under *Suboptions for use with legend(region())* below for specifying the inner margin between the border and contents. See [G] *marginstyle* for a list of margin choices.

textbox_options affect the rendition of the descriptive text associated with the keys. These are described in [G] ***textbox_options***. One of the most commonly specified *textbox_options* is size(*relativesize*), which specifies the size of font to be used for the descriptive text.

title_options allow placing titles, subtitles, notes, and captions on legends. For instance, leg-end(col(1) subtitle("Legend")) produces

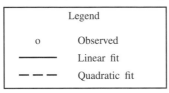

Note our use of subtitle() and not title(); title()s are nearly always too big. See [G] ***title_options***.

region(*roptions*) specifies the border and shading of the legend. You could remove the border around the legend by specifying legend(region(lstyle(none))) (thus doing away with the line) or legend(region(lcolor(none))) (thus making the line invisible). Alternatively, you could give the legend a gray background tint by specifying legend(region(fcolor(gs5))). See *Suboptions for use with legend(region())* below.

Suboptions for use with legend(region())

style(*areastyle*) specifies the overall style of the region in which the legend appears. The other suboptions allow you to change the region's attributes individually, but style() provides the starting point. See [G] ***areastyle*** for a list of choices.

color(*colorstyle*) specifies the color of the background of the legend and the line used to outline it. See [G] ***colorstyle*** for a list of color choices.

fcolor(*colorstyle*) specifies the background (fill) color for the legend. See [G] ***colorstyle*** for a list of color choices.

lstyle(*linestyle*) specifies the overall style of the line used to outline the legend, which includes its pattern (solid, dashed, etc.), its thickness, and its color. The other suboptions listed below allow you to change the line's attributes individually, but lstyle() is the starting point. See [G] ***linestyle*** for a list of choices.

lcolor(*colorstyle*) specifies the color of the line used to outline the legend. See [G] ***colorstyle*** for a list of color choices.

lwidth(*linewidthstyle*) specifies the thickness of the line used to outline the legend. See [G] ***linewidthstyle*** for a list of choices.

lpattern(*linepatternstyle*) specifies whether the line used to outline the legend is solid, dashed, etc. See [G] ***linepatternstyle*** for a list of choices.

margin(*marginstyle*) specifies the inner margin between the border and the contents of the legend. Also see bmargin() under *Content suboptions for use with legend()* above for specifying the outer margin around the legend. See [G] ***marginstyle*** for a list of margin choices.

Location suboptions for use with legend()

off and on determine whether the legend appears. The default is on when more than one symbol (meaning marker, line style, or color swatch) appears in the legend. In those cases, legend(off) will suppress the display of the legend.

position(*clockpos*) and ring(*ringpos*) override the default location of the legend, which is usually centered below the plot region. position() specifies a direction (*sic*) according to the hours on the dial of a 12-hour clock, and ring() specifies the distance from the plot region.

ring(0) is defined as being inside the plot region itself and allows you to place the legend inside the plot. ring(k), $k > 0$, specifies positions outside the plot region; the larger the ring() value, the farther away from the plot region is the legend. ring() values may be integers or nonintegers and are treated ordinally.

position(12) puts the legend directly above the plot region (assuming ring() > 0), position(3) directly to the right of the plot region, and so on.

See *Where legends appear* under *Remarks* below and see *Positioning of titles* in [G] ***title_options*** for more information on the position() and ring() suboptions.

span specifies that the legend is to be placed in an area spanning the entire width (or height) of the graph rather than an area spanning the plot region. This affects whether the legend is centered with respect to the plot region or the entire graph. See *Spanning* in [G] ***title_options*** for more information on span.

at(*#*) is for use only when the *twoway_option* by() is also specified. It specifies that the legend is to appear in the *#*th position of the $R \times C$ array of plots, using the same coding as by(..., holes()). See *Use of legends with by()* under *Remarks* below and see [G] ***by_option***.

Remarks

Remarks are presented under the headings

> *When legends appear*
> *The contents of legends*
> *Where legends appear*
> *Putting titles on legends*
> *Use of legends with by()*
> *Problems arising with or due to legends*

When legends appear

Legends appear on the graph whenever more than one symbol is used, where symbol is broadly defined to include markers, lines, and color swatches (such as those used to fill bars). When you draw a graph with only one symbol on it, such as

```
. sysuse uslifeexp, clear
(U.S. life expectancy, 1900-1999)
```

```
. line le year
```

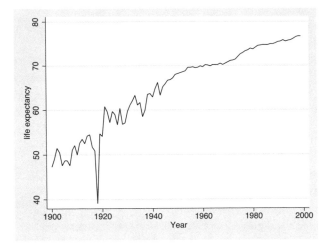

no legend appears. When there is more than one symbol, a legend is added:

```
. line le_m le_f year
```

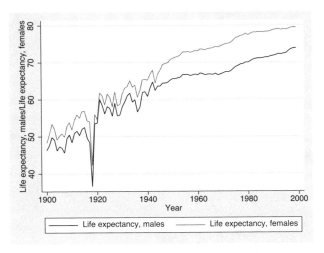

Even when there is only one symbol, a legend is constructed. It is merely not displayed. Specifying legend(on) will force the display of the legend:

```
. line le year, legend(on)
```

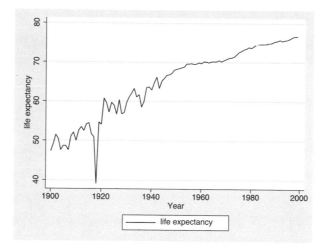

Similarly, when there is more than one symbol and you do not want the legend, you can specify legend(off) to suppress it:

```
. line le_m le_f year, legend(off)
```

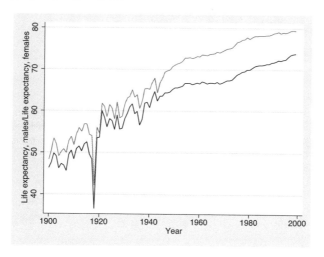

The contents of legends

By default, the descriptive text is obtained from the variable's variable label; see [R] **label**. If the variable has no variable label, the variable's name is used. In

```
. line le_m le_f year
```

the variable le_m had previously been labeled "Life expectancy, males" and the variable le_f had been labeled "Life expectancy, females". In the legend of this graph, repeating "life expectancy" is unnecessary. The graph would be improved if we changed the labels on the variables:

```
. label var le_m "Males"
. label var le_f "Females"
. line le_m le_f year
```

Alternatively, we can specify the `label()` suboption to change the descriptive text. We obtain the same visual result without relabeling our variables:

```
. line le_m le_f year, legend(label(1 "Males") label(2 "Females"))
```

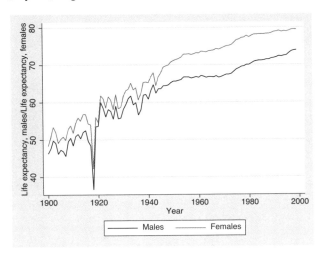

Where legends appear

By default, legends appear beneath the plot, centered, at what is technically referred to as `position(6) ring(3)`. Suboptions `position()` and `ring()` specify the location of the label. `position()` specifies on which side of the plot region the legend appears—`position(6)` means 6 o'clock—and `ring()` specifies the distance from the plot region—`ring(3)` means farther out than the *title_option* `b2title()` but inside the *title_option* `note()`; see [G] ***title_options***.

If we specify `legend(position(3))`, the legend will be moved to the 3 o'clock position:

```
. line le_m le_f year, legend(pos(3))
```

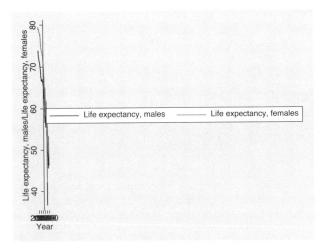

This may not be what we desired, but it is what we asked for. The legend was moved to be beside the graph and, given the size of the legend, the graph was squeezed to fit. When you move legends to the side, you invariably also want to specify the `col(1)` option:

```
. line le_m le_f year, legend(pos(3) col(1))
```

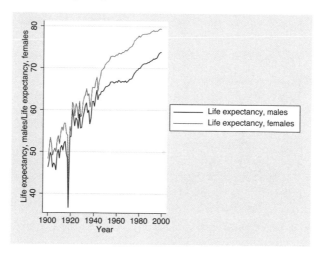

As a matter of syntax, we could have typed the above command with two `legend()` options

```
. line le_m le_f year, legend(pos(3)) legend(col(1))
```

instead of one combined: `legend(pos(3) col(1))`. We would obtain the same results either way.

Putting aside syntax, the above graph would look better with less descriptive-text,

```
. line le_m le_f year, legend(pos(3) col(1)
                 lab(1 "Males") lab(2 "Females"))
```

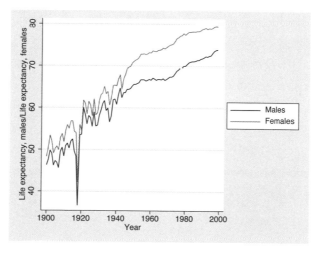

and we can further reduce the width required by the legend by specifying the `stack` suboption:

```
. line le_m le_f year, legend(pos(3) col(1)
                 lab(1 "Males") lab(2 "Females") stack)
```

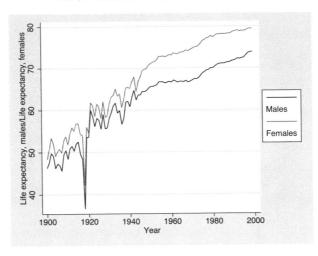

We can make this look better by placing a blank line between the first and second keys:

```
. line le_m le_f year, legend(pos(3) col(1)
                 lab(1 "Males") lab(2 "Females") stack
                     order(1 - " " 2))
```

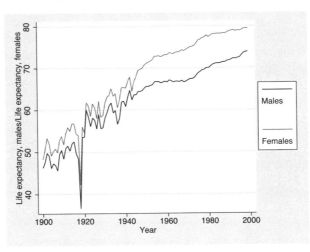

ring()—the suboption that specifies the distance from the plot region—is seldom specified but, when it is specified, ring(0) is the most useful. ring(0) specifies that the legend be moved inside the plot region:

(Continued on next page)

```
. line le_m le_f year, legend(pos(5) ring(0) col(1)
                 lab(1 "Males") lab(2 "Females"))
```

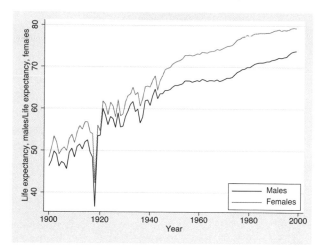

Our use of `position(5) ring(0)` put the legend inside the plot region, at 5 o'clock, which is to say, the bottom right corner. Had we specified `position(2) ring(0)`, the legend would have appeared in the top left corner.

We might now add some background color to the legend:

```
. line le_m le_f year, legend(pos(5) ring(0) col(1)
                 lab(1 "Males") lab(2 "Females")
                 region(fcolor(gs15)))
```

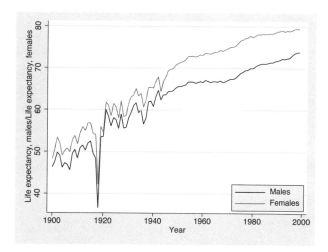

Putting titles on legends

Legends may include titles:

```
. line le_m le_f year, legend(pos(5) ring(0) col(1)
                          lab(1 "Males") lab(2 "Females")
                          region(fcolor(gs15)))
                        legend(subtitle("Legend"))
```

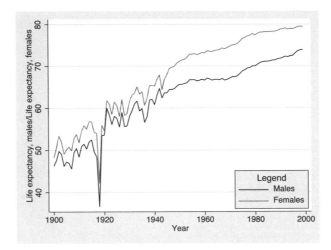

Above we specified `subtitle()` rather than `title()` because, when we tried `title()`, it seemed too big.

Legends may also contain `notes()` and `captions()`; see [G] *title_options*.

Use of legends with by()

If you want the legend located in the default location, no special action need be taken when you use `by()`:

```
. sysuse auto, clear
(1978 Automobile Data)
. scatter mpg weight || lfit mpg weight ||, by(foreign, total row(1))
```

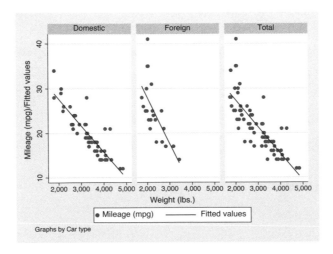

If, however, you wish to move the legend, you must distinguish between legend(*contents*) and legend(*location*). The former must appear outside the by(). The latter appears inside the by():

```
. scatter mpg weight || lfit mpg weight ||,
        legend(cols(1))
        by(foreign, total legend(pos(4)))
```

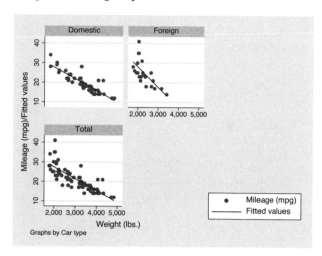

Note that legend(col(1)) was placed in the command just where we would place it had we not specified by(), but that legend(pos(4)) was moved to be inside the by() option. We did that because the cols() suboption is documented under *contents* in the syntax diagram whereas position() is documented under *location*. The logic is that, at the time the individual plots are constructed, they must know what style of key they are producing. The placement of the key, however, is something that happens when the overall graph is assembled, and so where the key is to be placed must be told to by(). Were we to forget this distinction and simply to type

```
. scatter mpg weight || lfit mpg weight ||,
        legend(cols(1) pos(4))
        by(foreign, total)
```

the cols(1) suboption would have been ignored.

An additional *location* suboption is provided for use in the case of by(): at(#). You specify this option to tell by() to place the legend inside the $R \times C$ array it creates:

```
. scatter mpg weight || lfit mpg weight ||,
        legend(cols(1))
        by(foreign, total legend(at(4) pos(0)))
```

(*Continued on next page*)

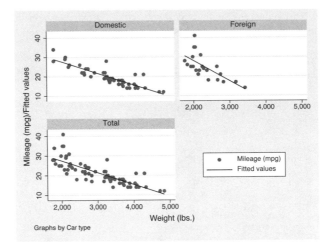

In the above, we specified at(4) to mean that the key was to appear in the 4th position of the 2×2 array, and we specified pos(0) to move the key to the middle (0 o'clock) position within the cell.

If you wish to suppress the legend altogether, you must specify the legend(off) inside the by() option:

```
. scatter mpg weight || lfit mpg weight ||,
        by(foreign, total legend(off))
```

Problems arising with or due to legends

There are two problems associated with legends:

1. Text may flow outside the border of the legend box.

2. The presence of the legend may cause the title of the y axis to run into the values labeled on the axis.

The first problem arises because Stata uses an approximation to obtain the width of a text line. The solution is to specify the width(*relativesize*) *textbox_option*:

```
. graph ..., ... legend(width(#))
```

See *Use of the textbox option width() in* [G] *added_text_option*.

The second problem arises when the key is in its default position(6) o'clock location and the descriptive text for one or more of the keys is very long. In position(6), the borders of the key are supposed to line up with the borders of the plot region. Usually the plot region is wider than the key and so the key is expanded to fit below it. When the key is wider than the plot region, however, it is the plot region that is widened. As the plot region expands, it will eat away at whatever is at it sides, namely the y axis labels and title. Margins will disappear. In extreme cases, the title will be printed on top of the labels and the labels themselves may end up on top of the axis!

The solution to this problem is to shorten the descriptive text, either by using fewer words or by breaking the long description into multiple lines. Use the legend(label(# "*text*")) option to modify the longest line of the descriptive text.

Also See

Complementary: [G] *title_options*

Title

> *legendstyle* — Choices for look of legends

Syntax

legendstyle may be

legendstyle	description
default	determined by scheme

Other *legendstyles* may be available; type

    . graph query legendstyle

to obtain the full list installed on your computer.

Description

legendstyle specifies the overall style of legends and is specified in the `legend(style())` option:

    . graph ..., legend(... style(*legendstyle*) ...)

Remarks

Remarks are presented under the headings

> *What is a legend?*
> *What is a legendstyle?*
> *You do not need to specify a legendstyle*

What is a legend?

A legend is a table that shows the symbols used in a graph along with text describing their meaning. Each symbol/text entry in a legend is called a key. See [G] ***legend_option*** for more information.

What is a legendstyle?

The look of a legend is defined by fourteen attributes:

1. The number of columns or rows of the table.

2. Whether, in a multicolumn table, the first, second, ..., keys appear across the rows or down the columns.

3. Whether the symbol/text of a key appears horizontally adjacent or vertically stacked.

4. The gap between lines of the legend.

5. The gap between columns of the legend.

6. How the symbol of a key is aligned and justified.

7. The gap between the symbol and text of a key.

8. The height to be allocated in the table for the symbol of the key.

9. The width to be allocated in the table for the symbol of the key.

10. The width to be allocated in the table for the text of the key.

11. Whether the above specified height, width, and width are to be dynamically adjusted according to contents of the keys.

12. The margin around the legend.

13. The color, size, etc. of the text of a key (17 attributes).

14. The look of any titles, subtitles, notes, and captions placed around the table (23 attributes each).

The *legendstyle* specifies all fourteen of these attributes.

You do not need to specify a legendstyle

The *legendstyle* is specified in the option

```
legend(style(legendstyle))
```

Correspondingly, option `legend()` has other suboptions that will allow you to specify the fourteen attributes individually; see [G] *legend_option*.

Specify the *legendstyle* when a style exists that is exactly what you desire or when another style would allow you to specify fewer changes to obtain what you want.

Also See

Complementary: [G] *legend_option*

Title

> *line_options* — Options for determining the look of lines

Syntax

The *line_options* are

line_options	description
lstyle(*linestyle*)	overall style of line
lpattern(*linepatternstyle*)	whether line solid, dashed, etc.
lwidth(*linewidthstyle*)	thickness of line
lcolor(*colorstyle*)	color of line

See [G] *linestyle*, [G] *linepatternstyle*, [G] *linewidthstyle*, and [G] *colorstyle*.

All options are *rightmost*; see [G] **repeated options**.

Description

The *line_options* determine the look of a line in some contexts.

Options

lstyle(*linestyle*) specifies the overall style of the line: its pattern, thickness, and color.

You need not specify lstyle() just because there is something you want to change about the look of the line. The other *line_options* will allow you to make changes. You specify lstyle() when another style exists that is exactly what you desire or when another style would allow you to specify fewer changes.

See [G] *linestyle* for a list of available line styles.

lpattern(*linepatternstyle*) specifies whether the line is solid, dashed, etc. See [G] *linepatternstyle* for a list of available patterns. lpattern() is not allowed with graph pie; see [G] **graph pie**.

lwidth(*linewidthstyle*) specifies the thickness of the line. See [G] *linewidthstyle* for a list of available thicknesses.

lcolor(*colorstyle*) specifies the color of the line. See [G] *colorstyle* for a list of available colors.

Remarks

Lines occur in many contexts and, in some of those contexts, the above options are used to determine the look of the line. For instance, in the dot plots drawn by graph dot, if you specify option linetype(line), lines are used in place of the dots and the *line_options* then determine the look of the lines. In this case, graph dot's syntax is

```
. graph dot ..., ... linetype(line) lines(line_options) ...
```

In this case, the *line_options* are specified inside option lines(). If you wanted to specify the *line_option* lcolor(green), you would type

> . graph dot ..., ... linetype(line) lines(lcolor(green))

Also See

Complementary: [G] **lines**; [G] **graph dot**

Title

linepatternstyle — Choices for whether lines are solid, dashed, etc.

Syntax

linepatternstyle may be

linepatternstyle	description
solid	solid line
dash	dashed line
dot	dotted line
dash_dot	
shortdash	
shortdash_dot	
longdash	
longdash_dot	
blank	invisible line
"*formula*"	e.g., "-." or "--.." etc.

A *formula* is composed of any combination of

l	solid line
_	(underscore) a long dash
-	(hyphen) a medium dash
.	short dash (almost a dot)
#	small amount of blank space

For a palette displaying each of the above named line styles, type

```
palette linepalette [ , scheme(schemename) ]
```

Other *linepatternstyles* may be available; type

```
. graph query linepatternstyle
```

to obtain the full list installed on your computer.

Description

A line's look is determined by its pattern, thickness, and color; see [G] **lines**. *linepatternstyle* specifies the pattern.

linepatternstyle is specified via options named

⟨*object*⟩⟨l or li or line⟩pattern()

For instance, in the case of connecting lines (the lines used to connect points in a plot) used by graph twoway function, the option is named clpattern():

```
. twoway function ..., clpattern(linepatternstyle) ...
```

390

In some cases, you will see that a *linepatternstylelist* is allowed:

. twoway line ..., clpattern(*linepatternstylelist*) ...

A *linepatternstylelist* is a sequence of *linepatterns* separated by spaces. Shorthands are allowed to make specifying the list easier; see [G] *stylelists*.

Remarks

While you may choose a prerecorded pattern (e.g., solid or dash), understand that you can build any pattern you wish by specifying a line-pattern formula. For example,

formula	Description
"1"	solid line, same as solid
"_"	a long dash
"_-"	a long dash followed by a short dash
"_--"	a long dash followed by two short dashes
"_--_#"	a long dash, two short dashes, a long dash, and a bit of space
etc.	

When you specify a formula, you must enclose it in double quotes.

Also See

Complementary: [G] **lines**; [G] *linestyle*; [G] *linewidthstyle*; [G] *colorstyle*, [G] *connectstyle*

Title

> **lines** — Concept definition: line

Syntax

The following affects how a line appears:

linestyle	overall style
linepatternstyle	whether solid, dashed, etc.
linewidthstyle	its thickness
colorstyle	its color

See [G] ***linestyle***, [G] ***linepatternstyle***, [G] ***linewidthstyle***, and [G] ***colorstyle***.

Description

Lines occur in many contexts—in borders, axes, the ticks on axes, the outline around symbols, the connecting of points in a plot, and more. *linestyle*, *linepatternstyle*, *linewidthstyle*, and *colorstyle* define the look of the line.

Remarks

linestyle, *linepatternstyle*, *linewidthstyle*, and *colorstyle* are specified inside options that control how the line is to appear. These options have names of the form

*⟨object⟩*lstyle (*linestyle*)
*⟨object⟩*lpattern (*linepatternstyle*)
*⟨object⟩*lwidth (*linewidthstyle*)
*⟨object⟩*lcolor (*linecolorstyle*)

For instance,

- The options to specify how points in a plot are to appear are specified by the options clstyle(), clpattern(), clwidth(), and clcolor(); see [G] ***connect_options***.

- The suboptions to specify how the border around a textbox such as a title are to appear are named blstyle(), blpattern(), blwidth(), and blcolor(); see [G] ***textbox_options***.

Wherever these options arise, they always come in a group of four and the four pieces have the same meaning.

linestyle

linestyle is specified inside the *⟨object⟩*lstyle() option.

linestyle specifies the overall style of the line—its pattern (solid, dashed), etc.—its thickness, and its color.

You need not specify the ⟨*object*⟩lstyle() option just because there is something you want to change about the look of the line and, in fact, most of the time you do not. You specify ⟨*object*⟩lstyle() when another style exists that is exactly what you desire or when another style would allow you to specify fewer changes to obtain what you want.

See [G] *linestyle* for the list of what may be specified inside the ⟨*object*⟩lstyle() option.

linepatternstyle

linepatternstyle is specified inside the ⟨*object*⟩lpattern() option.

linepatternstyle specifies whether the line is solid, dashed, etc.

See [G] *linepatternstyle* for the list of what may be specified inside the ⟨*object*⟩lpattern() option.

linewidthstyle

linewidthstyle is specified inside the ⟨*object*⟩lwidth() option.

linewidthstyle specifies the thickness of the line.

See [G] *linewidthstyle* for the list of what may be specified inside the ⟨*object*⟩lwidth() option.

colorstyle

colorstyle is specified inside the ⟨*object*⟩lcolor() option.

colorstyle specifies the color of the line.

See [G] *colorstyle* for the list of what may be specified inside the ⟨*object*⟩ lcolor() option.

Also See

Complementary: [G] *linestyle*, [G] *linepatternstyle*, [G] *linewidthstyle*, [G] *colorstyle*;
[G] *connect_options*

Title

linestyle — Choices for overall look of lines

Syntax

linestyle may be

linestyle	description
foreground	borders, axes, etc.; in foreground color
grid	grid lines
minor_grid	a lesser grid line or same as grid
major_grid	a bolder grid line or same as grid
refline	reference lines
yxline	yline() or xline()
none	nonexistent line
p1–p15	used by first plot, second plot, ...
p1solid–p15solid	same as p1–p15 but always solid; same color, same thickness

Other *linestyles* may be available; type

        . graph query linestyle

to obtain the full list installed on your computer.

Description

linestyle sets the overall pattern, thickness, and color of a line; see [G] **lines** for more information.

linestyle is specified via options named

⟨*object*⟩⟨l or li or line⟩style()

For instance, in the case of connecting lines (the lines used to connect points in a plot) used by graph twoway function, the option is named clstyle():

        . twoway function ..., clstyle(*linestyle*) ...

In some cases, you will see that a *linestylelist* is allowed:

        . twoway line ..., clstyle(*linestylelist*) ...

A *linestylelist* is a sequence of *linestyles* separated by spaces. Shorthands are allowed to make specifying the list easier; see [G] **stylelists**.

Remarks

Remarks are presented under the headings

> *What is a line?*
> *What is a linestyle?*
> *You do not need to specify a linestyle*
> *Specifying a linestyle can be convenient*
> *Suppressing lines*

What is a line?

Nearly everything that appears on a graph is a line, the exceptions being markers, fill areas, bars, and the like, and even they are outlined or bordered by a line.

What is a linestyle?

Lines are defined by three attributes:

1. The *linepattern*—whether it is solid, dashed, etc.; see [G] **linepatternstyle**.

2. The *linewidth*—how thick the line is; see [G] **linewidthstyle**.

3. The *linecolor*—the color of the line; see [G] **colorstyle**.

The *linestyle* specifies all three of these attributes.

You do not need to specify a linestyle

The *linestyle* is specified in options named

⟨*object*⟩⟨l or li or line⟩style(*linestyle*)

Correspondingly, you will always find that three other options are available:

⟨*object*⟩⟨l or li or line⟩pattern(*linepatternstyle*)

⟨*object*⟩⟨l or li or line⟩width(*linewidthstyle*)

⟨*object*⟩⟨l or li or line⟩color(*colorstyle*)

You specify the *linestyle* when a style exists that is exactly what you desire or when another style would allow you to specify fewer changes to obtain what you want.

Specifying a linestyle can be convenient

Consider the command

```
. line y1 y2 x
```

Assume you wanted the line for y2 versus x to be the same as y1 versus x. One way you could proceed would be to set the pattern, width, and color of the line for y1 versus x and then set the pattern, width, and color of the line for y2 versus x to be the same. Easier, however, would be to type

    . line y1 y2 x, clstyle(p1 p1)

clstyle() is the name of the option that specifies the style of connected lines. When you do not specify the clstyle() option, results are as if you specified

    clstyle(p1 p2 p3 p4 p5 p6 p7 p8 p9 p10 p11 p12 p13 p14 p15)

where the extra elements are ignored. In any case, p1 is one set of pattern, thickness, and color values, p2 is another set, and so on.

Say that you wanted y2 versus x to look like y1 versus x except that you wanted the line to be green; you could type

    . line y1 y2 x, clstyle(p1 p1) clcolor(. green)

There is nothing special about the *linestyles* p1, p2, . . . ; they merely specify sets of pattern, thickness, and color values just like any other named *linestyle*. Type

    . graph query linestyle

to find out what other line styles are available. You may find something pleasing and, if so, that is more easily specified than each of the individual options to modify the individual elements.

Also see *Styles and composite styles* in [G] **graph twoway scatter** for more information.

Suppressing lines

Sometimes you want to suppress lines. For instance, you might want to remove the border around the plot region. There are two ways to do this: by specifying

⟨*object*⟩⟨l or li or line⟩style(none)

or by specifying

⟨*object*⟩⟨l or li or line⟩color(*color*)

The first works well in most cases; see *Suppressing the axes* in [G] ***axis_scale_options*** for an example.

For the outlines of solid objects, however, remember that lines have a thickness and removing the outline by setting its linestyle to none sometimes makes the resulting object seem too small, especially when the object was small to begin with. In those cases, specify

⟨*object*⟩⟨l or li or line⟩color(*color*)

and set the outline color to be the same as the interior color.

Also See

Complementary: [G] **lines**; [G] *linepatternstyle*, [G] *linewidthstyle*, [G] *colorstyle*;
[G] *connectstyle*

Title

> *linewidthstyle* — Choices for thickness of lines

Syntax

linewidthstyle may be

linewidthstyle	description
none	line has zero width; it vanishes
vvthin	thinnest
vthin	
thin	
medthin	
medium	
medthick	
thick	
vthick	
vvthick	
vvvthick	thickest
relativesize	any size you want

See [G] *relativesize*.

Other *linewidthstyles* may be available; type

      . graph query linewidthstyle

to obtain the full list installed on your computer.

Description

A line's look is determined by its pattern, thickness, and color; see [G] **lines**. *linewidthstyle* specifies the line's thickness.

linewidthstyle is specified via options named

⟨*object*⟩⟨l or li or line⟩width()

For instance, in the case of connecting lines (the lines used to connect points in a plot) used by graph twoway function, the option is named clwidth():

      . twoway function ..., clwidth(*linewidthstyle*) ...

In some cases, you will see that a *linewidthstylelist* is allowed:

      . twoway line ..., clwidth(*linewidthstylelist*) ...

A *linewidthstylelist* is a sequence of *linewidths* separated by spaces. Shorthands are allowed to make specifying the list easier; see [G] **stylelists**.

Remarks

If you specify the line width as none, the line will vanish.

Also See

Complementary: [G] **lines**; [G] *linepatternstyle*, [G] *colorstyle*, [G] *linestyle*; [G] *connectstyle*

Title

marginstyle — Choices for size of margins

Syntax

marginstyle may be

marginstyle	description	
zero	no margin	
tiny	tiny margin, all four sides	(smallest)
vsmall		
small		
medsmall		
medium		
medlarge		
large		
vlarge	very large margin, all four sides	(largest)
bottom	medium on the bottom	
top	medium on the top	
top_bottom	medium on bottom and top	
left	medium on the left	
right	medium on the right	
sides	medium on left and right	
# # # #	specified margins; left, right, bottom, top	
marginexp	specified margin or margins	

where *marginexp* is one or more elements of the form

$$\{ l \mid r \mid b \mid t \} [\, space \,] [\, + \mid - \mid = \,] \#$$

such as

```
l=5
l=5  r=5
l+5
l+5  r=7.2   b-2  t+1
```

In both the # # # # syntax and the $\{ l \mid r \mid b \mid t \} [+ \mid - \mid =] \#$ syntax, # is interpreted as a percent of the minimum of the width and height of the graph. Thus, a distance of 5 is the same in both the vertical and horizontal directions.

When applying margins to rotated textboxes, note that the terms left, right, bottom, and top refer to the box before rotation; see [G] *textbox_options*.

Other *marginstyles* may be available; type

```
. graph query marginstyle
```

to obtain the full list installed on your computer. If other *marginstyles* do exist, they are merely names associated with # # # # margins.

Description

marginstyle is used to specify margins: areas to be left unused.

Remarks

marginstyle is used, for instance, in the `margin()` suboption of `title()`:

    . graph ..., title("My title", margin(*marginstyle*)) ...

In this case, *marginstyle* specifies the margin between the text and the borders of the textbox that will contain the text (which box will ultimately be placed on the graph). See [G] *title_options* and [G] *textbox_options*.

As another example, *marginstyle* is allowed by the `margin()` suboption of `graphregion()`:

    . graph ..., graphregion(margin(*marginstyle*)) ...

In this case, it allows you to put margins around the plot region within the graph. See *Controlling the aspect ratio* in [G] *region_options* for an example.

Also See

Complementary: [G] *textbox_options*, [G] *region_options*

Title

> *marker_label_options* — Options for specifying marker labels

Syntax

The *marker_label_options* are

marker_label_options	description
mlabel(*varname*)	specify marker variable
mlabstyle(*markerlabelstyle*)	overall style of label
mlabposition(*clockpos*)	where to locate the label
mlabvposition(*varname*)	where to locate the label 2
mlabgap(*relativesize*)	gap between marker and label
mlabangle(*anglestyle*)	angle of label
mlabtextstyle(*textstyle*)	overall style of text
mlabsize(*textsizestyle*)	size of label
mlabcolor(*colorstyle*)	color of label

See [G] *markerlabelstyle*, [G] *clockpos*, [G] *relativesize*, [G] *anglestyle*,
[G] *textstyle*, [G] *textsizestyle*, and [G] *colorstyle*.

All options are *rightmost*; see [G] **repeated options**.

In some cases—such as when used with scatter—lists are allowed inside the arguments. A list is a sequence of the elements separated by spaces. Shorthands are allowed to make specifying the list easier; see [G] *stylelists*. When lists are allowed, option mlabel() allows a *varlist* in place of a *varname*.

Description

Marker labels are labels that appear next to (or in place of) markers. Markers are the ink used to mark where points are on a plot.

Options

mlabel(*varname*) specifies the (usually string) variable to be used that provides, observation by observation, the marker "text". For instance, you might have

```
. sysuse auto, clear
(1978 Automobile Data)
```

(Continued on next page)

401

```
. list mpg weight make in 1/4
```

	mpg	weight	make
1.	22	2,930	AMC Concord
2.	17	3,350	AMC Pacer
3.	22	2,640	AMC Spirit
4.	20	3,250	Buick Century

Typing

```
. scatter mpg weight, mlabel(make)
```

would draw a scatter of mpg versus weight and label each point in the scatter according to its make. (Recommendation, include "in 1/10" on the above command. Marker labels only work well when there is a small amount of data.)

mlabstyle(*markerlabelstyle*) specifies the overall look of marker labels, including their position, their size, their text style, etc. The other options documented below allow you to change each attribute of the marker label, but mlabstyle() is the starting point.

You need not specify mlabstyle() just because there is something you want to change about the look of a marker and, in fact, most people seldom specify the mlabstyle() option. You specify mlabstyle() when another style exists that is exactly what you desire or when another style would allow you to specify fewer changes to obtain what you want.

mlabposition(*clockpos*) and mlabvposition(*varname*) specify where the label is to be located relative to the point. mlabposition() and mlabvposition() are alternatives; the first specifies a constant position for all points and the second specifies a variable that contains *clockpos* (a number 0 through 12) for each point. If both options are specified, mlabvposition() takes precedence.

If neither option is specified, the default is mlabposition(3) o'clock—meaning to the right of the point.

mlabposition(12) means above the point, mlabposition(1) means above and to the right of the point, and so on. mlabposition(0) o'clock is taken to mean that the label is to be put directly on top of the point (in which case remember to also specify the msymbol(i) option so that the marker does not also display; see [G] *marker_options*).

mlabvposition(*varname*) specifies a numeric variable containing values 0 through 12 which are used, observation-by-observation, to locate the labels relative to the points.

See [G] *clockpos* for more information on specifying *clockpos*.

mlabgap(*relativesize*) specifies how much space should be put between the marker and the label. See [G] *relativesize*.

mlabangle(*anglestyle*) specifies the angle of text. The default is usually mlabangle(horizontal). See [G] *anglestyle*.

mlabtextstyle(*textstyle*) specifies the overall look of text of the marker labels, which in this case means their size and color. When you see [G] *textstyle*, you will find that a textstyle defines much more, but all those other things are ignored in the case of marker labels. In any case, the mlabsize() and mlabcolor() options documented below allow you to change the size and color, but the mlabtextstyle is the starting point.

As with mlabstyle(), you need not specify mlabtextstyle() just because there is something you want to change. You specify mlabtextstyle() when another style exists that is exactly what you desire or when another style would allow you to specify fewer changes to obtain what you want.

mlabsize(*textsizestyle*) specifies the size of the text. See [G] *textsizestyle*.

mlabcolor(*colorstyle*) specifies the color of the text. See [G] *colorstyle*.

Remarks

Remarks are presented under the headings

> *Typical use*
> *Eliminating overprinting and overruns*
> *Advanced use*
> *Using marker labels in place of markers*

Typical use

Markers are the ink used to mark where points are on a plot and marker labels optionally appear beside the markers to identify the points. For instance, if you were plotting country data, marker labels would allow you to have "Argentina", "Bolivia", ..., appear next to each point. Marker labels visually work well when there is a small amount of data.

To obtain marker labels, you specify the mlabel(*varname*) option, such as mlabel(country). *varname* is the name of a variable that, observation by observation, specifies the text with which the point is to be labeled. *varname* may be a string or numeric variable, but usually it is a string. For instance, consider the South American subset of the life expectancy by country data:

```
. sysuse lifeexp, clear
(Life expectancy, 1998)

. list country lexp gnppc if region==2
```

	country	lexp	gnppc
45.	Canada	79	19170
46.	Cuba	76	.
47.	Dominican Republic	71	1770
48.	El Salvador	69	1850
49.	Guatemala	64	1640
50.	Haiti	54	410
51.	Honduras	69	740
52.	Jamaica	75	1740
53.	Mexico	72	3840
54.	Nicaragua	68	1896
55.	Panama	74	2990
56.	Puerto Rico	76	.
57.	Trinidad and Tobago	73	4520
58.	United States	77	29240

We might graph these data and use labels to indicate the country by typing

```
. scatter lexp gnppc if region==2, mlabel(country)
```

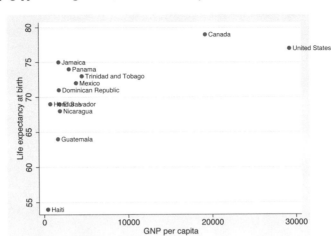

Eliminating overprinting and overruns

Note that in the graph the label "United States" runs off the right edge and that the labels for Honduras and El Salvador are overprinted. Problems like that invariably occur when using marker labels. The mlabposition() allows specifying where the labels appear, and we might try

```
. scatter lexp gnppc if region==2, mlabel(country) mlabpos(9)
```

to move the labels to the 9 o'clock position, which is to say, to being to the left of the point. In this case, however, that will introduce more problems than it will solve. You could try other clock positions around the point, but we could not find one that was satisfactory.

If our only problem were with "United States" running off the right, an adequate solution might be to widen the x axis so that there would be room for the label "United States" to fit:

```
. scatter lexp gnppc if region==2, mlabel(country)
              xscale(range(35000))
```

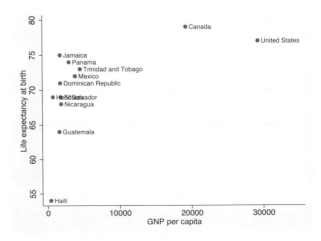

   That would solve one problem but will leave us with the overprinting problem. The way to solve that problem is to move the Honduras label to being to the left of its point, and the way to do that is to specify the option `mlabvposition(`*varname*`)` rather than `mlabposition(`*clockpos*`)`. We will create new variable `pos` stating where we want each label:

```
. generate pos = 3
```

```
. replace pos = 9 if country=="Honduras"
```

```
. scatter lexp gnppc if region==2, mlabel(country) mlabv(pos)
             xscale(range(35000))
```

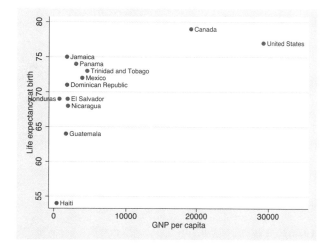

   We are near a solution: Honduras is running off the left edge of the graph, but we know how to fix that. You may be tempted to solve this problem just as we solved the problem with the United States label: expand the range, say to `range(-500 35000)`. That would be a fine solution.

   In this case, however, we will increase the margin between the left edge of the plot area and the y axis by adding the option `plotregion(margin(l+9))`; scc [G] *region_options*. `plotregion(margin(l+9))` says to increase the margin on the left by 9 percent and is really the "right" way to handle margin problems:

```
. scatter lexp gnppc if region==2, mlabel(country) mlabv(pos)
             xscale(range(35000))
             plotregion(margin(l+9))
```

(Continued on next page)

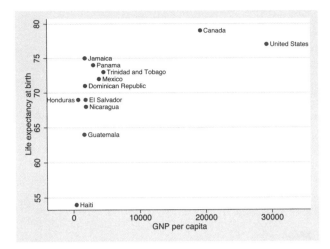

The overall result is adequate. Were we producing this graph for publication, we would move the label for United States to the left of its point, just as we did with Honduras, rather than widening the x axis.

Advanced use

Let us now consider properly graphing the life-expectancy data and graphing more of it. This time we will include South America as well as North and Central America, and we will graph the data on a log(GNP) scale.

```
. sysuse lifeexp, clear
(Life expectancy, 1998)

. keep if region==2 | region==3                                          (note 1)

. replace gnppc = gnppc / 1000
. label var gnppc "GNP per capita (thousands of dollars)"                 (note 2)

. generate lgnp = log(gnp)
. quietly reg lexp lgnp
. predict hat
. label var hat "Linear prediction"                                      (note 3)

. replace country = "Trinidad" if country=="Trinidad and Tobago"
. replace country = "Para" if country == "Paraguay"                      (note 4)

. generate pos = 3
. replace pos = 9 if lexp > hat                                          (note 5)

. replace pos = 3 if country == "Colombia"
. replace pos = 3 if country == "Para"
. replace pos = 3 if country == "Trinidad"
. replace pos = 9 if country == "United States"                          (note 6)

. twoway (scatter lexp gnppc, mlabel(country) mlablv(pos))
         (line hat gnppc, sort)
       , xscale(log) xlabel(.5 5 10 15 20 25 30, grid)
         legend(off)
         title("Life expectancy vs. GNP per capita")
         subtitle("North, Central, and South America")
         note("Data source:  World bank, 1998")
         ytitle("Life expectancy at birth (years)")
```

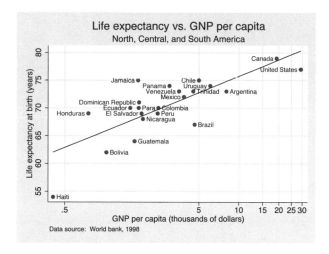

Notes:

1. In these data, region 2 is North and Central America and region 3 is South America.

2. We divide `gnppc` by 1,000 to keep the x axis labels from running into each other.

3. We add a linear regression prediction. We cannot use `graph twoway lfit` because we want the predictions based on a regression of log(GNP), not GNP.

4. The first time we graphed the results, we discovered that there was no way we could make the names of these two countries fit on our graph, so we shortened them.

5. We are going to place the marker labels to the left of the marker when life expectancy is above the regression line and to the right of the marker otherwise.

6. To keep labels from overprinting, for a few countries we need to override rule (5).

Also see [G] *scale_option* for another rendition of this graph. In that rendition, we specify one more option—`scale(1.1)`—to increase the size of the text and markers by 10%.

Using marker labels in place of markers

In addition to specifying where the marker label goes relative to the marker, you can specify that the marker label be used instead of the marker. `position(0)` o'clock is taken to mean that the label be centered where the marker would appear. To suppress the display of the marker as well, specify option `msymbol(i)`; see [G] *marker_options*.

Using the labels in place of the points tends to work well in analysis graphs where our interest is often in identifying the outliers. Below we graph the entire `lifeexp.dta` data.

```
. scatter lexp gnppc, xscale(log) mlab(country) m(i)
```

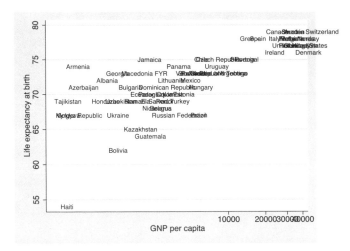

In the above graph, we also specified `xscale(log)` to convert the x axis to a log scale. A log x scale is more appropriate for these data but, had we used it earlier, the overprinting problem with Honduras and El Salvador would have disappeared, and we wanted to show how to handle the problem.

Also See

Complementary: [G] **graph twoway scatter**; [G] *anglestyle*, [G] *clockpos*, [G] *colorstyle*, [G] *markerlabelstyle*, [G] *relativesize*, [G] *textstyle*, [G] *textsizestyle*

Title

> *marker_options* — Options for specifying markers

Syntax

The *marker_options* are

marker_options	description
mstyle(*markerstyle*)	overall style of marker
msymbol(*symbolstyle*)	shape of marker
mcolor(*colorstyle*)	color of marker, inside and out
msize(*markersizestyle*)	size of marker
mfcolor(*colorstyle*)	inside or "fill" color
mlstyle(*linestyle*)	overall style of line used to outline marker's shape
mlpattern(*linepatternstyle*)	whether line is solid, dashed, ...
mlwidth(*linewidthstyle*)	thickness of line
mlcolor(*colorstyle*)	color of line

See [G] **markerstyle**, [G] **symbolstyle**, [G] **colorstyle**, [G] **markersizestyle**, [G] **linestyle**, [G] **linepatternstyle**, and [G] **linewidthstyle**.

All options are *rightmost*; see [G] **repeated options**.

One example of each of the above is

```
msymbol(O)        mfcolor(red)     mlcolor(olive)      mstyle(p1)
mcolor(green)                      mlwidth(thick)      mlstyle(p1)
msize(medium)                      mlpattern(solid)
```

It is sometimes allowed to specify a list of elements, with the first element applying to the first variable, the second to the second, and so on. See, for instance, [G] **graph twoway scatter**. One example would be

```
msymbol(O o p)
mcolor(green blue black)
msize(medium medium small)
mfcolor(red red none)
mlcolor(olive olive green)
mlwidth(thick thin thick)
mlpattern(solid solid solid)
mstyle(p1 p2 p3)
mlstyle(p1 p2 p3)
```

For information on specifying lists, see [G] *stylelists*.

Description

Markers are the ink used to mark where points are on a plot. The important options are

msymbol(*symbolstyle*)   (choice of symbol)
mcolor(*colorstyle*)   (choice of color)
msize(*markersizestyle*)   (choice of size)

Options

mstyle(*markerstyle*) specifies the overall look of markers, such as their shape, their color, etc. The other options allow you to change each attribute of the marker, but mstyle() is the starting point.

You need not specify mstyle() just because there is something you want to change about the look of the marker and, in fact, most people seldom specify the mstyle() option. You specify mstyle() when another style exists that is exactly what you desire or when another style would allow you to specify fewer changes to obtain what you want.

See [G] *markerstyle* for a list of available marker styles.

msymbol(*symbolstyle*) specifies the shape of the marker and is one of the more commonly specified options. See [G] *symbolstyle* for more information on this important option.

mcolor(*colorstyle*) specifies the color of the marker. This option sets both the color of the line used to outline the markers shape and the color of the inside of the marker. Also see options mfcolor() and mlcolor() below. See [G] *colorstyle* for a list of color choices.

msize(*markersizestyle*) specifies the size of the marker. See [G] *markersizestyle* for a list of size choices.

mfcolor(*colorstyle*) specifies the color of the inside of the marker. See [G] *colorstyle* for a list of color choices.

mlstyle(*linestyle*), mlpattern(*linepatternstyle*), mlwidth(*linewidthstyle*), and mlcolor(*colorstyle*) specify the look of the line used to outline the shape of the marker. See [G] **lines**.

Remarks

You will never need to specify all nine marker options, and seldom will you even need to specify more than one or two of them. Many people think that there is just one important marker option,

msymbol(*symbolstyle*)

msymbol() specifies the shape of the symbol; see [G] *symbolstyle* for choice of symbol. A few people would add to the important list a second option,

mcolor(*colorstyle*)

mcolor() specifies the marker's color; see [G] *colorstyle* for choice of color. Finally, a very few would add msize(*markersizestyle*)

msize() specifies the marker's size; see [G] *markersizestyle* for choice of sizes.

After that, we are really into the details. One of the remaining options, however, is of interest:

mstyle(*markerstyle*)

A marker has a set of characteristics:

{shape, color, size, inside details, outside details}

Each of the options other than mstyle() modifies something in that set. mstyle() sets the values of the entire set. It is from there that the changes you specify are made. See [G] *markerstyle*.

Also See

Complementary: [G] *symbolstyle*, [G] *colorstyle*, [G] *markersizestyle*, [G] *linewidthstyle*, [G] *linepatternstyle*, [G] *linestyle*; [G] *markerstyle*

Title

> *markerlabelstyle* — Choices for overall look of marker labels

Syntax

markerlabelstyle may be

markerlabelstyle	description
p1–p15	used by first plot, second plot, ...

Other *markerlabelstyles* may be available; type

        . graph query markerlabelstyle

to obtain the full list installed on your computer.

Description

markerlabelstyle defines the position, gap, angle, size, and color of the marker label. See [G] ***marker_label_options*** for more information.

markerlabelstyle is specified inside the `mlabstyle()` option,

        . graph ..., mlabstyle(*markerlabelstyle*) ...

In some cases (e.g., with `twoway scatter`), a *markerlabelstylelist* is allowed: A *markerlabelstylelist* is a sequence of *markerlabelstyles* separated by spaces. Shorthands are allowed to make specifying the list easier; see [G] ***stylelists***.

Remarks

Remarks are presented under the headings

> *What is a markerlabel?*
> *What is a markerlabelstyle?*
> *You do not need to specify a markerlabelstyle*
> *Specifying a markerlabelstyle can be convenient*

What is a markerlabel?

Marker labels are identifying text that appear next to (or in place of) markers. Markers are the ink used to mark where points are on a plot.

What is a markerlabelstyle?

The look of marker labels are defined by four attributes:

1. The marker label's position—where the marker is located relative to the point; see [G] *clockpos*.
2. The gap between the marker label and the point; see [G] *clockpos*.
3. The angle at which the identifying text is presented; see [G] *anglestyle*.
4. The overall style of the text; see [G] *textstyle*.
   a. The size of the text; see [G] *textsizestyle*.
   b. The color of the text; see [G] *colorstyle*.

The *markerlabelstyle* specifies all four of these attributes.

You do not need to specify a markerlabelstyle

The *markerlabelstyle* is specified by the option

mstyle(*markerlabelstyle*)

Correspondingly, you will find other options available:

mlabposition(*clockpos*)
mlabgap(*relativesize*)
mlabangle(*anglestyle*)
mlabtextstyle(*textstyle*)
mlabsize(*textstyle*)
mlabcolor(*colorstyle*)

You specify the *markerlabelstyle* when a style exists that is exactly what you desire or when another style would allow you to specify fewer changes to obtain what you want.

Specifying a markerlabelstyle can be convenient

Consider the command

. scatter y1 y2 x, mlabel(country country)

Assume that you want the marker labels for y2 versus x to appear the same as for y1 versus x. (An example of this can be found under *Eliminating overprinting and overruns* and under *Advanced use* in [G] *marker_label_options*.) One way you could proceed would be to set all the attributes for the marker labels for y1 versus x and then set all the attributes for y2 versus x to be the same. Easier, however, would be to type

. scatter y1 y2 x, mlabel(country country) mlabstyle(p1 p1)

When you do not specify mlabstyle(), results are as if you specified

mlabstyle(p1 p2 p3 p4 p5 p6 p7 p8 p9 p10 p11 p12 p13 p14 p15)

where the extra elements are ignored. In any case, p1 is one set of marker-label attributes, p2 is another set, and so on.

Say that you wanted y2 versus x to look like y1 versus x except that you wanted the line to be green; you could type

```
. scatter y1 y2 x, mlabel(country country) mlabstyle(p1 p1)
                    mlabcolor(. green)
```

There is nothing special about *markerlabelstyles* p1, p2, . . . ; they merely specify sets of marker-label attributes just like any other named *markerlabelstyle*. Type

```
. graph query markerlabelstyle
```

to find out what other marker-label styles are available.

Also see *Styles and composite styles* in [G] **graph twoway scatter** for more information.

Also See

Complementary: [G] *marker_label_options*

Title

> *markersizestyle* — Choices for the size of markers

Syntax

markersizestyle may be

markersizestyle	description
vtiny	the smallest
tiny	
vsmall	
small	
medsmall	
medium	
medlarge	
large	
vlarge	
huge	
vhuge	
ehuge	the largest
relativesize	any size you want including size modification

See [G] ***relativesize***.

Other *markersizestyles* may be available; type

    . graph query markersizestyle

to obtain the full list installed on your computer.

Description

Markers are the ink used to mark where points are on a plot; see [G] ***marker_options***. *markersizestyle* specifies the size of the markers.

Remarks

markersizestyle is specified inside the msize() option:

    . graph ..., msize(*markersizestyle*) ...

In some cases, you will see that a *markersizestylelist* is allowed:

    . scatter ..., msymbol(*markersizestylelist*) ...

A *markersizestylelist* is a sequence of *markersizestyles* separated by spaces. Shorthands are allowed to make specifying the list easier; see [G] ***stylelists***.

Also See

Complementary: [G] *marker_options*; [G] *symbolstyle*, [G] *colorstyle*, [G] *linewidthstyle*, [G] *linepatternstyle*, [G] *linestyle*, [G] *markerstyle*

Title

> *markerstyle* — Choices for overall look of markers

Syntax

markerstyle may be

markerstyle	description
p1–p15	used by first plot, second plot, ...

Other *markerstyles* may be available; type

    . graph query markerstyle

to obtain the full list installed on your computer.

Description

Markers are the ink used to mark where points are on a plot. *markerstyle* defines the symbol, size, and color of a marker. See [G] ***marker_options*** for more information.

markerstyle is specified inside the mstyle() option,

    . graph ..., mstyle(*markerstyle*) ...

In some cases, you will see that a *markerstylelist* is allowed:

    . twoway scatter ..., mstyle(*markerstylelist*) ...

A *markerstylelist* is a sequence of *markerstyles* separated by spaces. Shorthands are allowed to make specifying the list easier; see [G] ***stylelists***.

Remarks

Remarks are presented under the headings

> *What is a marker?*
> *What is a markerstyle?*
> *You do not have to specify a markerstyle*
> *Specifying a markerstyle can be convenient*

What is a marker?

Markers are the ink used to mark where points are on a plot. Some people use the word point or symbol, but a point is where the marker is placed and a symbol is merely one characteristic of a marker.

417

What is a markerstyle?

Markers are defined by five attributes:

1. The *symbol*—the shape of the marker; see [G] **symbolstyle**.

2. The *markersize*—the size of the marker; see [G] **markersizestyle**.

3. The overall color of the marker; see [G] **colorstyle**.

4. The interior (fill) color of the marker; see [G] **colorstyle**.

5. The line that outlines the shape of the marker:

    a. The overall style of the line; see [G] **linestyle**.

    b. The pattern of the line—whether solid, dashed, etc.; see [G] **linepatternstyle**.

    c. The thickness of the line; see [G] **linewidthstyle**.

    d. The color of the line; see [G] **colorstyle**.

The *markerstyle* defines all five (eight) of these attributes.

You do not have to specify a markerstyle

The *markerstyle* is specified via the

    mstyle(*markerstyle*)

option. Correspondingly, you will find eight other options available:

    msymbol(*symbolstyle*)
    msize(*markersizestyle*)
    mcolor(*colorstyle*)
    mfcolor(*colorstyle*)
    mlstyle(*linestyle*)
    mlpattern(*linepatternstyle*)
    mlwidth(*linewidthstyle*)
    mlcolor(*colorstyle*)

You specify the *markerstyle* when a style exists that is exactly what you desire or when another style would allow you to specify fewer changes to obtain what you want.

Specifying a markerstyle can be convenient

Consider the command

    . scatter y1var y2var xvar

Pretend that you wanted the markers for y2var versus xvar to be the same as y1var versus xvar. One way you could proceed would be to set all the characteristics of the marker for y1var versus xvar and then set all the characteristics of the marker for y2var versus xvar to be the same. Easier, however, would be to type

    . scatter y1var y2var xvar, mstyle(p1 p1)

mstyle() is the name of the option that specifies the overall style of the marker. When you do not specify the mstyle() option, results are as if you specified

> mstyle(p1 p2 p3 p4 p5 p6 p7 p8 p9 p10 p11 p12 p13 p14 p15)

where the extra elements are ignored. In any case, p1 is one set of marker characteristics, p2 another, and so on.

Say that you wanted y2var versus xvar to look like y1var versus xvar except that you wanted the symbols to be green; you could type

> . scatter y1var y2var xvar, mstyle(p1 p1) mcolor(. green)

There is nothing special about the *markerstyles* p1, p2, ...; they merely specify sets of marker attributes just like any other named *markerstyle*. Type

> . graph query markerstyle

to find out what other marker styles are available. You may find something pleasing and, if so, that is more easily specified than each of the individual options to modify the shape, color, size, ... elements.

Also See

Complementary: [G] *marker_options*; [G] *symbolstyle*, [G] *colorstyle*, [G] *markersizestyle*, [G] *linewidthstyle*, [G] *linepatternstyle*, [G] *markerstyle*, [G] *linestyle*; [G] *stylelists*

Title

> *mf_options* — Options for exporting to Windows Metafiles

Syntax

The *mf_options* when exporting to Windows Metafiles or Enhanced Metafiles are

mf_options	description
<u>font</u>face(*fontname*)	font to use

Current default values may be listed by typing

```
. graph set mf
```

and default values may be set by typing

```
. graph set mf name value
```

where *name* is the name of an *mf_option*, omitting the parentheses.

Description

These *mf_options* are used with `graph export` when creating Windows Metafiles or Windows Enhanced Metafiles; see [G] **graph export**.

Options

<u>font</u>face(*fontname*) specifies the name of the font to be used to render text. If *fontname* includes blanks, *fontname* must be enclosed in double quotes.

Remarks

Remarks are presented under the headings

> *Using the mf_options*
> *Setting defaults*

Using the mf_options

You have drawn a graph and wish to create an Enhanced Metafile for including in a document. You wish, however, to change the text from Ariel to Times New Roman:

```
. graph ...                              (draw a graph)
. graph export myfile.emf, fontface("Times New Roman")
```

Setting defaults

If you always wanted `graph export` to use Times New Roman when exporting to a Metafile, you could type

```
. graph set mf fontface "Times New Roman"
```

At a future date, you could type

```
. graph set mf fontface Ariel
```

to set it back. You can list the current *mf_option* settings for Metafiles by typing

```
. graph set mf
```

Also See

Complementary: [G] **graph export**

Title

name_option — Option for naming graph in memory

Syntax

The *name_option* is

name_option	description
name(*name*[, replace])	specify name

name() is *unique*; see [G] **repeated options**.

Description

Option name() specifies the name of the graph being created.

Options

name(*name*[, replace]) specifies the name of the graph. If name() is not specified, name(Graph, replace) is assumed.

In fact, name(Graph) has the same effect as name(Graph, replace) because replace is assumed when the name is Graph. For all other *names*, you must specify suboption replace if a graph under that name already exists.

Remarks

When you type, for instance,

    . scatter yvar xvar

that results in you seeing a graph. In addition, the graph is stored in memory. For instance, try the following: close the Graph window and then type

    . graph display

Your graph will reappear.

Every time you draw a graph, that previously remembered graph is discarded and the new graph replaces it.

You can have more than one graph stored in memory. When you do not specify the name under which the graph is to be remembered, it is remembered under the default name Graph. For instance, if you were now to type

    . scatter y2var xvar, name(g2)

You would now have two graphs stored in memory: Graph and g2. If you typed

    . graph display

or

    . graph display Graph

you would see your first graph. Type

    . graph display g2

and you will see your second graph.

Do not confuse Stata's storing of graphs in memory with the saving of graphs to disk. Were you now to exit Stata, the graphs you have saved in memory would be gone forever. If you want to save your graphs, you want to specify the saving() option (see [G] *saving_option*) or you want to use the graph save command (see [G] **graph save**); either result in the same outcome.

You can find out what graphs you have in memory using graph dir, drop them using graph drop, rename them using graph rename, and so on, and of course, you can redisplay them using graph display. See [G] **graph manipulation** for the details on all of those commands.

You can drop all graphs currently stored in memory using graph drop _all or discard; see [G] **graph drop**.

Also See

Complementary: [G] **graph display**, [G] **graph drop**; [G] *saving_option*,
                        [G] **graph save**; [G] **graph manipulation**

Title

> *nodraw_option* — Option for suppressing display of graph

Syntax

The *nodraw_option* is

nodraw_option	description
nodraw	suppress display of graph

nodraw is *unique*; see [G] **repeated options**.

Description

Option nodraw prevents the graph from being displayed.

Options

nodraw specifies the graph is not to be displayed.

Remarks

When you type, for instance,

```
. scatter yvar xvar, saving(mygraph)
```

that results in you seeing a graph and it being stored in file mygraph.gph. If you type

```
. scatter yvar xvar, saving(mygraph) nodraw
```

the graph will still be saved in file mygraph.gph, but it will not be displayed. The result is the same as if you typed

```
. set graphics off
. scatter yvar xvar, saving(mygraph)
. set graphics on
```

It is not necessary that you specifying saving() (see [G] *saving_option*) to use nodraw. You could type

```
. scatter yvar xvar, nodraw
```

and later type (or code in an ado-file)

```
. graph display Graph
```

See [G] **graph display**.

Also See

Complementary: [R] **set**

424

Title

orientationstyle — Choices for orientation of textboxes

Syntax

orientationstyle may be

orientationstyle	description
horizontal	text reads left to right
vertical	text reads bottom to top
rhorizontal	text reads left to right (upside down)
rvertical	text reads top to bottom

Other *orientationstyles* may be available; type

    . graph query orientationstyle

to obtain the full list installed on your computer.

Description

A textbox contains one or more lines of text. *orientationstyle* specifies whether the textbox is horizontal or vertical.

orientationstyle is specified in the orientation() option nested within another option, such as title():

    . graph ..., title("My title", orientation(*orientationstyle*)) ...

See [G] *textbox_options* for more information on textboxes.

Remarks

orientationstyle specifies whether the text and box are oriented horizontally or vertically, vertically including text reading from bottom to top or from top to bottom.

Also See

Complementary: [G] *textbox_options*

Title

> **palette** — Display palettes of available selections

Syntax

> palette color *colorstyle* [*colorstyle*] [, sch<u>eme</u> (*schemename*)]
>
> palette <u>line</u>palette [, sch<u>eme</u> (*schemename*)]
>
> palette <u>symbol</u>palette [, sch<u>eme</u> (*schemename*)]

Description

palette produces graphs showing various selections available.

palette color shows how a particular color looks and allows you to compare two colors; see [G] *colorstyle*.

palette linepalette shows you the different *linepatternstyles*; see [G] *linepatternstyle*.

palette symbolpalette shows you the different *symbolstyles*; see [G] *symbolstyle*.

Options

scheme(*schemename*) specifies the scheme to be used to draw the graph. With this command, scheme() is rarely specified. We recommend specifying scheme(color) if you plan to print the graph on a color printer; see [G] **schemes**.

Remarks

The palette command is more a part of the documentation of graph than a useful command in its own right.

Also See

Complementary: [G] **graph**, [G] **graph query**

Title

pict_options — Options for exporting or printing to Macintosh PICT

Syntax

The *pict_options* when exporting to PICT are

pict_options	description
mag(#)	magnification/shrinkage factor; default = 100
fontface(*fontname*)	font to use

Current default values may be listed by typing

. graph set pict

and default values may be set by typing

. graph set pict *name value*

where *name* is the name of a *pict_option*, omitting the parentheses.

Description

These *pict_options* are used with graph export when creating a PICT file; see [G] **graph export**.

Options

mag(#) specifies that the graph is to be drawn smaller or larger than ordinarily. mag(100) is the default, meaning ordinary size. mag(110) would make the graph 10% larger than usual and mag(90) would make the graph 10% smaller than usual. # must be an integer.

fontface(*fontname*) specifies the name of the PostScript font to be used to render text. If *fontname* contains blanks, *fontname* must be enclosed in double quotes.

Remarks

Remarks are presented under the headings

Using the pict_options
Setting defaults

427

Using the pict_options

You have drawn a graph and wish to create a PICT file. You wish, however, to change the text from helvetica to roman, which is "Times" in PICT jargon:

```
. graph ...                                    (draw a graph)
. graph export myfile.pict, fontface(Times)
```

Setting defaults

If you always wanted `graph export` to use Times when exporting to PICT files, you could type

```
. graph set pict fontface Times
```

At a future date, you could type

```
. graph set pict fontface Geneva
```

to set it back. You can list the *pict_option* settings for PICT by typing

```
. graph set pict
```

Also See

Complementary: [G] **graph export**

Title

> *plot_option* — Option for adding additional twoway plots to command

Syntax

command ... $\big[$, ... `plot`(*plot* ... $\big[$ || *plot* ... $\big[$ ... $\big]$ $\big]$) ... $\big]$

where *plot* may be any subcommand of `graph twoway` such as `scatter`, `line`, `histogram`, etc.

Description

Some commands that draw graphs are documented in the other *Reference* manuals; documented there are those commands that do not start with the word `graph`. Many of those commands allow the `plot()` option. This option allows them to overlay their results on top of `graph twoway` plots.

Options

`plot`(*plot*) specifies the rest of the `graph twoway` subcommand(s) to be added to the `graph twoway` command issued by *command*.

Remarks

Remarks are presented under the headings

> *Commands that allow the plot() option*
> *Advantages of graph twoway commands*
> *Advantages of graphic commands implemented outside of graph twoway*
> *Use of the plot() option*

Commands that allow the plot() option

`graph` commands never allow the `plot()` option. The `plot()` option is allowed by commands outside of `graph` that are implemented in terms of `graph twoway`.

For instance, the `histogram` command—see [R] **histogram**—allows `plot()`. `graph twoway histogram`—see [G] **graph twoway histogram**—does not.

Advantages of graph twoway commands

The advantage of `graph twoway` commands is that they can be overlaid one on top of the other. For instance, you can type

   . `graph twoway scatter` *yvar xvar* || `lfit` *yvar xvar*

and the separate graphs `scatter` and `lfit` produced are combined into one. The variables to which each refers need not even be the same:

   . `graph twoway scatter` *yvar xvar* || `lfit` *y2var x2var*

Advantages of graphic commands implemented outside of graph twoway

Graphic commands implemented outside of `graph twoway` can have simpler syntax. For instance, the `histogram` command has an option `normal` that will overlay a normal curve on top of the histogram:

```
. histogram myvar, normal
```

That is easier than typing

```
. summarize myvar
. graph twoway histogram myvar ||
    function normden(x,'r(mean)','r(sd)'), range(myvar)
```

which is the `graph twoway` way of getting the same thing.

Thus, the trade off between `graph` and non-`graph` commands is one of greater flexibility versus easier use.

Use of the plot() option

The `plot()` option attempts to give back flexibility to non-`graph` graphic commands. Such commands are, in fact, implemented in terms of `graph twoway`. For instance, when you type

```
. histogram ...
```

or you type

```
. sts graph ...
```

the result is that those commands construct a complicated `graph twoway` command:

> → `graph twoway` *something_complicated*

and they run that for you. When you specify the plot option, such as

```
. histogram ..., plot(your_contribution)
```

or

```
. sts graph, plot(your_contribution)
```

the result is that the commands construct

> → `graph twoway` *something_complicated* || *your_contribution*

Let us assume you have survival data and wish to visually compare the Kaplan-Meier, i.e., the empirical survivor function, to the function that would be predicted if the survival times were assumed to be exponentially distributed. Simply typing

```
. sysuse cancer, clear
(Patient Survival in Drug Trial)
. quietly stset studytime, fail(died)
```

```
. sts graph
```

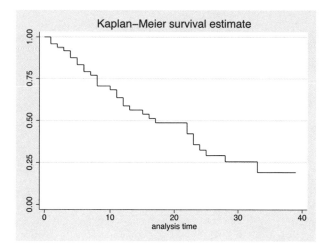

will obtain a graph of the empirical estimate. To obtain the exponential estimate, you might type

```
. quietly streg, distribution(exponential)
. predict S, surv
. graph twoway line S _t, sort
```

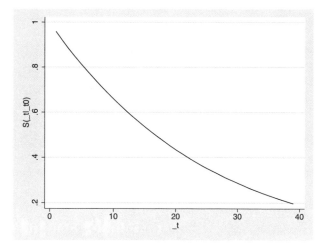

To put these two graphs together, we can type

(Continued on next page)

```
. sts graph, plot(line S _t, sort)
```

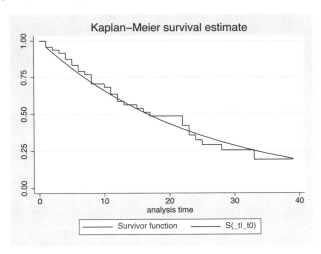

The result is just as if we typed

```
. sts graph || line S _t, sort
```

if only that were allowed.

Also See

Complementary: [G] **graph twoway**

Title

> *pr_options* — Options for use with graph print

Syntax

The *pr_options* are

pr_options	description
<u>t</u>margin(#)	top margin, inches, $0 \leq \# \leq 20$
<u>l</u>margin(#)	left margin, inches, $0 \leq \# \leq 20$
logo(on | off)	whether to display Stata logo

Current default values may be listed by typing

    . graph set print

The defaults may be changed by typing

    . graph set print *name value*

where *name* is the name of a *pr_option*, omitting the parentheses.

Description

The *pr_options* are used with graph print; see [G] **graph print**.

Options

tmargin(#) and lmargin(#) set the top and left page margins—the distance from the edge of the page to the start of the graph. # is specified in inches, must be between 0 and 20, and may be fractional.

logo(on) and logo(off) specify whether the Stata logo should be included at the bottom of the graph.

Remarks

Remarks are presented under the headings

> *Using the pr_options*
> *Setting defaults*
> *Note for Unix users*

Using the pr_options

You have drawn a graph and wish to print it. You wish, however, to suppress the Stata logo (although we cannot imagine why you would want to do that):

    . graph ...   (draw a graph)

    . graph print, logo(off)

Setting defaults

If you always wanted `graph print` to suppress the Stata logo, you could type

    . graph set print logo off

At a future date, you could type

    . graph set print logo on

to set it back. You can determine your default *pr_options* settings by typing

    . graph set print

Note for Unix users

In addition to the options documented above, there are additional options you may specify. Under Stata for Unix, the *pr_options* are in fact *ps_options*; see [G] **ps_options**.

Also See

Complementary: [G] **graph print**

Title

ps_options — Options for exporting or printing to PostScript

Syntax

The *ps_options* when exporting to PostScript are

ps_options	description
tmargin(#)	top margin in inches
lmargin(#)	left margin in inches
logo(on | off)	whether to include Stata logo
mag(#)	magnification/shrinkage factor; default = 100
fontface(fontname)	font to use
orientation(portrait | landscape)	whether vertical or horizontal
pagesize(letter | legal | executive | A4 | custom)	size of page
pageheight(#)	inches; relevant only if pagesize(custom)
pagewidth(#)	inches; relevant only if pagesize(custom)

Current default values may be listed by typing

. graph set ps

and default values may be set by typing

. graph set ps *name value*

where *name* is the name of a *ps_option*, omitting the parentheses.

Description

These *ps_options* are used with `graph export` when creating a PostScript file; see [G] **graph export**.

In addition, under Stata for Unix, these options are used with `graph print`; see [G] **graph print**.

Options

tmargin(#) and lmargin(#) set the top and left page margins—the distance from the edge of the page to the start of the graph. # is specified in inches, must be between 0 and 20, and may be fractional.

logo(on) and logo(off) specify whether the Stata logo should be included at the bottom of the graph.

mag(#) specifies that the graph is to be drawn smaller or larger than ordinarily. mag(100) is the default, meaning ordinary size. mag(110) would make the graph 10% larger than usual and mag(90) would make the graph 10% smaller than usual. # must be an integer.

`fontface`(*fontname*) specifies the name of the PostScript font to be used to render text. If *fontname* contains blanks, *fontname* must be enclosed in double quotes.

`orientation`(`portrait`) and `orientation`(`landscape`) specify whether the graph is to be presented vertically or horizontally.

`pagesize`() specifies the size of the page. `pagesize`(`letter`), `pagesize`(`legal`), `page-size`(`executive`), and `pagesize`(`A4`) are prerecorded sizes. `pagesize`(`custom`) specifies that you wish to explicitly specify the size of the page using the `pageheight`() and `pagewidth`() options.

`pageheight`(`##`) and `pagewidth`(`##`) are relevant only if `pagesize`(`custom`) is specified. They specify the height and width of the page in inches. # is specified in inches, must be between 0 and 20, and may be fractional.

Remarks

Remarks are presented under the headings

> *Using the ps_options*
> *Setting defaults*
> *Note for Unix users*

Using the ps_options

You have drawn a graph and wish to create a PostScript file. You wish, however, to change the text from helvetica to roman, which is "Times" in PostScript jargon:

```
. graph ...                              (draw a graph)
. graph export myfile.ps, fontface(Times)
```

Setting defaults

If you always wanted **graph export** to use Times when exporting to PostScript files, you could type

```
. graph set ps fontface Times
```

At a future date, you could type

```
. graph set ps fontface Helvetica
```

to set it back. You can list the current *ps_option* settings for PostScript by typing

```
. graph set ps
```

Note for Unix users

The PostScript settings are used not only by **graph export** when creating a PostScript file, but also by **graph print**. In [G] *pr_options*, you are told that you may list and set defaults by typing

```
graph set print ...
```

That is true, but under Unix, `print` is a synonym for `ps`, so whether you type **graph set print** or **graph set ps** makes no difference.

Also See

Complementary: [G] **graph export**

Title

pstyle — Choices for overall look of plot

Syntax

pstyle may be

pstyle	description
p1–p15	used by first plot, second plot, ...

Other *pstyles* may be available; type

```
. graph query pstyle
```

to obtain the full list installed on your computer.

Description

A *pstyle*—always specified inside option pstyle(*pstyle*)—specifies the overall style of a plot and is a composite of *markerstyle*, *markerlabelstyle*, *areastyle*, and, concerning connected lines, *linestyle*, *connectstyle*, and the *connect_option* cmissing().

Remarks

Remarks are presented under the headings

> *What is a plot?*
> *What is a pstyle?*
> *You do not need to specify a pstyle*
> *Specifying a pstyle can be convenient*

What is a plot?

When you type

```
. scatter y x
```

y versus x is called a plot. When you type

```
. scatter y1 x || scatter y2 x
```

or

```
. scatter y1 y2 x
```

y1 versus x is the first plot and y2 versus x is the second.

A plot is a single presentation of a data on a graph.

438

What is a pstyle?

The overall look of a plot—the *pstyle*—is defined by the attributes:

1. The way markers look, which includes their shape, color, size, etc.; see [G] *markerstyle*.

2. The way marker labels look, which includes the position, angle, size, color, etc.; see [G] *markerlabelstyle*.

3. The way lines look that are used to connect points, which includes their color, width, and whether solid, dashed, etc.; see [G] *linestyle*.

4. The way points are connected by lines (straight lines, stair step, etc.) if they are connected; see [G] *connectstyle*.

5. Whether missing values are ignored or cause lines to be broken when they points are being connected.

6. The way areas such as bars or beneath or between curves are filled, colored, or shaded; which includes whether and how they are outlined; see [G] *areastyle*.

7. The look of the "dots" in dot plots.

The *pstyle* specifies these seven attributes.

You do not need to specify a pstyle

The *pstyle* is specified by the option

pstyle(*pstyle*)

Correspondingly, you will always find that three other options are available to control each of the attributes; see, for instance, [G] **graph twoway scatter**.

You specify the *pstyle* when a style exists that is exactly what you desire or when another style would allow you to specify fewer changes to obtain what you want.

Specifying a pstyle can be convenient

Consider the command

    . scatter y1 y2 x, ...

and further, assume that lots of options are specified. Now imagine that you want to make the plot of y1 versus x look just like the plot of y2 versus x: you want the same marker symbols used, the same colors, the same style of connecting lines (if they are connecting), etc. Whatever attributes there are, you want them treated the same.

One solution would be to track down every little detail of how the things that are displayed appear, and specify options to make sure that they are specified the same. Easier, however, would be to type

    . scatter y1 y2 x, ... pstyle(p1 p1)

When you do not specify the pstyle() option, results are as if you specified

pstyle(p1 p2 p3 p4 p5 p6 p7 p8 p9 p10 p11 p12 p13 p14 p15)

where the extra elements are ignored. In any case, p1 is one set of plot-appearance values, p2 is another set, and so on. And so by typing

    . scatter y1 y2 x, ... pstyle(p1 p1)

all the appearance values used for y2 versus x are the same as those used for y1 versus x.

Say that you wanted y2 versus x to look like y1 versus x except that you wanted the markers to be green; you could type

```
. scatter y1 y2 x, ... pstyle(p1 p1) mcolor(. green)
```

There is nothing special about the *pstyles* p1, p2, . . . ; they merely specify sets of plot-appearance values just like any other. Type

```
. graph query pstyle
```

to find out what other plot styles are available.

Also see *Styles and composite styles* in [G] **graph twoway scatter** for more information.

Also See

Complementary: [G] **graph twoway scatter**; [G] *markerstyle*, [G] *markerlabelstyle*,
[G] *areastyle*, [G] *linestyle*, [G] *connectstyle*, [G] *connect_options*

Title

> *region_options* — Options for shading and outlining regions and controlling aspect

Syntax

The *region_options* are

region_options	description
ysize(*#*)	height of *available area* (in inches)
xsize(*#*)	width of *available area* (in inches)
graphregion(*suboptions*)	attributes of *graph region*
plotregion(*suboptions*)	attributes of *plot region*

Options ysize() and xsize() are *rightmost*; options graphregion() and plotregion() are *merged-implicit*; see [G] **repeated options**.

where *suboptions* are

suboptions	description
style(*areastyle*)	overall style of outer region
color(*colorstyle*)	line + fill color of outer region
fcolor(*colorstyle*)	fill color of outer region
lstyle(*linestyle*)	overall style of outline
lcolor(*colorstyle*)	color of outline
lwidth(*linewidthstyle*)	thickness of outline
lpattern(*linepatternstyle*)	whether outline is solid, dashed, etc.
istyle(*areastyle*)	overall style of inner region
icolor(*colorstyle*)	line + fill color of inner region
ifcolor(*colorstyle*)	fill color of inner region
ilstyle(*linestyle*)	overall style of outline
ilcolor(*colorstyle*)	color of outline
ilwidth(*linewidthstyle*)	thickness of outline
ilpattern(*linepatternstyle*)	whether outline is solid, dashed, etc.
margin(*marginstyle*)	margin between inner and outer regions

See [G] **areastyle**, [G] **colorstyle**, [G] **linestyle**, [G] **linewidthstyle**, [G] **linepatternstyle**, and [G] **marginstyle**.

The *available area*, *graph region*, and *plot region* are defined

441

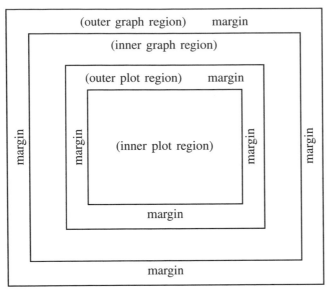

titles appear outside the borders of outer plot region

axes appear on the borders of the outer plot region

plot appears in inner plot region

Note: *what are called the "graph region" and the "plot region" are sometimes the inner and sometimes the outer regions.*

The available area and outer graph region are almost coincident; they differ only by the width of the border.

The borders of the outer plot or graph region are sometimes called the outer borders of the plot or graph region.

Description

The *region_options* set the size, margins, and color of the area in which the graph appears.

Options

ysize(*#*) and xsize(*#*) specify in inches the height and width of the *available area*. The defaults are usually ysize(4) and xsize(5), but this, of course, is controlled by the scheme; see [G] **schemes**. These two options can be used indirectly to control the aspect ratio. See *Controlling the aspect ratio* below.

graphregion(*suboptions*) and plotregion(*suboptions*) specify attributes for the *graph region* and *plot region*.

Suboptions

style(*areastyle*) and istyle(*areastyle*) specify the overall style of the outer and inner regions. The other suboptions allow you to change the region's attributes individually, but style() and istyle() provide the starting points. See [G] *areastyle* for a list of choices.

color(*colorstyle*) and icolor(*colorstyle*) specify the color of the line used to outline the outer and inner regions; see [G] *colorstyle* for a list of choices.

fcolor(*colorstyle*) and ifcolor(*colorstyle*) specify the fill color for the outer and inner regions; see [G] *colorstyle* for a list of choices.

lstyle(*linestyle*) and ilstyle(*linestyle*) specify the overall style of the line used to outline the outer and inner regions, which includes its pattern (solid, dashed, etc.), its thickness, and its color. The other suboptions listed below allow you to change the line's attributes individually, but lstyle() and ilstyle() are the starting points. See [G] *linestyle* for a list of choices.

lcolor(*colorstyle*) and ilcolor(*colorstyle*) specify the color of the line used to outline the outer and inner regions; see [G] *colorstyle* for a list of choices.

lwidth(*linewidthstyle*) and ilwidth(*linewidthstyle*) specify the thickness of the line used to outline the outer and inner regions; see [G] *linewidthstyle* for a list of choices.

lpattern(*linepatternstyle*) and ilpattern(*linepatternstyle*) specify whether the line used to outline the outer and inner regions is solid, dashed, etc.; see [G] *linepatternstyle* for a list of choices.

margin(*marginstyle*) specifies the margin between the outer and inner regions.

Remarks

Remarks are presented under the headings

> *Setting the offset between the axes and the plot region*
> *Controlling the aspect ratio*
> *Suppressing the border around the plot region*
> *Setting background and fill colors*
> *How graphs are constructed*

Setting the offset between the axes and the plot region

By default, most schemes (see [G] **schemes**) offset the axes from the region in which the data are plotted. This offset is specified plotregion(margin(*marginstyle*)); see [G] *marginstyle*.

If you do not want the axes offset from the contents of the plot, specify plotregion(margin(zero)). Compare

```
. sysuse auto, clear
(1978 Automobile Data)

. scatter price mpg
```

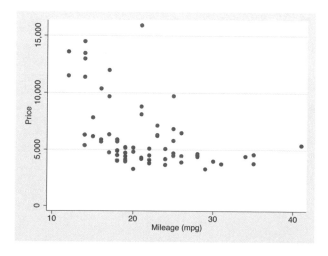

with

```
. scatter price mpg, plotr(m(zero))
```

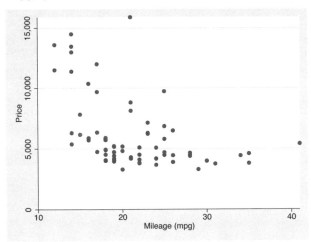

Controlling the aspect ratio

The best way to control the aspect ratio is by specifying xsize() or ysize() options. For instance, you draw a graph and find that the plot region is too wide given its height. To address the problem, either increase ysize() or decrease xsize(). The usual defaults (which of course are determined by the scheme; see [G] **schemes**) are ysize(4) and xsize(5.5), so you might try

```
. graph ..., ... ysize(5)
```

or

```
. graph ..., ... xsize(4.5)
```

For instance, compare

```
. scatter mpg weight
```

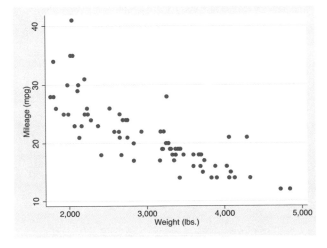

```
. scatter mpg weight, ysize(5)
```

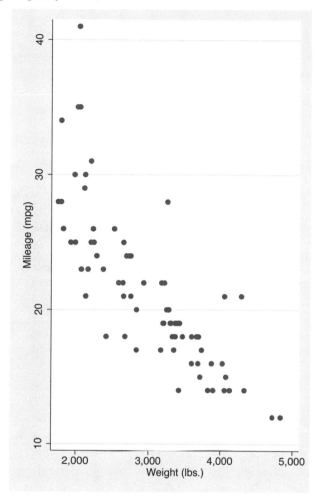

Another way to control the aspect ratio is to add to the outer margin of the *graph area*. This will keep the overall size of the graph the same while using less of the *available area*. For instance,

(Continued on next page)

```
. scatter mpg weight, graphregion(margin(l+10 r+10))
```

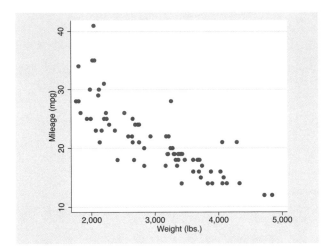

This method is especially useful when using `graph, by()`, but remember to specify the `graphre-gion(margin())` option inside the `by()` so that it affects the entire graph:

```
. scatter mpg weight, by(foreign, total graphr(m(l+10 r+10)))
```

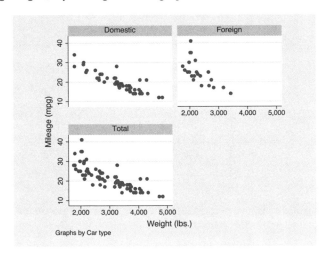

Compare the above with

(Continued on next page)

. scatter mpg weight, by(foreign, total)

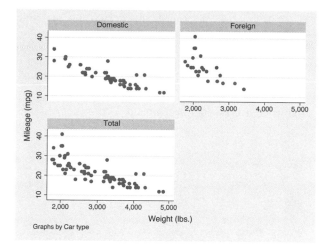

You do not have to get the aspect ratio or size right the first time you draw a graph; using graph display you can change the aspect ratio of an already drawn graph—even a graph saved in a .gph file. See *Changing the size and aspect ratio* in [G] **graph display**.

Suppressing the border around the plot region

To eliminate the border around the plot region, specify plotregion(style(none)):

. sysuse auto, clear
(1978 Automobile Data)
. scatter mpg weight, plotregion(style(none))

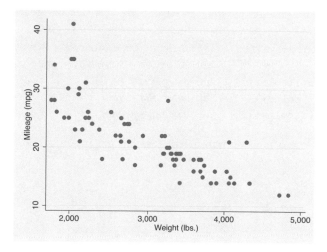

Setting background and fill colors

The background color of a graph is determined by default by the scheme you choose—see [G] **schemes**—and is usually black or white, perhaps with a tint. Option graphregion(fcolor(*colorstyle*)) allows you to override the scheme's selection. When doing this, choose a light background color for schemes that are naturally white and a dark background color for schemes that are naturally black, or you will have to type lots and lots of options to make your graph look good.

Below we draw a graph using a light gray background:

```
. sysuse auto, clear
(1978 Automobile Data)

. scatter mpg weight, graphregion(fcolor(gs13))
```

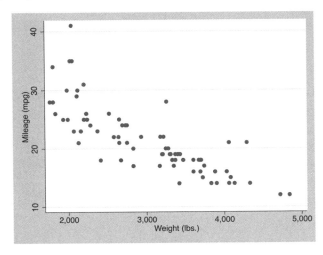

See [G] *colorstyle* for information on what may be specified inside the graphregion(fcolor()) option.

In addition to graphregion(fcolor()), there are three other fill-color options:

graphregion(ifcolor())	fills *inner graph region*	←*of little use*
plotregion(fcolor())	fills *outer plot region*	←*useful*
plotregion(ifcolor())	fills *inner plot region*	←*could be useful*

plotregion(fcolor()) is worth remembering. Below we make the plot region a light gray:

(Continued on next page)

. scatter mpg weight, plotr(fcolor(gs13))

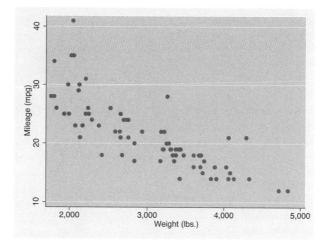

The other two options—graphregion(ifcolor()) and plotregion(ifcolor())—fill the *inner graph region* and *inner plot region*. Filling the *inner graph region* serves little purpose. Filling the *inner plot region*—which is the same as the *outer plot region* except that it omits the margin between the *inner plot region* and the axes—generally makes graphs appear too busy.

How graphs are constructed

graph works from the outside in, with the result that the dimensions of the *plot region* are what are left over.

graph begins with the *available area*, the size of which is determined by the xsize() and ysize() options. graph indents on all four sides by graphregion(margin()), and so defines the outer border of the *graph region*, the interior of which is the *inner graph region*.

Overall titles (if any) are now placed on the graph and, on each of the four sides, those titles are allocated whatever space they require. Next are placed any axis titles and labels, and they too are allocated whatever space necessary. That then determines the outer border of the *plot region* (or, more properly, the border of the *outer plot region*).

The axis (if any) is placed right on top of that border. graph now indents on all four sides by plotregion(margin()) and that determines the inner border of the plot region, which is to say, the border of the (inner) *plot region*. It is inside this that the data are plotted.

An implication of the above is that, if plotregion(margin(zero)), the axes are not offset from the region in which the data are plotted.

Now let us consider the lines used to outline the regions and the fill colors used to shade their interiors.

Starting once again with the *available area*, graph outlines its borders using graphregion(lstyle())—which is usually graphregion(lstyle(none))—and fills the area with the graphregion(fcolor()).

graph now moves to the inner border of the *graph region*, outlines it using graphregion(ilstyle()) and fills the *graph region* with graphregion(ifcolor()).

graph moves to the outer border of the *plot region*, outlines it using `plotregion(lstyle())` and fills the *outer plot region* with `plotregion(fcolor())`.

Finally, graph moves to the inner border of the *plot region*, using `plotregion(ilstyle())` and fills the (*inner*) *plot region* with `plotregion(ifcolor())`.

Also See

Complementary: [G] *areastyle*, [G] *colorstyle*, [G] *linestyle*, [G] *linewidthstyle*, [G] *linepatternstyle*, [G] *marginstyle*

Title

relativesize — Choices for sizes of objects

Syntax

relativesize	description
#	specify size; size 100 = minimum of width and height of graph; # required to be ≥ 0 depending on context
*#	specify size change via multiplication; *1 means no change, *2 twice as large, *.5 half; # required to be ≥ 0 depending on context

Negative sizes are allowed in certain contexts such as for gaps; in other cases such as the size of symbol, the size must be nonnegative, and negative sizes, if specified, are ignored.

Examples:

example	description
msize(2)	make marker diameter 2% of g
msize(1.5)	make marker diameter 1.5% of g
msize(.5)	make marker diameter .5% of g
msize(*2)	make marker size twice as large as default
msize(*1.5)	make marker size 1.5 times as large as default
msize(*.5)	make marker size half as large as default
xsca(titlegap(2))	make gap 2% of g
xsca(titlegap(.5))	make gap .5% of g
xsca(titlegap(-2))	make gap −2% of g
xsca(titlegap(-.5))	make gap −.5% of g
xsca(titlegap(*2))	make gap twice as large as default
xsca(titlegap(*.5))	make gap half as large as default
xsca(titlegap(*-2))	make gap −2 times as large as default
xsca(titlegap(*-.5))	make gap −.5 times as large as default

where g = min(width of graph, height of graph)

Description

A *relativesize* specifies a size relative to the graph (or subgraph) being drawn. Thus, as the size of the graph changes, so does the size of the object.

451

Remarks

relativesize is allowed, for instance, as a *textsizestyle* or a *markersizestyle*—see [G] **textsizestyle** and [G] **markersizestyle**—and as the size of many other things as well.

Note that relative sizes are not restricted to being integers; relative sizes of .5, 1.25, 15.1, etc. are allowed.

Also See

Complementary: [G] **markersizestyle**, [G] **textsizestyle**

Title

> **repeated options** — Interpretation of repeated options

Syntax

Options allowed with `graph` are categorized as being

> *unique*
> *rightmost*
> *merged-implicit*
> *merged-explicit*

What this means is described below.

Remarks

It may surprise you to learn that most `graph` options can be repeated within the same `graph` command. For instance, you can type

    . graph twoway scatter mpg weight, msymbol(Oh) msymbol(O)

and rather then getting an error, you will get back the same graph as if you omitted typing the `msymbol(Oh)` option. `msymbol()` is said to be a *rightmost* option.

`graph` allows that because so many other commands are implemented in terms of `graph`. Imagine an ado-file that constructs the "`scatter mpg weight, msymbol(Oh)`" part and you come along and use that ado-file, and you specify to it the option "`msymbol(O)`". The result is that the ado-file constructs

    . graph twoway scatter mpg weight, msymbol(Oh) msymbol(O)

and, because `graph` is willing to ignore all but the rightmost specification of the `msymbol()` option, the command works and does what you expect.

Options in fact come in three flavors, which are

1. *rightmost*: take the rightmost occurrence,

2. *merged*: merge the repeated instances together,

3. *unique*: the option may be specified only once; specifying it more than once is an error.

You will always find options categorized one of these three ways; typically that is done in the syntax diagram, but sometimes the categorization appears in the description of the option.

`msymbol()` is an example of a *rightmost* option. An example of a *unique* option is `saving()`; it may be specified only once.

Concerning *merged* options, they are broken into two subcategories:

2a. *merged-implicit*: always merge repeated instances together,

2b. *merged-explicit*: treat as *rightmost* unless an option within the option is specified, in which case it is merged.

`merged` can only apply to options that take arguments because otherwise there would be nothing to merge. Sometimes those options themselves take suboptions. For instance, the syntax of the `title()` option (the option that puts titles on the graph) is title("*string*" ["*string*" [...]] [, *suboptions*])

title() has suboptions that specify how the title is to look and among them is, for instance, color(); see [G] *title_options*. title() also has two other suboptions, prefix and suffix, that specify how repeated instances of the title() option are to be merged. For instance, specify

   ...title("My title") ...title("Second line", suffix)

and the result will be as if you specified

   ...title("My title" "Second line")

at the outset. Specify

   ...title("My title") ...title("New line", prefix)

and the result will be as if you specified

   ...title("New line" "My title")

at the outset. The prefix and suffix options specify exactly how repeated instance of the option are to be merged. If you do not specify one of those options,

   ...title("My title") ...title("New title")

the result will be as if you never specified the first option:

   ...title("New title")

title() is an example of a *merged-explicit* option. The suboption names for handling *merged-explicit* are not always prefix and suffix, but anytime an option is designated as *merged-explicit*, it will be documented under the heading *Interpretation of repeated options* exactly what and how the merge options work.

❏ Technical Note

Even when an option is *merged-explicit* and its merge suboptions are not specified, its other suboptions are merged. For instance, consider

   ...title("My title", color(red)) ...title("New title")

title() is *merged-explicit* but, since we did not specify one of its merge options, it is being treated as *rightmost*. Actually, it is almost being treated as rightmost because, rather than the title() being exactly what we typed, it will be

   ...title("New title", color(red))

This makes ado-files work as you would expect. Say you run the ado-file xyz.ado that constructs some graph and it constructs the command

   graph ..., ...title("Std. title", color(red)) ...

You specify an option to xyz.ado to change the title:

   . xyz ..., ...title("My title")

The overall result will be just as you expect: your title will substitute but the color of the title (and its size, position, etc.) will not change. If you wanted to change those things, you would have specified the appropriate suboptions in your title() option.

❏

Also See

Complementary: [G] **graph**

Title

> *ringpos* — Choices for location: distance from plot region

Syntax

ringpos is

#	$0 \leq \# \leq 100$, # real

Description

ringpos is specified inside options such as `ring()` and is typically used in conjunction with *clockpos* (see [G] *clockpos*) to specify a position for titles, subtitles, etc.

Remarks

See *Positioning of titles* under *Remarks* of [G] *title_options*.

Also See

Complementary: [G] *title_options*, [G] *clockpos*

Title

saving_option — Option for saving graph to disk

Syntax

The *saving_option* is

saving_option	description
saving(*filename* [, *suboptions*])	save graph to disk

saving() is *unique*; see [G] **repeated options**.

where *suboptions* are

suboptions	description
asis	freeze graph and save as-is
replace	okay to replace existing *filename*

Description

Option saving() saves the graph to disk.

Options

saving(*filename* [, *suboptions*]) specifies the name of the diskfile to be created or replaced. If *filename* is specified without an extension, .gph will be assumed.

Suboptions

asis specifies the graph is to be frozen and saved just as it is. The alternative—and the default if asis is not specified—is known as live format. In live format, the graph can continue to be edited in future sessions and, in addition, the overall look of the graph continues to be controlled by the chosen scheme (see [G] **schemes**).

Pretend you type

    . scatter yvar xvar, ... saving(mygraph)

That will create file mygraph.gph. Now pretend you send that file to a colleague. The way the graph appears on your colleague's computer might be different than it appears on yours. Perhaps you display titles on the top and your colleague has set his scheme to display titles on the bottom. Or perhaps your colleague prefers *y* axis on the right rather than the left. Understand that it will still be the same graph, but it might have a different look.

Or perhaps you just file away `mygraph.gph` for use later. Stored in the default live format, you can come back to it later and change the way it looks by specifying a different scheme and you can edit it.

If, on the other hand, you specify `asis`, the graph will look forever just as it looked the instant it was saved. You cannot edit it; you cannot change the scheme. If you send the as-is graph to colleagues, they will see it in exactly the form you see it.

Whether a graph is saved as-is or live makes no difference in terms of printing. As-is graphs usually require fewer bytes to store and they generally display more quickly, but that is all.

`replace` specifies that the file may be replaced if it already exists.

Remarks

To save a graph permanently, you add `saving()` to the end of the `graph` command (or any place among the options):

```
. graph ..., ... saving(myfile) ...
(file myfile.gph saved)
```

Alternatively, you can achieve the same result in two steps:

```
. graph ..., ...
. graph save myfile
(file myfile.gph saved)
```

The advantage of the two-part construction is that you can edit the graph between the time you first draw it and save it. The advantage of the one-part construction is that you will not forget to save it.

Also See

Complementary: [G] **graph save**, [G] **gph files**, [G] **graph manipulation**

Title

scale_option — Option for resizing text and markers

Syntax

The *scale_option* is

scale_option	description
scale(#)	specify scale; scale(1) default

scale() is *unique*; see [G] **repeated options**.

Description

Option scale() makes all the text and markers on a graph larger or smaller.

Options

scale(#) specifies a multiplier that affects the size of all text and markers on a graph. scale(1) is the default.

To increase the size of all text and markers on a graph by 20%, specify scale(1.2). To reduce the size of all text and markers on a graph by 20%, specify scale(.8).

Remarks

Under *Advanced use* in [G] **marker_label_options** we showed the following graph,

```
. twoway (scatter lexp gnppc, mlabel(country) mlablv(pos))
         (line hat gnppc, sort)
       , xsca(log) xlabel(.5 5 10 15 20 25 30, grid)
         legend(off)
         title("Life expectancy vs. GNP per capita")
         subtitle("North, Central, and South America")
         note("Data source:  World bank, 1998")
         ytitle("Life expectancy at birth (years)")
```

(Continued on next page)

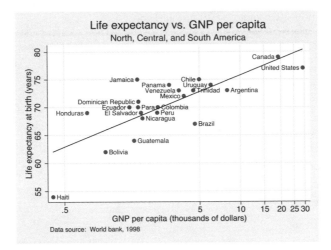

Here is the same graph with the size of all text and markers increased by 10%:

```
. twoway (scatter lexp gnppc, mlabel(country) mlablv(pos))
         (line hat gnppc, sort)
       , xsca(log) xlabel(.5 5 10 15 20 25 30, grid)
         legend(off)
         title("Life expectancy vs. GNP per capita")
         subtitle("North, Central, and South America")
         note("Data source:  World bank, 1998")
         ytitle("Life expectancy at birth (years)")
         scale(1.1)                                              (new)
```

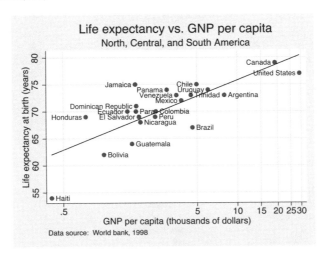

All we did was add the option `scale(1.1)` to the original command.

Also See

Complementary: [G] **graph**

Title

scheme economist — Scheme description: economist

Syntax

The economist scheme is

schemename	foreground	background	description
economist	color	white	*The Economist* magazine

For instance, you might type

    . graph ..., ... scheme(economist)

    . set scheme economist [, permanently]

See [G] *scheme_option* and [G] **set scheme**.

Description

Schemes determine the overall look of a graph; see [G] **schemes**.

Scheme economist specifies a look similar to that used by *The Economist* magazine.

Remarks

The Economist magazine (*http://www.economist.com*) employs a rather unique and clean graphics style that is both worthy of emulation and different enough from the usual as to provide an excellent example of just how much difference the scheme can make.

Among other things, *The Economist* puts the y axis on the right rather than the left of scatterplots.

Also See

Complementary: [G] **schemes**; [G] *scheme_option*, [G] **set scheme**

Title

> *scheme_option* — Option for specifying scheme

Syntax

The *scheme_option* is

scheme_option	description
scheme(*schemename*)	specify scheme to be used

scheme() is *unique*; see [G] **repeated options**.

For a list of available *schemenames*, see [G] **schemes**.

Description

Option scheme() specifies the graphics scheme to be used. The scheme specifies the overall look of the graph.

Options

scheme(*schemename*) specifies the scheme to be used.

Remarks

See [G] **schemes**.

Also See

Complementary: [G] **schemes**, [G] **set scheme**

Title

> **scheme s1** — Scheme description: s1 family

Syntax

The s1 family of schemes is

s1 family	foreground	background	description
s1rcolor	color	black	color on black
s1color	color	white	color on white
s1mono	monochrome	white	gray on white
s1manual	monochrome	white	s1mono, but smaller; used in some Stata manuals

For instance, you might type

    . graph ..., ... scheme(s1color)

    . set scheme s1rcolor [, permanently]

See [G] *scheme_option* and [G] **set scheme**.

Description

Schemes determine the overall look of a graph; see [G] **schemes**.

The s1 family of schemes is very similar to the s2 family—see [G] **scheme s2**—except that s1 uses a plain background, meaning no tint is applied to any part of the background.

Remarks

s1 is a conservative family of schemes which some people prefer to s2.

Of special interest is s1rcolor, which displays graphs on a black background. Because of pixel-bleeding, monitors have higher resolution when backgrounds are black rather than white. In addition, many users experience less eye strain viewing graphs on a monitor when the background is black. Scheme s1rcolor looks good when printed, but other schemes look better.

Schemes s1color and s1mono are derived from s1rcolor. Either of these schemes will deliver a better printed result. The important difference between s1color and s1mono is that s1color uses solid lines of different colors to connect points, while s1mono varies the line-pattern style.

Scheme s1manual is used in some of the Stata manuals, although it is not used in this one. s1manual is the same as s1mono, but presents graphs at a smaller overall size.

Also See

Complementary: [G] **schemes**; [G] *scheme_option*

Title

scheme s2 — Scheme description: s2 family

Syntax

The s2 family of schemes is

schemename	foreground	background	description
s2color	color	white	factory setting
s2mono	monochrome	white	s2color in monochrome
s2manual	monochrome	white	used in this manual

For instance, you might type

    . graph ..., ... scheme(s1mono)

    . set scheme s2mono [, permanently]

See [G] *scheme_option* and [G] **set scheme**.

Description

Schemes determine the overall look of a graph; see [G] **schemes**.

The s2 family of schemes is Stata's default scheme.

Remarks

s2 is the family of schemes that we like for displaying data. It provides a light background tint to give the graph better definition and make it visually more appealing. On the other hand, if you feel the tinting distracts from the graph, see [G] **scheme s1**; the s1 family is nearly identical to s2 but does away with the extra tinting.

In particular, we recommend you consider scheme s1rcolor; see [G] **scheme s1**. s1rcolor uses a black background and, for working at the monitor, it is difficult to find a better choice.

In any case, scheme s2color is Stata's default scheme. It look good on the screen, good when printed on a color printer, and more than adequate when printed on a monochrome printer.

Scheme s2mono has been optimized for printing on monochrome printers. In addition, rather than using the same symbol over and over and varying the color, s2mono will vary the symbol's shape and, in connecting points, s2mono varies the line pattern (s2color varies the color).

Scheme s2manual is the scheme used in printing this manual. It is basically s2mono, smaller, and in addition, the line pattern used to connect points is never varied.

Also See

Complementary: [G] **schemes**; [G] *scheme_option*

Title

scheme sj — Scheme description: sj

Syntax

The sj scheme is

schemename	foreground	background	description
sj	monochrome	white	*Stata Journal*

For instance, you might type

```
. graph ..., ... scheme(sj)
. set scheme sj [ , permanently ]
```

See [G] *scheme_option* and [G] **set scheme**.

Description

Schemes determine the overall look of a graph; see [G] **schemes**.

Scheme sj is the official scheme of the *Stata Journal*; see [R] **sj**.

Remarks

When submitting articles to the *Stata Journal*, graphs should be drawn using the scheme sj.

Before drawing graphs for inclusion with submissions, make sure scheme sj is up-to-date. Schemes are updated along with all the rest of Stata, so you just need to type

```
. update query
```

and follow any instructions given; see [R] **update**.

Also visit the *Stata Journal* web site for any special instructions. Point your browser to *http://www.stata-journal.com*.

Also See

Complementary: [G] **schemes**; [G] *scheme_option*, [G] **set scheme**;
[R] **sj**

Title

> **schemes** — Concept definition: schemes

Syntax

set scheme *schemename* $\left[\,,\; \underline{\text{permanently}}\,\right]$

<u>gr</u>aph ... $\left[\,,\; ...\,\text{scheme}(schemename)\; ...\,\right]$

where *schemename* may be

schemename	foreground	background	description
s1rcolor	color	black	a plain look on back background
s1color	color	white	a plain look
s1mono	monochrome	white	a plain look in monochrome
s1manual	monochrome	white	a plain look, but smaller; used in some Stata manuals
s2color	color	white	**factory setting**
s2mono	monochrome	white	s2color in monochrome
s2manual	monochrome	white	used in this manual
economist	color	white	*The Economist* magazine
sj	monochrome	white	*Stata Journal*

See [G] **scheme s1**, [G] **scheme s2**, [G] **scheme economist**, and [G] **scheme sj**.

Other *schemenames* may be available; type

. graph query, schemes

to obtain the full list installed on your computer.

Description

A scheme specifies the overall look of the graph.

set scheme sets the default scheme; see [G] **set scheme** for the details of this command.

Option scheme() specifies the graphics scheme to be used with this particular graph command without changing the default.

Remarks

Remarks are presented under the headings

The role of schemes
Finding out about other schemes
Setting your default scheme
The scheme is applied at display time
Background color
Foreground color
Obtaining new schemes
Making your own scheme

The role of schemes

When you type, for instance,

    . scatter yvar xvar

results are as if you typed

    . scatter yvar xvar, scheme(*your_default_scheme*)

Assuming you have not used the `set scheme` command to change your default scheme, *your_default_scheme* is `s2color`.

The scheme specifies the overall look for the graph, and by that we mean just about everything you can imagine. It determines whether y axes are on the left or the right. It determines how many values are by default labeled on the axes. It determines the colors that are used, if any. In fact, almost every statement made in other parts of this manual stating how something appears, or the relationship between how things appear, must not be taken too literally. How things appear is in fact controlled by the scheme:

- In [G] *symbolstyle*, we state that markers—the ink that denote the position of points on a plot—have a default size of `msize(medium)` and that small symbols have a size of `msize(small)`. That is in general true, but the size of the markers is in fact set by the scheme and a scheme might specify different default sizes.

- In [G] *axis_selection_options*, we state that when there is one y *axis*, it appears on the left and when there are two, the second appears on the right. What is in fact true is that where axes appear is controlled by the scheme and that most schemes work the way described. Another scheme named `economist`, however, displays things differently.

- In [G] *title_options*, we state where the titles, subtitles, etc. appear and we provide a diagram so that there can be no confusion. Except there can be confusion because where the titles, subtitles, etc., appear is in fact controlled by the scheme, and what we have described is what is true for the scheme named `s2color`.

The list goes on and on. If it has to do with the look of the result, it is controlled by the scheme.

To obtain some appreciation of just how much difference the scheme can make, you should type

    . scatter yvar xvar, scheme(economist)

`scheme(economist)` specifies a look similar to that used by *The Economist* magazine (*http://www.economist.com*) (whose graphs are worthy of emulation in our opinion). By comparison to the `s2color` scheme, the `economist` scheme moves y axes to the right, makes titles left justified, defaults grid lines to be on, sets a background color, and moves the note to the top right and expects it to be a number.

Finding out about other schemes

A list of schemes is provided in the syntax diagram above, but do not rely on the list being up to date. Instead, type

    . graph query, schemes

To obtain the full list installed on your computer.

Try drawing a few graphs with each:

    . graph ..., ... scheme(*schemename*)

Setting your default scheme

If you want to set your default scheme to, say, economist, type

    . set scheme economist

Scheme economist will now be your default scheme for the rest of this session, but the next time you use Stata, you will be back to using your old default scheme. If you type

    . set scheme economist, permanently

then economist will become your default scheme both the remainder of this session and in future sessions.

If you want to change your scheme back to s2color—the default scheme as Stata was originally shipped—type

    . set scheme s2, permanently

See [G] **set scheme**.

The scheme is applied at display time

Say you type

    . graph mpg weight, saving(mygraph)

to create and to save the file mygraph.gph (see [G] *saving_option*). If later you redisplay the graph by typing

    . graph use mygraph

the graph will reappear as you originally drew it. It will be displayed using the same scheme with which it was originally drawn, regardless of your current set scheme setting. If you type

    . graph use mygraph, scheme(economist)

the graph will be displayed using the economist scheme. It will be the same graph but will look different. You can change the scheme with which a graph is drawn either beforehand, on the original graph command, or later.

Background color

In the table at the beginning of the entry, we categorized the background color as being white or black, although actually what we mean is light or dark, because some of the schemes set background tinting. What we do mean is that "white" background schemes are suitable for printing. Printers (both the mechanical ones and the human ones) prefer you avoid dark backgrounds because of the large amounts of ink required and the corresponding problems with bleed through. On the other hand, dark backgrounds look very good on monitors.

In any case, You may change the background color of a scheme using the *region_options* `graphregion(fcolor())`, `graphregion(ifcolor())`, `plotregion(fcolor())`, and `plotregion(ifcolor())`; see [G] *region_options*. When overriding the background color, choose light colors for schemes that naturally have white backgrounds and dark colors for regions that have naturally black backgrounds.

Schemes that naturally have a black background are by default printed in monochrome. See [G] **set printcolor** if you wish to override this.

If you are producing graphs for printing on white paper, we suggest you choose a scheme with a naturally white background.

Foreground color

In the table at the beginning of this entry, we categorized the foreground as being color or monochrome. This refers to whether lines, markers, fills, etc. are presented by default in color or monochrome. Regardless of the scheme you choose, you can specify options such as `mcolor()`, `clcolor()`, etc., to control the color for each item on the graph.

Just because we categorized the foreground as monochrome does not mean you cannot specify colors in the options.

Obtaining new schemes

There may have already been installed in your copy of Stata schemes other than those documented in this manual. To find out, type

    . graph query, schemes

In addition, new schemes are added and existing schemes updated along with all the rest of Stata, so if you are connected to the Internet, type

    . update query

and follow any instructions given; see [R] **update**.

Finally, other users may have created schemes that could be of interest to you. To search the Internet, type

    . findit scheme

From there, you will be able to click to install any schemes that interest you; see [R] **search**.

Once a scheme is installed, which can be determined by verifying it shows in the list shown by

    . graph query, schemes

you can use it with the `scheme()` option

    . graph ..., ... scheme(*newscheme*)

or you can set it as your default, temporarily

. set scheme *newscheme*

or permanently,

. set scheme *newscheme*, permanently

Making your own scheme

Scheme *schemename* is stored in file *schemename*.scheme. For example, scheme s2color is stored in file s2color.scheme. You can locate where a scheme file is located by typing

. which *schemename*.scheme

If you copy a scheme file from its official place to your PERSONAL directory (see [P] **sysdir**), and if you change its name, you will have a personal new scheme of that new name.

For example, we have previously typed sysdir and determined that our PERSONAL directory is c:\ado\personal. We now type

. which s2color.scheme
c:\stata\ado\base\d\s2color.scheme
. copy c:\stata\ado\base\d\s2color.scheme c:\ado\personal\mine.scheme

We now have a new scheme called mine. We can edit file mine.scheme and change it how we wish. If you look at the file, you will find the lines very readable, especially if you compare what you are seeing with the contents of the other official scheme files.

Also See

Complementary: [G] *scheme_option*, [G] **set scheme**; [G] **scheme economist**,
[G] **scheme s1**, [G] **scheme s2**, [G] **scheme sj**

Title

set graphics — Set whether graphs are displayed

Syntax

<u>q</u>uery <u>graphics</u>

set <u>graphics</u> { on | off }

Description

`query graphics` shows the graphics settings.

`set graphics` allows you to change whether graphs are displayed.

Remarks

If you type

    . set graphics off

when you type a `graph` command, such as

    . scatter yvar xvar, saving(mygraph)

the graph will be "drawn" and saved in file `mygraph.gph`, but it will not be displayed. If you type

    . set graphics on

graphs will return to being displayed once again.

Drawing graphs secretly is sometimes useful in programming contexts, although in such contexts, it is better to specify the `nodraw` option; see [G] *nodraw_option*. Typing

    . scatter yvar xvar, saving(mygraph) nodraw

has the same effect as typing

    . set graphics off
    . scatter yvar xvar, saving(mygraph)
    . set graphics on

The advantage of the former is not only does it require less typing, but if the user should press **Break**, `set graphics` will not be left `off`.

Also See

Complementary: [G] *nodraw_option*

Title

set printcolor — Set how colors are treated when graphs are printed

Syntax

query graphics

set printcolor { automatic | asis | gs1 | gs2 | gs3 } [, permanently]

Description

query graphics shows the graphics settings.

set printcolor determines how colors are handled when graphs are printed.

Options

permanently specifies that the setting you make should be remembered across sessions.

Remarks

printcolor can be set one of five ways: automatic, asis, and gs1, gs2, or gs3. Four of the settings—asis and gs1, gs2, and gs3—specify how colors should be rendered when graphs are printed. The remaining setting—automatic—specifies that Stata should determine by context whether asis or gs1 is used.

Remarks are presented under the headings

> *What set printcolor affects*
> *The problem set printcolor solves*
> *set printcolor automatic*
> *set printcolor asis*
> *set printcolor gs1, gs2, and gs3*
> *The scheme matters, not the background color you set*

What set printcolor affects

set printcolor affects how graphs are printed when you pull down File and choose *Print graph* or when you use the graph print command; see [G] **graph print**.

set printcolor also affects the behavior of the graph export command when you use it to translate .gph files into another format, such as PostScript; see [R] **graph export**.

We will refer to all of the above in what follows as the "printing of graphs" or, equivalently, as the "rendering of graphs".

The problem set printcolor solves

If you should choose a scheme with a black background—see [G] **schemes**—and if you were then to print that graph, do you really want black ink poured onto the page so that what you get is exactly what you saw? Probably not. The purpose of `set printcolor` is to avoid such results.

set printcolor automatic

`set printcolor`'s default setting—`automatic`—looks at the graph to be printed and determines whether it should be rendered exactly as you see it on the screen or if instead the colors should be reversed and the graph printed in a monochrome gray scale.

`set printcolor automatic` bases its decision on the background color used by the scheme. If it is white (or light), the graph is printed `asis`. If it is black (or dark), the graph is printed `grayscale`.

set printcolor asis

If you specify `set printcolor asis`, all graphs will be rendered just as you see them on the screen, regardless of the background color of the scheme.

set printcolor gs1, gs2, and gs3

If you specify `set printcolor` gs1, gs2, or gs3, all graphs will be rendered according to a gray scale. If the scheme sets a black or dark background, the gray scale will be reversed (black becomes white and white becomes black).

gs1, gs2, and gs3 vary how colors are mapped to grays. gs1 bases its mapping on the average RGB value, gs2 on "true grayscale", and gs3 on the maximum RGB value. In theory, true grayscale should work best, but we have found that average generally works better with Stata graphs.

The scheme matters, not the background color you set

In all of the above, the background color you set using the *region_options* bgcolor(), graphre-gion(fcolor()) and plotregion(fcolor()) play no role in the decision that is made. Decisions are made based exclusively on whether the scheme naturally has a light or dark background.

You may set background colors but remember to start with the appropriate scheme. Set light background colors with light-background schemes and dark background colors with dark-background schemes.

Also See

Complementary: [G] **graph export**, [G] **graph print**

Title

---
set scheme — Set default scheme

Syntax

query graphics

set scheme *schemename* [, permanently]

For a list of available *schemenames*, see [G] **schemes**.

Description

query graphics shows the graphics settings, which includes the graphics scheme.

set scheme allows you to set the graphics scheme to be used.

Options

permanently specifies that the setting you make should be remembered across sessions.

Remarks

The graphics scheme specifies the overall look for the graph. You can specify the scheme to be used for an individual graph by specifying the scheme() option on the graph command, or you can specify the scheme once and for all using set scheme.

See [G] **schemes** for a description of schemes and a list of available *schemenames*.

One of the available *schemenames* is economist, which roughly corresponds to the style used by *The Economist* magazine. If you wanted to make the economist scheme the default for the rest of this session, you could type

    . set scheme economist

and if you wanted to make economist your default even in subsequent sessions, you could type

    . set scheme economist, permanently

Also See

Complementary: [G] **schemes**, [G] *scheme_option*

Title

> *std_options* — Options for use with graph construction commands

Syntax

The *std_options* are

std_options	description
title_options	titles, subtitles, notes, captions
scale(*#*)	resize text and markers
region_options	outlining, shading, aspect ratio, size
scheme(*schemename*)	overall look
nodraw	suppress display of graph
name(*name*, ...)	specify name for graph
saving(*filename*, ...)	save graph in file

See [G] *title_options*, [G] *scale_option*, [G] *region_options*, [G] *nodraw_option*, [G] *name_option*, and [G] *saving_option*.

Description

The above options are allowed with

command	manual entry
graph bar and graph hbar	[G] **graph bar**
graph dot	[G] **graph dot**
graph box	[G] **graph box**
graph pie	[G] **graph pie**

See [G] *twoway_options* for the standard options allowed with graph twoway.

Options

title_options specify what titles, subtitles, notes, and captions should be placed on the graph. See [G] *title_options*.

scale(*#*) specifies a multiplier that affects the size of all text and markers in a graph. scale(1) is the default, and scale(1.2) would make all text and markers 20% larger. See [G] *scale_option*.

region_options allow outlining the plot region (such as the placing or the suppressing of a border around the graph), specifying a background shading for the region, and the controlling of the aspect ratio. See [G] *region_options*.

475

scheme(*schemename*) specifies the overall look of the graph; see [G] *scheme_option*.

nodraw causes the graph to be constructed but not displayed; see [G] *nodraw_option*.

name(*name*[, replace]) specifies the name of the graph. name(Graph, replace) is the default. See [G] *name_option*.

saving(*filename*[, asis replace]) specifies the graph should be saved as *filename*. If *filename* is specified without an extension, .gph is assumed. asis specifies that the graph be saved just as it is. replace specifies that, if the file already exists, it is okay to replace it. See [G] *saving_option*.

Remarks

The above options may be used with any of the graph commands listed above.

Also See

Complementary: [G] *name_option*, [G] *nodraw_option*, [G] *region_options*, [G] *saving_option*, [G] *scale_option*, [G] *title_options*

Title

stylelists — Lists of style elements and shorthands

Syntax

A *stylelist* is a generic; specific examples of *stylelists* include *symbolstylelist*, *colorstylelist*, etc.

A *stylelist* is

$$el \left[\, el \left[\, \ldots \right] \right]$$

where each *el* may be

el	description
as_defined_by_style	what *symbolstyle*, *colorstyle*, . . . allows
"*as defined by style*"	must quote *els* containing spaces
' "*as* "*defined*" *by style*" '	compound quote *els* containing quotes
.	specifies the "default"
=	repeat previous *el*
. .	repeat previous *el* until end
. . .	same as . .

If the list ends prematurely, it is as if the list were padded out with . (meaning default for the remaining elements).

If the list has more elements than required, extra elements are ignored.

= in the first element is taken to mean . (period).

If the list allows numbers including missing value, if missing value is not the default, and if you want to specify missing value for an element, you must enclose the period in quotes: ".".

Examples:

```
. ..., ... msymbol(O d p o) ...
. ..., ... msymbol(O . p) ...
. ..., ... mcolor(blue . green green) ...
. ..., ... mcolor(blue . green =) ...
. ..., ... mcolor(blue blue blue blue) ...
. ..., ... mcolor(blue = = =) ...
. ..., ... mcolor(blue ...) ...
```

Description

Sometimes an option takes not a *colorstyle* but a *colorstylelist*, or not a *symbolstyle* but a *symbolstylelist*. *colorstyle* and *symbolstyle* are just two examples; there are lots of styles. Whether an option allows a list is documented in its syntax diagram. For instance, you might see graph matrix ... $\Big[$, ... mcolor(*colorstyle*) ... $\Big]$

in one place and

graph twoway scatter ... $\Big[$, ... mcolor(*colorstylelist*) ... $\Big]$

in another. In either case, to learn about *colorstyles*, you would see [G] ***colorstyle***. Here we have discussed how you would generalize a *colorstyle* into a *colorstylelist*, or a *symbolstyle* into a *symbolstylelist*, etc.

Also See

Complementary: [G] **graph twoway**

Title

> *symbolstyle* — Choices for the shape of markers

Syntax

symbolstyle may be

symbolstyle	synonym (if any)	description
circle	O	solid
diamond	D	solid
triangle	T	solid
square	S	solid
plus	+	
x	X	
smcircle	o	solid
smdiamond	d	solid
smsquare	s	solid
smtriangle	t	solid
smplus smx	x	
circle_hollow	Oh	hollow
diamond_hollow	Dh	hollow
triangle_hollow	Th	hollow
square_hollow	Sh	hollow
smcircle_hollow	oh	hollow
smdiamond_hollow	dh	hollow
smtriangle_hollow	th	hollow
smsquare_hollow	sh	hollow
point	p	a very small dot
none	i	a symbol that is invisible

For a symbol palette displaying each of the above symbols, type

    palette symbolpalette [, scheme(*schemename*)]

Other *symbolstyles* may be available; type

    . graph query symbolstyle

to obtain the full list installed on your computer.

Description

Markers are the ink used to mark where points are on a plot; see [G] ***marker_options***. *symbolstyle* specifies the shape of the marker.

You specify the *symbolstyle* inside the `msymbol()` option allowed with many of the `graph` commands:

        . graph twoway ... , msymbol(*symbolstyle*) ...

In some cases, you will see that a *symbolstylelist* is allowed:

        . scatter ... , msymbol(*symbolstylelist*) ...

A *symbolstylelist* is a sequence of *symbolstyles* separated by spaces. Shorthands are allowed to make specifying the list easier; see [G] ***stylelists***.

Remarks

Remarks are presented under the headings

> *Typical use*
> *Filled and hollow symbols*
> *Size of symbols*

Typical use

`msymbol(`*symbolstyle*`)` is one of the more commonly specified options. For instance, you may not be satisfied with the default rendition of

        . scatter mpg weight if foreign ||
          scatter mpg weight if !foreign

and prefer

        . scatter mpg weight if foreign, msymbol(oh) ||
          scatter mpg weight if !foreign, msymbol(x)

When you are graphing multiple y variables in the same plot, you can specify a list of *symbolstyles* inside the `msymbol()` option:

        . scatter mpg1 mpg2 weight, msymbol(oh x)

The result is the same as typing

        . scatter mpg1 weight, msymbol(oh) ||
          scatter mpg2 weight, msymbol(x)

Also note that in the above, we specified the symbol-style synonyms. Whether you type

        . scatter mpg1 weight, msymbol(oh) ||
          scatter mpg2 weight, msymbol(x)

or

        . scatter mpg1 weight, msymbol(smcircle_hollow) ||
          scatter mpg2 weight, msymbol(smx)

makes no difference.

Filled and hollow symbols

The *symbolstyle* specifies the *shape* of the symbol and, in that sense, one of the styles `circle` and `hcircle`—and `diamond` and `hdiamond`, etc.—are unnecessary in that each is a different rendition of the same shape. The option `mfcolor(`*colorstyle*`)` (see [G] *marker_options* specifies how the inside is of the symbol is filled. `hcircle()`, `hdiamond`, etc., are included for convenience and are equivalent to specifying

`msymbol(Oh)`: `msymbol(O) mfcolor(none)`

`msymbol(dh)`: `msymbol(d) mfcolor(none)`

etc.

Using `mfcolor()` to fill the inside of a symbol with different colors in some cases creates what are effectively new symbols. For instance, if you take `msymbol(O)` and fill its interior with a lighter shade of the same color used to outline the shape, you obtain a very pleasing result. For instance, you might try

`msymbol(O) mlcolor(yellow) mfcolor(.5*yellow)`

or

`msymbol(O) mlcolor(gs5) mfcolor(gs12)`

as in

    . scatter mpg weight, msymbol(O) mlcolor(gs5) mfcolor(gs14)

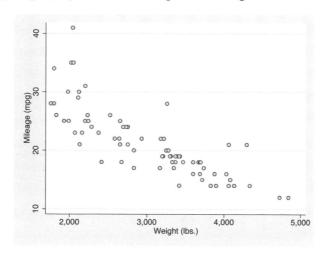

Size of symbols

Just as `msymbol(O)` and `msymbol(Oh)` differ only in `mfcolor()`, so it is that `msymbol(O)` and `msymbol(o)`—symbols `circle` and `smcircle`—differ only in `msize()`. In particular,

`msymbol(O)`: `msymbol(O) msize(medium)`

`msymbol(o)`: `msymbol(O) msize(small)`

and the same is true for all the other large and small symbol pairs.

`msize()` is interpreted as being relative to the size of the graph region (see [G] *region_options*) and so the same symbol size will in fact be a little different in

```
. scatter mpg weight
```

and

```
. scatter mpg weight, by(foreign total)
```

Also See

Complementary: [G] *marker_options*; [G] *markersizestyle*, [G] *colorstyle*,
                        [G] *linepatternstyle*, [G] *linewidthstyle*, [G] *linestyle*,
                        [G] *markerstyle*

Title

---

textbox_options — Options for textboxes and concept definition

---

Syntax

Textboxes contain one or more lines of text. The appearance of textboxes are controlled by the options

textbox_options	description
tstyle(*textboxstyle*)	overall style
orientation(*orientationstyle*)	whether vertical or horizontal
size(*textsizestyle*)	size of text
color(*colorstyle*)	color of text
justification(*justificationstyle*)	text left, centered, right justified
alignment(*alignmentstyle*)	text top, middle, bottom baseline
margin(*marginstyle*)	margin from text to border
linegap(*relativesize*)	space between lines
width(*relativesize*)	width of textbox override
height(*relativesize*)	height of textbox override
box or nobox	whether border drawn around
bcolor(*colorstyle*)	color of background and border
bfcolor(*colorstyle*)	color of background
blstyle(*linestyle*)	overall style of border
blpattern(*linepatternstyle*)	line pattern of border
blwidth(*linewidthstyle*)	thickness of border
blcolor(*colorstyle*)	color of border
bmargin(*marginstyle*)	margin from border outwards
bexpand	expand box in direction of text
placement(*compassdirstyle*)	location of textbox override

See [G] *textboxstyle*, [G] *orientationstyle*, [G] *textsizestyle*, [G] *colorstyle*, [G] *justificationstyle*, [G] *alignmentstyle*, [G] *marginstyle*, [G] *relativesize*, [G] *linestyle*, [G] *linepatternstyle*, [G] *linewidthstyle*, and [G] *compassdirstyle*.

The above options invariably occur inside other options. For instance, the syntax of title() (see [G] *title_options*) is

$$\texttt{title("}string\texttt{" } \big[\texttt{"}string\texttt{" } [\dots]\big] \big[\texttt{, } title\_options \; textboxoptions \big]\texttt{)}$$

and so any of the options above can appear inside the title() option:

```
. graph ..., ... title("My title", color(green) box) ...
```

483

Description

A textbox contains one or more lines of text. The textbox options listed above specify how the text and textbox should appear.

Options

tstyle(*textboxstyle*) specifies the overall style of the textbox. Think of a textbox as a set of characteristics that include, in addition to the text, the size of font, the color, whether lines are drawn around the box, etc. The *textboxstyle* you choose specifies all of those things and it is from there that the changes you make by specifying the other operations take effect.

The default is determined by the overall context of the text (such as whether it is due to title(), subtitle(), etc.), and that in turn is specified by the scheme (see [G] **schemes**). All of which is to say, identifying the name of the default style in a context is virtually impossible.

Option tstyle() is rarely specified. Usually, one simply lets the overall style be whatever it is and specify the other textbox options to modify it. Do not, however, dismiss the idea of looking for a better overall style that more closely matches your desires.

See [G] ***textboxstyle***.

orientation(*orientationstyle*) specifies whether the text and box are oriented horizontally or vertically (text reading from bottom to top or text reading from top to bottom). See [G] ***orientationstyle***.

size(*textsizestyle*) specifies the size of the text that appears inside the textbox. See [G] ***textsizestyle***.

color(*colorstyle*) specifies the color of the text that appears inside the textbox. See [G] ***colorstyle***.

justification(*justificationstyle*) specifies how the text is "horizontally" aligned in the box. Choices include left, right, and center. Think of the textbox as being horizontal even if it is vertical when specifying this option. See [G] ***justificationstyle***.

alignment(*alignmentstyle*) specifies how the text is "vertically" aligned in the box. Choices include baseline, middle, and top. Think of the textbox as being horizontal even if it is vertical when specifying this option. See [G] ***alignmentstyle***.

margin(*marginstyle*) specifies the margin around the text, the distance from the text to the borders of the box. The text that appears in a box, plus margin(), determine the overall size of the box. See [G] ***marginstyle***.

When dealing with rotated textboxes, textboxes for which orientation(vertical) or orientation(rvertical) has been specified, note that the margins for the left, right, bottom, and top refer to the margins before rotation.

linegap(*relativesize*) specifies the distance between lines.

width(*relativesize*) and height(*relativesize*) override Stata's usual determination of the width and height of the textbox based on its contents. See *Width and height* under *Remarks* below.

box and nobox specify whether a box should be drawn outlining the border of the textbox. The default is determined by the tstyle() which in turn is determined by context, etc. In general, the default is not to outline boxes and so the option to outline boxes is box. If an outline appears by default, then nobox is the option to suppress the outlining of the border. No matter what the default, you can specify box or nobox.

bcolor(*colorstyle*) specifies the color of both the background of the box and the color of the outlined border. This option is typically not specified because it results in the border disappearing into the background of the textbox; see options bfcolor() and blcolor() below for alternatives. The color only matters if `box` is also specified; otherwise, `bcolor()` is ignored. See [G] *colorstyle* for a list of color choices.

`bfcolor`(*colorstyle*) specifies the color of the background of the box. The background of the box is filled with the `bfcolor()` only if `box` is also specificd; otherwise, `bfcolor()` is ignored. See [G] *colorstyle* for a list of color choices.

`blstyle`(*linestyle*) specifies the overall style of the line used to outline the border. The style includes the line's pattern (solid, dashed, etc.), its thickness, and its color.

You need not specify `blstyle()` just because there is something you want to change about the look of the line. Options `blpattern`, `blwidth()`, and `blcolor()` will allow you to change the attributes individually. You specify `blstyle()` when there is a style that is exactly what you desire or when another style would allow you to specify fewer changes.

See [G] *linestyle* for a list of style choices and see [G] **lines** for a discussion of lines in general.

`blpattern`(*linepatternstyle*) specifies the pattern of the line outlining the border. See [G] *linepatternstyle*. Also see [G] **lines** for a discussion of lines in general.

`blwidth`(*linewidthstyle*) specifies the thickness of the line outlining the border. See [G] *linewidthstyle*. Also see [G] **lines** for a discussion of lines in general.

`blcolor`(*colorstyle*) specifies the color of the border of the box. The border color only matters if `box` is also specified; otherwise the `blcolor()` is ignored. See [G] *colorstyle* for a list of color choices.

`bmargin`(*marginstyle*) specifies the margin between the border and the containing box. See [G] *marginstyle*.

`bexpand` specifies that the textbox is to be expanded in the direction of the text, made wider if the text is horizontal and made longer if the text is vertical. It is expanded to the borders of its containing box. See [G] *title_options* for a demonstration of this option.

`placement`(*compassdirstyle*) overrides default placement; see *Appendix: Overriding default or context-specified positioning* below. See [G] *compassdirstyle* for argument choices.

Remarks

Remarks are presented under the headings

> *Definition of a textbox*
> *Position*
> *Justification*
> *Position and justification combined*
> *Margins*
> *Width and height*
> *Appendix: Overriding default or context-specified positioning*

Definition of a textbox

A textbox is one or more lines of text

single-line textbox

| 1st line of multiple-line textbox |
| 2nd linc of multiple-line textbox |

for which the borders may or may not be visible (controlled by the `box`/`nobox` option). Textboxes can be horizontal or vertical

| in an `orientation(vertical)` textbox, letters are rotated 90 degrees counterclockwise; `orientation(vertical)` reads bottom to top | in an `orientation(rvertical)` textbox, letters are rotated 90 degrees clockwise; `orientation(rvertical)` reads top to bottom |

Even in vertical textboxes, options are stated in horizontal terms of left-and-right. Think horizontally and imagine the rotation as being performed at the end.

Position

Textboxes are first formed and second positioned on the graph. The *textbox_options* affect the construction of the textbox, not its positioning. The options that control its positioning are provided by the context in which the textbox is used. For instance, the syntax of the `title()` option—see [G] *title_options*—is

title("*string*" ... [, position(...) ring(...) span(...) ... *textbox_options*]

It is `title()`'s `position()`, `ring()`, and `span()` options that determine where the title (i.e., textbox) is positioned. Once the textbox is formed, its contents no longer matter; it is just a box to be positioned on the graph.

Textboxes are positioned inside other boxes. For instance, the textbox might be

| title |

and, due to the `position()`, `ring()`, and `span()` options specified, it might be `title()`'s desire to position that box somewhere on the top "line":

There are many ways the smaller box could be fit into the larger box, which is the usual case, and forgive us for combining two discussions into one: how boxes fit into each other and the controlling of placement. If you specified `title()`'s `position(11)` o'clock option, the result would be

If you specified `title()`'s `position(12)` o'clock option, the result would be

If you specified `title()`'s `position(1)` o'clock option, the result would be

Justification

An implication of the above is that it is not the textbox option `justification()` that determines whether the title is centered; it is `title()`'s `position()` option.

Remember, textbox options describe the construction of textboxes, not their use. `justification(left | right | center)` determines how text is placed in multiple-line textboxes:

```
┌─────────────────────────────────┐
│ Example of multiple-line textbox │
│ justification(left)              │
└─────────────────────────────────┘
┌─────────────────────────────────┐
│ Example of multiple-line textbox │
│        justification(right)      │
└─────────────────────────────────┘
┌─────────────────────────────────┐
│ Example of multiple-line textbox │
│    justification(center)         │
└─────────────────────────────────┘
```

Textboxes are no wider than the text of their longest line. justification() determines how lines shorter than the longest are placed inside the box. In a single-line textbox,

```
┌──────────────────┐
│ single-line textbox │
└──────────────────┘
```

it does not matter how the text is justified.

Position and justification combined

With positioning options provided by the context in which the textbox is being used, and the justification() option, you can create many different effects in the presentation of multi-line textboxes. For instance, considering title(), you could produce

```
┌──────────────────┬─────────────────┬──────────────────────┐
│                  │ First line of title │                   │   (1)
│                  │ Second line      │                      │
└──────────────────┴─────────────────┴──────────────────────┘
```

or

```
┌──────────────────┬─────────────────┬──────────────────────┐
│                  │ First line of title │                   │   (2)
│                  │   Second line    │                      │
└──────────────────┴─────────────────┴──────────────────────┘
```

or

```
┌──────────────────┬─────────────────┬──────────────────────┐
│                  │ First line of title │                   │   (3)
│                  │      Second line │                      │
└──────────────────┴─────────────────┴──────────────────────┘
```

or

```
┌──────────────────────────────────┬─────────────────┐
│                                   │ First line of title │   (4)
│                                   │ Second line      │
└──────────────────────────────────┴─────────────────┘
```

or

```
┌──────────────────────────────────┬─────────────────┐
│                                   │ First line of title │   (5)
│                                   │   Second line    │
└──────────────────────────────────┴─────────────────┘
```

or

```
┌──────────────────────────────────┬─────────────────┐
│                                   │ First line of title │   (6)
│                                   │      Second line │
└──────────────────────────────────┴─────────────────┘
```

or many others. The commands would be

```
. graph ..., title("First line of title" "Second line",      (1)
                            position(12) justification(left))
. graph ..., title("First line of title" "Second line",      (2)
                            position(12) justification(center))
. graph ..., title("First line of title" "Second line",      (3)
                            position(12) justification(right))
. graph ..., title("First line of title" "Second line",      (4)
                            position(1) justification(left))
```

```
. graph ... , title("First line of title" "Second line",       (5)
                         position(1) justification(center))
. graph ... , title("First line of title" "Second line",       (6)
                         position(1) justification(right))
```

Margins

There are two margins: `margin()` and `bmargin()`. `margin()` specifies the margin between the text and the border. `bmargin()` specifies the margin between the border and the containing box.

By default, textboxes are the smallest rectangle that will just contain the text. If you specify `margin()`, you add space between the text and the borders of the bounding rectangle:

`margin(zero) textbox`

textbox with lots of margin on all four sides

`margin(`*marginstyle*`)` allows different amounts of padding to be specified above, below, left, and right of the text; see [G] *marginstyle*. `margin()` margins make the textbox look better when the border is outlined via the `box` option and/or the box is shaded via the `bcolor()` or `bfcolor()` options.

`bmargin()` margins are used to move the textbox a little or a lot when the available positioning options are inadequate. Consider specifying the `caption()` option (see [G] *title_options*) so that it is inside the plot region:

```
. graph ... , caption("My caption", ring(0) position(7))
```

Seeing the result, you decide you want to shift the box up and to the right a bit:

```
. graph ... , caption("My caption", ring(0) position(7)
              bmargin("2 0 2 0"))
```

The `bmargin()` numbers (and `margin()` numbers) are the top, bottom, left, and right amounts, and the amounts are specified as relative sizes (see [G] *relativesize*). We specified a 2% bottom margin and a 2left margin, thus pushing the caption box up and to the right.

Width and height

The width and the height of a textbox are determined by its contents (the text width and number of lines) plus the margins just discussed. The width calculation, however, is based on an approximation, with the result that the textbox that should look like this

Stata approximates the width of textboxes

can end up looking like this

Stata approximates the width of textboxes

or like this

Stata approximates the width of textboxes

You will not notice this problem unless the borders are being drawn (option `box`), because without borders, in all three cases you would see

Stata approximates the width of textboxes

For an example of this problem and the solution, see *Use of the textbox option width()* in [G] ***added_text_option***. If the problem arises, use width(*relativesize*) to work around it. Getting the width() right is a matter of trial and error. The correct width will nearly always be between 0 and 100.

Corresponding to width(*relativesize*), there is height(*relativesize*). This option is of less use because Stata never gets the height incorrect.

Appendix: Overriding default or context-specified positioning

What follows is really a footnote. We said previously that where a textbox is located is determined by the context in which it is used and by the positioning options provided by that context. There may be cases where you wish to override that default or where the context does not provide such control. In such cases, the option placement() allows you to take control.

Let us begin by correcting a misconception we introduced. We previously said that textboxes are fit inside other boxes when they are positioned. That is not exactly true. For instance, what happens when the textbox is bigger than the box into which it is being placed? Say we have the textbox

and we need to put it "in" the box

The way things work, textboxes are not put inside other boxes; they are merely positioned so that they align a certain way with the pre-existing box. Those alignment rules are such that, if the pre-existing box is larger than the textbox, the result will be what is commonly meant by "inside". The alignment rules are either to align one of the four corners or to align and center on one of the four edges.

In the example just given, the textbox could be positioned so that its northwest corner is coincident with the northwest corner of the pre-existing box,

placement(nw)

or so that their northeast corners are coincident,

placement(ne)

or so that their southwest corners are coincident,

placement(sw)

or so that their southeast corners are coincident,

placement(se)

or so that the midpoint of the top edges are the same,

placement(n)

or so that the midpoint of the left edges are the same,

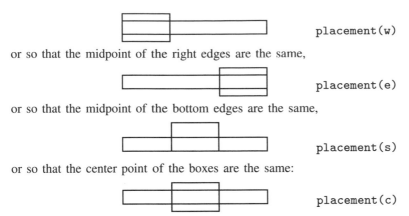

or so that the midpoint of the right edges are the same,

or so that the midpoint of the bottom edges are the same,

or so that the center point of the boxes are the same:

If you have trouble seeing any of the above, consider what you would obtain if the pre-existing box were larger than the textbox. Below we show the pre-existing box with 8 different textboxes:

placement(nw)	placement(n)	placement(ne)
placement(w)	placement(c)	placement(e)
placement(sw)	placement(s)	placement(se)

Also See

Complementary: [G] *alignmentstyle*; [G] *colorstyle*, [G] *compassdirstyle*, [G] *justificationstyle*, [G] *linepatternstyle*, [G] *linewidthstyle*, [G] *marginstyle*, [G] *relativesize*, [G] *orientationstyle*, [G] *textboxstyle*, [G] *textsizestyle*; [G] *title_options*

Title

textboxstyle — Choices for the overall look of text including border

Syntax

textboxstyle may be

textboxstyle	description
heading	large text suitable for headings
subheading	medium text suitable for subheadings
body	medium text
smbody	small text
p1–p15	default used to label points

Other *textboxstyles* may be available; type

    . graph query textboxstyle

to obtain the full list installed on your computer.

Description

A textbox contains one or more lines of text. *textboxstyle* specifies the overall style of the textbox.

textboxstyle is specified in the style() option nested within another option, such as title():

    . graph ..., title("My title", style(*textboxstyle*)) ...

See [G] **textbox_options** for more information on textboxes.

In some cases, you will see that a *textboxstylelist* is allowed. A *textboxstylelist* is a sequence of *textboxstyles* separated by spaces. Shorthands are allowed to make specifying the list easier; see [G] **stylelists**.

Remarks

Remarks are presented under the headings

> *What is a textbox?*
> *What is a textboxstyle?*
> *You do not need to specify a textboxstyle*

What is a textbox?

A textbox is one or more lines of text that may or may not have a border around it.

What is a textboxstyle?

Textboxes are defined by eleven attributes:

1. Whether the textbox is vertical or horizontal; see [G] *orientationstyle*.

2. The size of the text; see [G] *textsizestyle*.

3. The color of the text; see [G] *colorstyle*.

4. Whether the text is left justified, centered, or right justified; see [G] *justificationstyle*.

5. How the text aligns with the baseline; see [G] *alignmentstyle*.

6. The margin from the text to the border; see [G] *marginstyle*.

7. The gap between lines; see [G] *relativesize*.

8. Whether a border is drawn around the box and, if so

   a. The color of the background; see [G] *colorstyle*.

   b. The overall style of the line used to draw the border, which includes its color, width, and whether solid or dashed, etc.; see [G] *linestyle*.

9. The margin from the border outwards; see [G] *marginstyle*.

10. Whether the box expands to fill the box in which it is placed.

11. Whether the box is to be shifted when placed on the graph; see [G] *compassdirstyle*.

The *textboxstyle* specifies all eleven of these attributes.

You do not need to specify a textboxstyle

The *textboxstyle* is specified in option

```
tstyle(textboxstyle)
```

Correspondingly you will find other options are available for setting each of the individual attributes above; see [G] *textbox_options*.

You specify the *textboxstyle* when a style exists that is exactly what you desire or when another style would allow you to specify fewer changes to obtain what you want.

Also See

Complementary: [G] *textbox_options*; [G] *textstyle*

Title

> *textsizestyle* — Choices for the size of text

Syntax

textsizestyle may be

textsizestyle	description
zero	no size whatsoever, vanishingly small
minuscule	smallest
quarter_tiny	
third_tiny	
half_tiny	
tiny	
vsmall	
small	
medsmall	
medium	
medlarge	
large	
vlarge	
huge	
vhuge	largest
tenth	one-tenth the size of the graph
quarter	one-fourth the size of the graph
third	one-third the size of the graph
half	one-half the size of the graph
full	text the size of the graph
relativesize	any size you want

See [G] **relativesize**.

Other *textsizestyles* may be available; type

```
. graph query textsizestyle
```

to obtain the full list installed on your computer.

Description

textsizestyle specifies the size of the text.

textsizestyle is specified inside options such as the `size()` suboption of `title()` (see [G] *title_options*):

```
. graph ..., title("My title", size(textsizestyle)) ...
```

493

Also see [G] *textbox_options* for information on other characteristics of text.

Also See

Complementary: [G] *marker_label_options*, [G] *textbox_options*

Title

> *textstyle* — Choices for the overall look of text

Syntax

textstyle may be

textstyle	description
title	default used by `title()`
subtitle	default used by `subtitle()`
note	default used by `note()`
caption	default used by `caption()`
axis_title	default for axis titles
smbody	small text
body	medium size text
heading	large text suitable for headings
subheading	medium text suitable for subheadings
label	text suitable for labeling
keylabel	default used to label keys in legends
smlabel	default used to label points
ticklabel	default used to label major ticks
mticklabel	default used to label minor ticks

Other *textstyles* may be available; type

        . graph query textboxstyle   *(sic)*

to obtain the full list installed on your computer. The *textstyle* list is the same as the *textboxstyle* list.

Description

textstyle specifies the overall look of single lines of text. *textstyle* is specified inside options such as the marker-label option `mltextstyle()` (see [G] ***marker_label_options***):

        . twoway scatter ..., mlabel(...) mltextstyle(*textstylelist*) ...

In the example above, a *textstylelist* is allowed. A *textstylelist* is a sequence of *textstyles* separated by spaces. Shorthands are allowed to make specifying the list easier; see [G] ***stylelists***.

A *textstyle* is in fact a *textboxstyle*, it is just that only a subset of the attributes of the textbox matter; see [G] ***textboxstyle***.

Remarks

Remarks are presented under the headings

What is text?
What is a textstyle?
You do not need to specify a textstyle
Relationship between textstyles and textboxstyles

What is text?

Text is a single line of text.

What is a textstyle?

How text appears is defined by five attributes:

1. Whether the text is vertical or horizontal; see [G] *orientationstyle*.

2. The size of the text; see [G] *textsizestyle*.

3. The color of the text; see [G] *colorstyle*.

4. Whether the text is left justified, centered, or right justified; see [G] *justificationstyle*.

5. How the text aligns with the baseline; see [G] *alignmentstyle*.

The *textstyle* specifies these five attributes.

You do not need to specify a textstyle

The *textstyle* is specified in options such as

    mltextstyle(*textstyle*)

Correspondingly you will find other options are available for setting each of the individual attributes above; see [G] *marker_label_options*.

You specify the *textstyle* when a style exists that is exactly what you desire or when another style would allow you to specify fewer changes to obtain what you want.

Relationship between textstyles and textboxstyles

textstyles are in fact a subset of the attributes of *textboxstyles*; see [G] *textboxstyle*. A textbox is a heavyweight concept—it allows multiple lines, has an optional border around it, has a background color, and more. By comparison, text is a lightweight: it is just a line of text, and *textstyle* is the overall style of that single line.

Most textual graphical elements are textboxes, but there are a few simple graphical elements that are merely text, such as the marker labels mentioned above. The mltextstyle(*textstyle*) option really should be documented as mltextstyle(*textboxstyle*) because it is in fact a *textboxstyle* that mltextstyle() accepts. When mltextstyle() processes the *textboxstyle*, however, it looks only at the five attributes listed above and ignores the other attributes *textboxstyle* defines.

Also See

Complementary: [G] *marker_label_options*; [G] *textboxstyle*

Title

> *tickstyle* — Choices for the overall look of axis ticks and axis tick labels

Syntax

tickstyle may be

tickstyle	description
major	major tick and major tick label
major_nolabel	major tick with no tick label
major_notick	major tick label with no tick
minor	minor tick and minor tick label
minor_nolabel	minor tick with no tick label
minor_notick	minor tick label with no tick
none	no tick, no tick label

Other *tickstyles* may be available; type

```
. graph query tickstyle
```

to obtain the full list installed on your computer.

Description

Ticks are the marks that appear on axes. *tickstyle* specifies the overall look of ticks. See [G] *axis_label_options*.

Remarks

Remarks are presented under the headings

> *What is a tick? What is a tick label?*
> *What is a tickstyle?*
> *You do not need to specify a tickstyle*
> *Suppressing ticks and/or tick labels*

What is a tick? What is a tick label?

A tick is the small line that extends or crosses an axis and next to which, sometimes, numbers are placed.

A tick label is the text (typically a number) that optionally appears beside the tick.

What is a tickstyle?

tickstyle is really misnamed; it ought to be called a *tick_and_tick_label_style* in that it controls both the look of ticks and their labels.

Ticks are defined by three attributes:

1. The length of the tick; see [G] *relativesize*.

2. Whether the tick extends out, in, or crosses the axis.

3. The line style of the tick which includes its thickness, color, and whether solid, dashed, etc.; see [G] *linestyle*.

Labels are defined by two attributes:

1. The size of the text.

2. The color of the text.

Ticks and tick labels share one more attribute:

1. The gap between the tick and the tick label.

The *tickstyle* specifies all six of these attributes.

You do not need to specify a tickstyle

The *tickstyle* is specified in the options named

{ y | x }{ label | tick | mlabel | mtick } (tstyle(*tickstyle*))

Correspondingly, there are other { y | x }{ label | tick | mlabel | mtick } () suboptions that allow you to specify the individual attributes; see [G] *axis_label_options*.

You specify the *tickstyle* when a style exists that is exactly what you desire or when another style would allow you to specify fewer changes to obtain what you want.

Suppressing ticks and/or tick labels

If you wish to suppress the ticks that usually appear, specify one of the styles

tickstyle	description
major_nolabel	major tick with no tick label
major_notick	major tick label with no tick
minor_nolabel	minor tick with no tick label
minor_notick	minor tick label with no tick
none	no tick, no tick label

For instance, you might type

```
. scatter ..., ylabel(,tstyle(major_notick))
```

Suppressing the ticks can be useful when creating special effects. For instance, consider a case where you wish to add grid lines to a graph at $y = 10$, 20, 30, and 40, but you do not want ticks or labels at those values. Moreover, you do not want even to interfere with the ordinary ticking or labeling of the graph. The solution is

```
. scatter ..., ymtick(10(10)40, grid tstyle(none))
```

We "borrowed" the `ymtick()` option and changed it so that it did not output ticks. We could just as well have borrowed the `ytick()` option. See [G] *axis_label_options*.

Also See

Complementary: [G] *axis_label_options*

Title

---

title_options — Options for specifying titles

Syntax

title_options	description
title(*tinfo*)	overall title
subtitle(*tinfo*)	subtitle of title
note(*tinfo*)	note about graph
caption(*tinfo*)	explanation of graph
t1title(*tinfo*) t2title(*tinfo*)	rarely used
b1title(*tinfo*) b2title(*tinfo*)	rarely used
l1title(*tinfo*) l2title(*tinfo*)	vertical text
r1title(*tinfo*) r2title(*tinfo*)	vertical text

The above options are *merged-explicit*; see [G] **repeated options**.

$\{t|b|l|r\}\{1|2\}$title() are allowed with graph twoway only.

where *tinfo* is

> "*string*" ["*string*" [. . .]] [, *suboptions*]

suboptions	description
prefix and suffix	add to title text
position(*clockpos*)	position of title—side
ring(*ringpos*)	position of title—distance
span	"centering" of title
textbox_options	rendition of title

See [G] ***textbox_options*** for a description of *textbox_options*.
Option position() is not allowed with $\{t|b|l|r\}\{1|2\}$title().

Examples include

```
title("My graph")
note('"includes both "high" and "low" priced items"')

title("First line" "Second line")
title("Third line", suffix)
title("Fourth line" "Fifth line", suffix)
```

The definition of *ringpos* and the default positioning of titles is

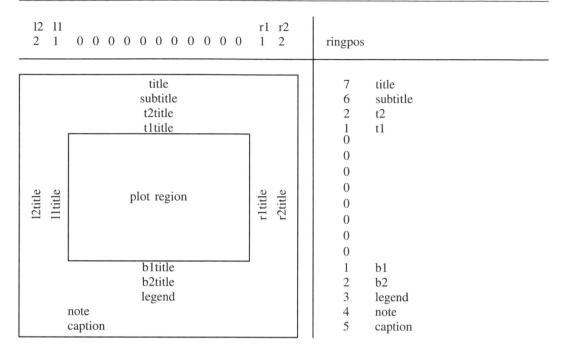

l2	l1											r1	r2	ringpos
2	1	0	0	0	0	0	0	0	0	0	0	1	2	

ringpos	
7	title
6	subtitle
2	t2
1	t1
0	
0	
0	
0	
0	
0	
0	
0	
0	
1	b1
2	b2
3	legend
4	note
5	caption

Description

Titles are the adornment around a graph that explains its purpose.

Options

title(*tinfo*) specifies the overall title of the graph. The title usually appears centered at the top of the graph. It is sometimes desirable to specify the span suboption when specifying the title, as in

. graph ..., ... title("Life expectancy", span)

See *Spanning* under *Remarks* below.

subtitle(*tinfo*) specifies the subtitle of the graph. The subtitle appears near the title (usually directly under it) and is presented in a slightly smaller font. subtitle() is used in conjunction with title(), and subtitle() is used by itself when the title() seems too big. For instance, you might type

. graph ..., ... title("Life expectancy") subtitle("1900-1999")

or

. graph ..., ... subtitle("Life expectancy" "1900-1999")

If subtitle() is used in conjunction with title() and you specify suboption span with title(), remember also to specify span with subtitle().

note(*tinfo*) specifies notes to be displayed with the graph. Notes are usually displayed in a small font placed at the bottom-left corner of the graph. By default, the left edge of the note will align with the left edge of the plot region. Specify suboption span if you wish the note moved all the way left; see *Spanning* under *Remarks* below.

caption(*tinfo*) specifies an explanation to accompany the graph. Captions are usually displayed at the very bottom of the graph, below the note(), in a font slightly larger than used for the note(). By default, the left edge of the caption will align with the left edge of the plot region. Specify suboption span if you wish the caption moved all the way left; see *Spanning* under *Remarks* below.

$\{t|b|l|r\}\{1|2\}$title() are rarely specified. It is generally better to specify the *axis_title_options* ytitle() or xtitle(); see [G] *axis_title_options*. The $\{t|b|l|r\}\{1|2\}$title() options are included for backwards compatibility with previous versions of Stata.

Suboptions

prefix and suffix specify that the specified text is to be added as separate lines either before or after any existing title of the specified type. See *Interpretation of repeated options* below.

position(*clockpos*) and ring(*ringpos*) override the default location of the title. position() specifies a direction *(sic)* according to the hours on the dial of a 12-hour clock and ring() specifies how far from the plot region the title is to appear.

ring(0) is defined as inside the plot region and is for the special case when you are placing a title directly on top of the plot. ring(k), $k>0$, specifies positions outside the plot region; the larger the ring() value, the farther away from the plot region. ring() values may be integer or noninteger and are treated ordinally.

position(12) puts the title directly above the plot region (assuming ring()>0), position(3) puts the title directly to the right of the plot region, and so on.

span specifies that the title is to be placed in an area spanning the entire width (or height) of the graph rather than an area spanning the plot region. See *Spanning* under *Remarks* below.

textbox_options are any of the options allowed with a textbox. Important options include

justification(left | center | right): determines how the text is centered;

orientation(horizontal | vertical): determines whether the text in the box reads from left to right or from bottom to top (there are other alternatives as well);

color(): determines the color of the text;

box: determines whether a box is drawn around the text;

width(*relativesize*): overrides the calculated width of the text box and is used in cases when text flows outside the box or when there is too much space between the text and the right border of the box; see *Width and height* under [G] *textbox_options*.

See [G] *textbox_options* for a description of each of the above options.

Remarks

Titles is the generic term we will use for titles, subtitles, keys, etc., and title options is the generic term we will use for title(), subtitle(), note(), caption(), and $\{t|b|l|r\}\{1|2\}$title(). Titles and title options all work the same way. In our examples, we will most often use the title() option, but understand that we could equally well use any of the title options.

Remarks are presented under the headings

> *Multiple-line titles*
> *Interpretation of repeated options*
> *Positioning of titles*
> *Alignment of titles*
> *Spanning*
> *Use of the textbox options box and bexpand*

Multiple-line titles

Titles can have multiple lines:

```
. graph ..., title("My title") ...
```

specifies a single-line title,

```
. graph ..., title("My title" "Second line") ...
```

specifies a two-line title, and

```
. graph ..., title("My title" "Second line" "Third line") ...
```

specifies a three-line title. You may have as many lines in your titles as you wish.

Interpretation of repeated options

Each of the title options can be specified more than once in the same command. For instance,

```
. graph ..., title("One") ... title("Two") ...
```

This does not produce a two-line title. Rather, when you specify multiple title options, it is the rightmost option that is operative and the earlier options are ignored. The title in the above command will be "Two".

That is, the earlier options will be ignored unless you specify `prefix` or `suffix`. In

```
. graph ..., title("One") ... title("Two", suffix) ...
```

the title will consist of two lines, the first line being "One" and the second, "Two". In

```
. graph ..., title("One") ... title("Two", prefix) ...
```

the first line will be "Two" and the second line "One".

Repeatedly specifying title options may seem silly, but it is easier to do than you might expect. Consider the command

```
. twoway (sc y1 x1, title("x1 vs. y1")) (sc y2 x2, title("x2 vs. y2"))
```

`title()` is in fact an option of `twoway`, not `scatter`, and graphs have only one `title()` (although it might consist of multiple lines). Thus, the above is probably not what the user intended. Had the user typed

```
. twoway (sc y1 x1) (sc y2 x2), title("x1 vs. y1") title("x2 vs. y2")
```

he would have seen his mistake. It is, however, okay to put `title()` options inside the `scatters`; `twoway` knows to pull them out. Nevertheless, only the rightmost one will be honored (because neither `prefix` nor `suffix` were specified) and thus the title of this graph will be "x2 vs. y2".

Multiple title options arise usefully when you are using a command that draws graphs that itself is written in terms of **graph**. For instance, the command **sts graph** (see [ST] **sts**) will graph the Kaplan–Meier survivor function. When you type

     . sts graph

with the appropriate data in memory, a graph will appear and that graph will have a **title()**. Yet, if you type

     . sts graph, title("Survivor function for treatment 1")

your title will override **sts graph**'s default. Inside the code of **sts graph**, both **title()** options appear on the **graph** command. First appears the default and second appears the one that you specified. This programming detail is worth understanding because, as an implication, if you type

     . sts graph, title("for treatment 1", suffix)

your title will be suffixed to the default. Most commands work this way, so if you use some command and it produces a title you do not like, specify **title()** (or **subtitle()**, ...) to override it, or specify **title(...**, **suffix)** (or **subtitle(...**, **suffix)**, ...) to add to it.

❑ Technical Note

Title options before the rightmost are not completely ignored. Their options are merged and honored so, if a title is moved or the color changed early on, the title will continue to be moved or the color changed. You can always specify the options to change it back.

❑

Positioning of titles

Where titles appear is determined by the scheme that you choose; see [G] **schemes**. Options **position(***clockpos***)** and **ring(***ringpos***)** override that location and let you place the title where you want it.

position() specifies a direction *(sic)* according to the hours of a 12-hour clock and **ring()** specifies how far from the plot region the title is to appear.

	11	12	1	
10	10 or 11	12	1 or 2	2
9	9	0	3	3
8	7 or 8	6	4 or 5	4
	7	6	5	

position() has two interpretations, one for **ring(0)** and another for **ring(***k***)**, $k > 0$. **ring(0)** is for the special case when you are placing a title directly on top of the plot. Put that case aside; titles usually appear outside the plot region.

A title directly above the plot region is at **position(12)** o'clock, a title to the right at **position(3)** o'clock, and so on. If you put your title at **position(1)** o'clock, it will end up above and to the right of the plot region.

Now consider two titles—say `title()` and `subtitle()`—both located at `position(12)`. Which is to appear first? That is determined by their respective `ring()` values. `ring()` specifies ordinally how far a title is from the plot region. The title with the larger `ring()` value is placed farther out. `ring()` values may be integer or noninteger.

For instance, `legend()` (see [G] *legend_option*) is closer to the plot region than `caption()` because, by default, `legend()` has a `ring()` value of 4 and `caption()` a `ring()` value of 5. Because both appear at `position(7)` o'clock, both appear below the plot region and because 4 < 5, the `legend()` appears above the `caption()`. These defaults assume you are using the default scheme.

If you wanted to put your legend below the caption, you could specify

. graph ... , legend(... ring(5.5)) caption("My caption")

or

. graph ... , legend(...) caption("My caption", ring(3.5))

The plot region itself is defined as `ring(0)` and, if you specified that, the title would appear inside the plot region, right on top of what is plotted! You can specify where inside the plot region you want the title using `position()`, and the title will put itself on the corresponding edge of the plot region. In `ring(0)`, the clock positions 1 and 2, 4 and 5, 7 and 8, and 10 and 11 are treated as being the same. In addition, `position(0)` designates the center of the plot region.

Within the plot region—within `ring(0)`—given a `position()`, you can further shift the title up or down or left or right by specifying the title's `margin()` *textbox_option*. For instance, you might specify

. graph ... , caption(... , ring(0) pos(7)) ...

and then discover that the caption needed to be moved up and right a little, and so change the `caption()` option to read

. graph ... , caption(... , ring(0) pos(7) margin(medium)) ...

See [G] *textbox_options* and [G] *marginstyle* for more information on the `margin()` option.

Alignment of titles

How should the text be placed in the textbox: left justified, centered, or right justified? The defaults that have been set vary according to title type:

title type	default justification
`title()`	centered
`subtitle()`	centered
$\{$ t $|$ b $|$ l $|$ r $\}\{$ 1 $|$ 2 $\}$`title()`	centered
`note()`	left-justified
`caption()`	left-justified

Actually, how a title is justified is by default determined by the scheme, and in the above, we assume you are using a default scheme.

You can change the justification using the *textbox_option* `justification(left | center | right)`. For instance,

. graph ... , title("My title", justification(left)) ...

See [G] *textbox_options*.

Spanning

Option `span` specifies that the title is to be placed in an area spanning the entire width (or height) of the graph rather than an area spanning the plot region. That is,

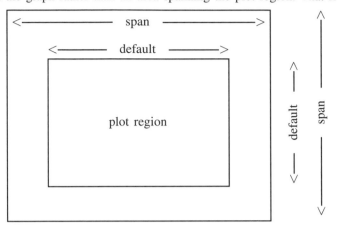

For instance, the `title()` is usually centered at the top of the graph. Is it to be centered above the plot region (the default) or between the borders of the entire available area (`title(..., span)` specified)? The `note()` is usually presented left justified below the plot region. Is it left justified to align with the border of the plot region (the default) or left justified to the entire available area (`note(..., span)` specified)?

Do not confuse `span` with the *textbox* option `justification(left—center—right)` which places the text left justified, centered, or right justified in whatever area is being spanned; see *Alignment of titles* above.

Use of the textbox options box and bexpand

The *textbox_options* box and bexpand—see [G] *textbox_options*—can be put to effective use with titles. We need you to look at three graphs:

```
. scatter mpg weight, title("Mileage and weight")
```

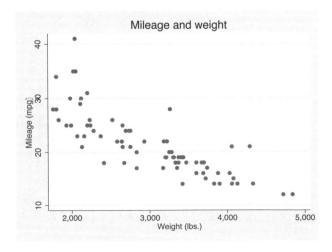

```
. scatter mpg weight, title("Mileage and weight", box)
```

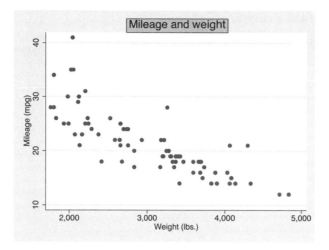

```
. scatter mpg weight, title("Mileage and weight", box bexpand)
```

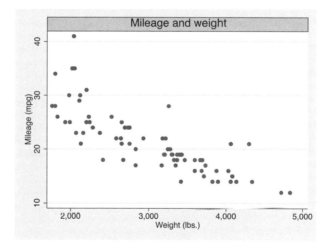

We want to direct your attention to the treatment of the title, which will be

Mileage and weight

Without options, the title appeared as is.

The textbox option `box` drew a box around the title.

The textbox options `bexpand` expanded the box to line up with the plot region and drew a box around the expanded title.

In both the second and third examples, in the graphs you will also note that the background of the textbox was shaded. That is because most schemes set the textbox option `bfcolor()`, but `bfcolor()` only becomes effective when the textbox is boxed.

Also See

Complementary: [G] *legend_option*; [G] *textbox_options*, [G] **schemes**

Title

twoway_options — Options for twoway graphs

Syntax

The *twoway_options* allowed with all `twoway` graphs are

twoway_options	description
added_line_options	draw lines at specified y or x values
added_text_option	display text at specified (y,x) value
axis_options	labels, ticks, grids, log scales
title_options	titles, subtitles, notes, captions
legend_option	legend explaining what means what
scale(#)	resize text and markers
region_options	outlining, shading, aspect ratio, size
scheme(*schemename*)	overall look
by(*varlist*, ...)	repeat for subgroups
nodraw	suppress display of graph
name(*name*, ...)	specify name for graph
saving(*filename*, ...)	save graph in file
advanced_options	difficult to explain

See [G] *added_line_options*, [G] *added_text_option*, [G] *axis_options*,
[G] *title_options*, [G] *legend_option*, [G] *scale_option*, [G] *region_options*,
[G] *by_option*, [G] *nodraw_option*, [G] *name_option*, [G] *saving_option*,
[G] *advanced_options*

Description

The above options are allowed with all `twoway` plots, `scatter`, `line`, ...

Options

added_line_options specify that horizontal or vertical lines be drawn on the graph; see
[G] *added_line_options*. If your interest is in drawing grid lines through the plot region, see
axis_options below.

added_text_option specifies text to be displayed on the graph (inside the plot region); see
[G] *added_text_option*.

axis_options specify how the axes are to look including values to be labeled or ticked on the axes.
These options also allow you to obtain logarithmic scales and grid lines. See [G] *axis_options*.

title_options specify what titles, subtitles, notes, and captions should be placed on the graph. See
[G] *title_options*.

legend_option specifies whether a legend is to appear and allows you to modify the legend's contents. See [G] *legend_option*.

scale(#) specifies a multiplier that affects the size of all text and markers in a graph. scale(1) is the default and scale(1.2) would make all text and markers 20% larger. See [G] *scale_option*.

region_options allow outlining the plot region (such as the placing or the suppressing of a border around the graph), specifying a background shading for the region, and the controlling of the aspect ratio. See [G] *region_options*.

scheme(*schemename*) specifies the overall look of the graph; see [G] *scheme_option*.

by(*varlist*, ...) specifies that the plot(s) should be repeated for each set of values of *varlist*; see [G] *by_option*.

nodraw causes the graph to be constructed but not displayed; see [G] *nodraw_option*.

name(*name*[, replace]) specifies the name of the graph. name(Graph, replace) is the default. See [G] *name_option*.

saving(*filename*[, asis replace]) specifies the graph should be saved as *filename*. If *filename* is specified without an extension, .gph is assumed. asis specifies that the graph be saved just as it is. replace specifies that, if the file already exists, it is okay to replace it. See [G] *saving_option*.

advanced_options are not so much advanced as they are difficult to explain and are rarely used. They are also invaluable when you need them; see [G] *advanced_options*.

Remarks

The above options may be used with any of the twoway plottypes—see [G] **graph twoway**—for instance,

```
. twoway scatter mpg weight, by(foreign)
. twoway line le year, xlabel(,grid) saving(myfile, replace)
```

Understand that the above options are options of twoway, meaning that they affect the entire twoway graph and not just one or the other of the plots on it. For instance, in

```
. twoway lfitci  mpg weight, stdf ||
         scatter mpg weight, ms(O) by(foreign, total row(1))
```

the by() option applies to the entire graph and in theory you should type

```
. twoway lfitci  mpg weight, stdf  ||
         scatter mpg weight, ms(O) ||, by(foreign, total row(1))
```

or

```
. twoway (lfitci  mpg weight, stdf)
         (scatter mpg weight, ms(O)), by(foreign, total row(1))
```

to demonstrate your understanding of that fact. You need not do that, however, and in fact it does not matter to which plot you attach the *twoway options*. You could even type

```
. twoway lfitci  mpg weight, stdf by(foreign, total row(1)) ||
         scatter mpg weight, ms(O)
```

and, when specifying multiple *twoway_options*, you could even attach some to one plot and the others to another:

```
. twoway lfitci  mpg weight, stdf by(foreign, total row(1)) ||
         scatter mpg weight, ms(O) saving(myfile)
```

Also See

Complementary: [G] **graph twoway**; [G] *axis_options*, [G] *title_options*, [G] *legend_option*, [G] *scale_option*, [G] *region_options*, [G] *scheme_option*, [G] *by_option*, [G] *nodraw_option*, [G] *name_option*, [G] *saving_option*, [G] *advanced_options*

Subject and author index

This is the subject and author index for the *Stata Graphics Reference Manual*. Readers interested in topics other than graphics should see the combined subject index at the end of Volume 4 of the *Stata Base Reference Manual*, which indexes the *Stata Base Reference Manual*, the *Stata User's Guide*, the *Stata Programming Reference Manual*, the *Stata Cluster Analysis Reference Manual*, the *Stata Cross-Sectional Time-Series Reference Manual*, the *Stata Survey Data Reference Manual*, the *Stata Survival Analysis & Epidemiological Tables Reference Manual*, and the *Stata Time-Series Reference Manual*.

Semicolons set off the most important entries from the rest. Sometimes no entry will be set off with semicolons; this means all entries are equally important.